P9-AFV-210

# DATE DUE

| | | | |
|---|---|---|---|
| MY 26 98 | | | |
| NO 13 02 | | | |
| | | | |
| | | | |
| | | | |
| | | | |
| | | | |
| | | | |
| | | | |
| | | | |
| | | | |
| | | | |
| | | | |
| | | | |
| | | | |
| | | | |
| | | | |

DEMCO 38-296

"Return from the Bear Hunt." William Hahn, 1882. Scene probably near The Geysers, Sonoma County, California. Oil painting on canvas, 54¾ × 88 in. Collection of Miss Rosina Hahn. By permission of M. H. de Young Memorial Museum, San Francisco.

R

# CALIFORNIA GRIZZLY

*By Tracy I. Storer and Lloyd P. Tevis, Jr.*

Foreword by Rick Bass

## UNIVERSITY OF CALIFORNIA PRESS

BERKELEY   LOS ANGELES   LONDON

Riverside Community College
Library
4800 Magnolia Avenue
Riverside, California 92506

DEC '97

QL 737 .C27 S73 1996

Storer, Tracy Irwin, 1889-

California grizzly

University of California Press
Berkeley and Los Angeles, California

University of California Press, Ltd.
London, England

First California Paperback Printing 1996

COPYRIGHT, 1955, BY THE REGENTS OF THE UNIVERSITY OF CALIFORNIA

© renewed 1983 Ruth Storer, Lloyd P. Tevis, Jr.
Foreword © 1996 Rick Bass

Library of Congress Cataloging-in-Publication Data
Storer, Tracy Irwin, 1889–1973
Califoria grizzly / by Tracy I. Storer and Lloyd P. Tevis, Jr.
p. cm.
Includes bibliographical references (p. 305) and index.
ISBN 0-520-20520-0 (alk. paper)
1. Grizzly bear.   2. Grizzly bear—California.   I. Tevis, Lloyd
P. (Lloyd Pacheco), 1916–    . II. Title.
QL737 .C27S73
599.74'446—dc20                                          96-21929
CIP

Library of Congress Catalogue Card Number: 55-9884

Printed in the United States of America
1  2  3  4  5  6  7  8  9

Designed by Marion Jackson

The paper used in this publication meets the minimum requirements
of American National Standard for Information Sciences—
Permanence of Paper for Printed Library Materials,
ANSI Z39.48-1984. ♾

# Preface

In 1860 Theodore H. Hittell published a volume on *James Capen Adams, Mountaineer and Grizzly Bear Hunter of California.* A century has passed since the subject of that book was roaming the mountains and valleys of the state, shooting or trapping bears. It is almost as long since reporter Hittell began to write down the experiences of hunter Adams. Before and after the days of Adams and Hittell a great deal was recorded about the grizzly, in private diaries, in books on California and the West, and in newspapers. A few bits of scientific knowledge were added but the total in this field is small. The present account, ninety-four years after the one by Hittell, is an attempt to bring these scraps of information into a continuity—to describe the bear itself, the manner of life it had in successive contacts with different human contemporaries, the imprint it made on their lives and activities, and its ultimate effect on the civilization that developed in California. Hittell's book recounted the experiences of one man with many bears. The present volume tells of the relations between many bears and a great host of people.

The grizzly—emblem of California—never had and never can have an adequate biography. In the days of the Indians, in the Spanish and Mexican periods, and in the gold rush, this huge beast was common, indeed abundant, over much of the lowlands and foothills of the state, except in the deserts. It was a constant and fearful element in the lives of the Indians; it competed with them for food and, because of its prowess, it became the motif for a certain class of shamans. Through the era of the missions and ranchos the grizzly ravaged the herds of livestock, and men captured it alive for bear-and-bull fights. In the days of '49 and

v

later, as pioneers and settlers from the East swarmed into California, it became a source of meat and robes, a hazard to people, and a menace to the cattle, sheep, and hogs on pioneer farms. Hunters, miners, and ranchers were vigorous in their attempts to destroy it—a rugged and dangerous pursuit in which many persons were maimed or killed. As a result, in a few years hardly a grizzly remained near any of the early communities. Elsewhere in the state a gradual but persistent decline followed, and by the 1920's the grizzly was extinct in California.

This account must be historical and mainly from the written and published record, since the people who had firsthand experiences with the grizzly have also gone. It is a patchwork, deficient in many respects, but the only sort practicable in the circumstances.

Abbreviations used in the text are as follows: MVZ, Museum of Vertebrate Zoölogy, University of California; USNM, United States National Museum, Washington, D.C. References to articles in newspapers are cited thus: (N 34), and so forth. A numbered list of newspaper articles follows the Bibliography.

We are grateful to the following copyright owners for permission to quote from copyrighted material: Academy Library Guild, Fresno, _California Vaquero_, by A. J. Rojas, 1953; Columbia University Press, _Gold Rush_, by J. Goldsborough Bruff, edited by G. W. Read and Ruth Gaines, 1949; Doubleday & Company, Inc., _Lives of Game Animals_, by Ernest Thompson Seton, 1929; E. P. Dutton & Co., Inc., _California Cowboys_, by Dane Coolidge, 1939; Franciscan Fathers, Old Mission, Santa Barbara, _San Buenaventura_, 1930, _Mission Santa Ines_, 1932, _Mission San Luis Obispo_, 1933, and _Mission San Carlos Borromeo_, 1934, by Zephyrin Engelhardt; Wallace Hebberd, _Reminiscences of a Ranger_, by Horace Bell, 1927; John Howell, _Seventy-five Years in California_, by William Heath Davis, 1929; The Huntington Library, San Marino, California, _The Cattle on a Thousand Hills_, by Robert Glass Cleland, 1951, and _The Life and Adventures of Don Agustín Janssens, 1834–1856_, edited by William H. Ellison and Francis Price, 1953; Alfred A. Knopf, Inc., _From Wilderness to Empire_, by R. G. Cleland, 1944; William Morrow & Company, Inc., _On the Old West Coast_,

*Being Further Reminiscences of a Ranger, Major Horace Bell,* edited by Lanier Bartlett, 1930; Newell & Chamberlain, Gazette Press, Mariposa, *The Call of Gold,* by N. D. Chamberlain, 1936; Marco R. Newmark, *Sixty Years in Southern California, 1853–1913,* by Harris Newmark, 1916, 1926, and 1930; Rinehart & Company, Inc., *The Salinas,* by Anne B. Fisher, 1945; Harry L. Fraser, Administrator of Saunders Press, *Bound for Sacramento,* by Carl Meyer, translated by Ruth Frey Axe, 1938; The Wistar Institute of Anatomy and Biology, *The Autobiography of Isaac J. Wistar,* 1937.

We are also grateful to The Huntington Library for permission to reproduce the cover of *Life of J. C. Adams, Known as Old Adams,* New York, 1860, and to quote from the manuscript by William Heath Davis, HM DA–2(228).

TRACY I. STORER

LLOYD P. TEVIS, JR.

*Davis, California*
*May, 1955*

# Contents

# *Illustrations*

# *Foreword*

It is dusk outside my cabin in the Yaak Valley of extreme northwest Montana, right up on the Canadian border, and a soft gold light is falling in long shadows and slants of light across the marsh at which I am staring. A lone bittern is calling stoically, steadily, plaintively—one lone bird—and I cannot help but imagine, at this haunted time of day, and by the repetitive, unceasing nature of his calling, that he is singing to something that has gone away: that he is lamenting, trying to speak back across the years to a thing—though what that thing might be, I could not begin to say; and almost anyone will tell you that birds sing only for territoriality and the here and now.

Within the watershed of my valley—fifty miles north-to-south, and about that far, east-to-west—there may be as few as a dozen grizzlies left, hiding out between clearcuts, or possibly as many as thirty. Either way, it's not very many. As the clearcuts, seed-tree cuts, and re-gen cuts—all the same, for biological or as we also say practical purposes—continue unabated, and increase —as the last cores and sanctuaries of grizzly country are entered with roads and the fire-trails of noxious weeds that follow these roads—I find myself wondering what it would be like to see grizzlies go extinct in my valley, as other generations have watched them go extinct in theirs, throughout the country. I wonder if I could stand that kind of loss. This is a wild valley and I wonder if we, humans, could stand that kind of humiliation.

I suspect that a lot of us could. We've had a fair amount of practice.

We don't begin, as scientists, to know more than one relative iota about the fit that grizzlies (or anything else, really) have in this world. We're surprised when, once wolves are re-established to an area, one of the grizzlies skips hibernation (another thing about which we know nothing) and prowls the snowy woods behind the wolf-packs, cleaning up on the leavings, though sometimes commandeering the spoils. Revelations concerning the bears squeeze forth from science each year in small batches—the bears' role in seed dispersal, with their seasonal migrations up and down the mountains—etc., etc. It always stuns us, the increasing complexity we discover concerning the way things fit in the world. It's a miracle! we cry. Each small thing we discover about these incredible animals raises two new questions, questions we didn't even previously know how to ask. They like to live in untouched, unharmed areas—in places of mystery—and this is a commodity in ever-diminishing supply on the public lands of the West these days.

We are on somewhat firmer ground in understanding the grizzly's place, with regard to humans, in the realm of art—the dwelling of story and culture. Here the bears enrich our imagination as they have enriched the imagination of all mankind, for as long as bears and humans have been in the world together. All cultures, up to and miraculously even still including the present one, have been built on the backs of the great bear. Oral as well as written stories of place look always to the other inhabitants of that place—to the rest of the animal world from which we now seem pathologically intent on excluding ourselves. These are the means by which we used to define the country in which we lived and tell one another about the shape it was in, and the things that went on in that place—about the needs and requirements and habits and workings of a place. All animals had a role in these stories, for all animals are part of the workings of a place, but the grizzly has almost always been the dominant creature in each landscape it has inhabited, and so its stories have assumed the extra weight and import of that dominance. Simply put, when you went for a walk in the woods or told a story about that country and your relation to it, *the grizzly used to matter.* It used to be the key factor. It is with fear, or lamentation, that I consider the consequences for our own culture if the grizzly sinks from sight beneath us: the foundation upon which we have built all that we are now.

This book then, though sad in that it is about a thing no longer—or not presently—with us, arouses in me a great wonderment and an excitement of the imagination, when I read of the old days in California, when California was so truly rich, not yet 150 years gone by. The richness of those times is hard to imagine compared to the impoverishments of today. The map in this book of the original distribution of grizzlies in California is startling for its saturation: nearly all of the state, save for the salt-desert, was once hospitable to them. It's hard to imagine or construct, even in the imagination, a country that was better suited for them: thousands of square miles of rich grass that was higher than their backs, in which they could graze like horses; rich mast crops, and shady oak groves, in which they could gorge like, well, grizzlies; the beaches, washing in with each tide the high-protein wrack of whales, fish, crabs, seals, dolphins—a fertility of wildness, and mild, easy weather, unequalled perhaps anywhere in the world. I do not think it is too far from correct to use words like Eden and Genesis and Paradise; and surely if any blasphemy is involved in imagining or reading about the natural history of those days, it is not of uttering the closeness these times must have had to Godliness, but of letting it—*all of it*—be lost.

Estimates of 10,000 California grizzlies rest next to those of 130,000 Indians, making for an interesting bear-to-man ratio. We've gone from 13-to-1 to a billion-to-zero. And we wonder why some days we feel a little hollow inside: as if something's missing, or something's not quite right.

What stories, what tales are held, still in the shallow past, but fast on its way to becoming the deep past, of the California grizzlies and the intersection of their fate with that of "civilized" man: stories of grizzlies burying or "caching" people alive, and the night-woods alive with the sound of grizzlies crunching and cracking bones outside settlers' windows in the evenings—sweet dreams! The grizzly feasts that were had, for a short while, as the tender stock was brought into this country—cattle and horses, chickens and pigs—so much easier to catch than deer—as if the bears had fallen into heaven, is how it must have seemed to them at first, though of course it was not heaven at all, but some other place.

It is with great sadness that I learn of yet another accounting in which the seemingly-ineradicable human gene for sadism is manifested, and not exclusively in the white-devil stock of European Anglos that crossed over to this continent 200 years ago, but which was already firm in place in the Spaniards of the 1600s—as, I suspect, it always has been, and may always be, in all of us (p. 131):

"The animals were lassoed by the throat and also by the hind leg, a horseman at each end, and the two pulling in opposite directions till the poor beast succumbed. The fun was kept up until about daylight, and when they got through they were completely exhausted . . ."

What fun! But of course it was not simply the Spaniards who had lost their senses, lost their minds, lost their reverence for the land and its inhabitants; we should not allow to pass unobserved the antics of the Sacramentoites who, lacking a bull to put in an arena with a grizzly, would sometimes put in the bull's place a donkey or a burro, to watch the bear bite off the head or leg of the creature. Storer and Tevis write (p. 161):

"Finally, somebody with a perverted sense of humor conceived the idea of letting hundreds of city rats loose into a well-closed arena, where they tormented the grizzly to distraction by swarming over him and crawling under his fur."

All around me, I am too often in the cynic's habit, the scientists's or artist's habit, of looking at what is marvelous today, but which will not be here tomorrow. Will the next disappearing thing be something as ancient and complex-in-this-world as cedar fronds, or white pine forests? What of the beloved old growth tamarack forests, the autumnal gold-needled giants that seem to be in some rare niche, halfway between the worlds of deciduous and coniferous trees?

It is not just in the extreme northwest corner of Montana,

the Yaak, where grizzlies flirt with—or hell, are forced to embrace—extinction. In northern Washington, in southern Colorado, in northern Idaho, they are down to only single- or double-digit populations; you could wad them all into a large service elevator—and even in our showpiece, Yellowstone, they are threatened, down to a population of about thirty females.

Rich country, rich soil, yields great and rich animals, which yield great and rich stories—which yield great cultures. Wolves, grizzlies, mountain lions, jaguars, buffalo, elk, sea lions, condors—these giants of both reality and the imagination rose out of California's soil, in the not-yet-distant past. We had a thing once that was holy. How could a country's soil *not* be holy when it was once plowed by bears in moonlight—the great grizzly feeding nocturnally in that rich soil, digging for acorns and roots, raking and furrowing that black night earth with his shining claws? How could great stories—and a great culture—*not* arise from such a natural history, such a legacy?

And is it any wonder that, having allowed such magic, such grace, to be lost—such *prodigy* to be pissed away—we now find ourselves in such shambles, and so bereft of heart?

For the record, the grizzly's last years in California were around the early 1920s. The story of their disappearance from California is heartbreakingly familiar to that of their disappearance from all the other places in the West: receding lastly to the high mountains, but unable to hide safely even in those uncontested crags of so much rock and ice. The intensity of Storer's and Tevis's account—and the richness of it—leaves a reader feeling robbed: as if reading of some subterranean (or high-aloft) place of glory that lies just beneath our feet, or just above our heads, but which we cannot reach.

William Kelly, in 1851, writes of the phenomenon of grizzly springs: places where the bears would scratch at the earth, digging for fresh damp bulbs or roots, and in so doing, would excavate little underground seeps and springs which would then form puddles and ponds, around which willows and other vegetation would grow—small gardens or oases, as if the bears had shaped from clay a miniature, life-giving world, around which

other animals would come and drink. I have heard of similar instances of coyotes doing this in the desert, and while we still
have our coyotes in the world, what does it bode for us to be losing grizzlies-as-creators? If we run out of creation legends, and
creation models, will we turn instead to legends of destruction,
and models of destruction?

Where are our creation myths now? What a diminishment
of results: where once the grizzlies clawed life from the earth, we
now scratch tiny scribblings of ink in paper in wan attempts to
convince our politicians—or whomever they belong to—to protect these valuable places. I wish California well in protecting its
remaining wild character, and in reclaiming its injured areas. I
hope that with their great political power, Californians will work
to demand the protection of our country's other remaining wild
places.

All things are connected, and we are all complicit in the accounting of loss. The valley next to mine, the ghost of the Ural
Valley, once home to wild grizzlies, lies beneath hundreds of feet
of water that was once the Kootenai River, so that power can be
provided through a dam that sends the juice through the Bon-
neville Power Administration all the way down to California.
This lake, besides drowning the Ural Valley almost thirty years
ago, has helped isolate and further fragment and endanger the
genetic vigor of the handful of grizzlies remaining in the mountains outside my cabin tonight, up in those snowy vestigial corners. I say this not with judgment or rancor, only to illustrate
how every wild place that is not yet gone is in immediate peril of
being crushed by the so-ponderous weight of our sheer mass
upon the world. The very least we can do is commit these last
places—any wild and good country—to the future, as wilderness
reserves, so that they and the miracles within them will forever
be in this world; and that perhaps too future generations
brighter and more committed can then take these last little
cores, or gems, that we had the awkward grace to protect, and
begin then the true and hard work of reconnecting them so that
grizzlies and wolves could once more exist in the world with us,
and bring back with them more legends, more stories of creation
and regeneration.

Isaac Wistar describes finding, in 1849 or 1850, the haunting scene of a skeleton of a grizzly-in-repose, down in the "deep, gloomy bottom" of a dark and secluded canyon, lying in a manner in which other bears have also been found across the West—the death-pose (p. 75):

"The position was one not unfrequently assumed by the animal in death, that is, prone of all fours, the head resting on the forepaws, something like a dog which waits impatiently for his master."

Do I wait for, and dream of, a day when grizzlies could hole up in, or wander unmolested—which is to say, to not be thrust with swords for sport, or used as target practice by Spanish missionaries—the wild places in California? Certainly. It seems to me that our first job is to hang on to the wild cores that we've got—to reverse the downward trend in ecosystems such as Yellowstone, and the Yaak, the Bitterroots, the Salmon-Selway, and elsewhere.

With the grizzly gone—for now—from the state of California, at least let the image on the state's flag be honored by a renewed and reinvigorated commitment to protecting country, good country, in California and elsewhere in the grizzly's range. None of us like to be reminded of it but we have debts, huge debts, in almost every direction we look. Our debts to the environment and wild nature are not our only obligations to the world, here at century's end, but they are appallingly immediate, and grow ever more staggering and perilous to both ourselves and the coming generations with each passing year of the continued diminishment, near unto an absence, of the wild.

There is no law that says our stories must turn to loss and destruction. The soil is still rich, still desires the legends and stories of creation and rebirth. We need only to begin digging.

RICK BASS
*Yaak Valley 1996*

# CALIFORNIA GRIZZLY

The grizzly bear is the largest of the carnivora and one of the most formidable animals in the world. He grows to a height of four feet and a length of seven or eight and attains a weight of two thousand pounds. His hair is coarse, usually gray but sometimes brownish and dark brown on the legs. His strength is tremendous, being able to knock down a bull or carry off a horse. It is difficult to kill him and, even when pierced to the heart with a rifle-ball, he often lives for some time. It is seldom that he attacks man; but when wounded he is ferocious. He usually dens in the chaparral and rarely attempts to climb a tree. He lives chiefly upon grass, clover, berries, acorns and roots, but is fond of beef, veal, venison and especially pork, when he can get them. He has been known to lie on his back in a grassy plain and throw his huge legs in the air to attract too inquisitive cattle close enough to enable him to seize them and, having seized a prize, to carry it off bodily. He will break into a corral and carry off a calf, sheep or pig as a cat carries off a mouse. Cases have been known of sagacious old bears that have decimated herds and flocks and have been the terror of neighborhoods for years.

In early times grizzly bears were very plentiful all over the country and did great damage to the cattle and gardens of the first settlers. In 1799 the troops of Purísima made a regular campaign against the bears in that region. In July, 1801, Raymundo Carrillo wrote from Monterey that the vaqueros in that neighborhood had within the year killed thirty-eight bears, but that the depredations by others continued unabated; and he proposed an ambuscade by the troops at a certain place where the carcasses of a few old mares should be exposed. Notwithstanding repeated expeditions against them, bears continued to be plentiful down to the time when the American hunters and trappers came to the country; but they then began to be thinned out. At the time of the American occupation there were still many, but, as the country filled up, they became scarcer and are now only found in remote places. As a cub the grizzly is very clumsy but at the same time very playful. If taken at an early age, he can be easily tamed and becomes kind and affectionate. In 1855, a hunter named Adams brought to San Francisco and for several years exhibited, among others, two old grizzlies, which he had tamed and trained to accompany him in his hunting excursions and to pack his blankets on their backs.

T. H. HITTELL, History of California (1898), 2 : 560–561

# Introduction

Bears are inhabitants of the Northern Hemisphere—Europe, Asia, North America, and the Arctic islands—except for the spectacled bear (*Tremarctos*), in the Andes of South America (see p. 2). They comprise a rather homogeneous group that is characterized by medium to large size, stocky form, very short tail, somewhat lengthened muzzle, and stout legs with five toes on each broad foot. The animals are plantigrade, like man, walking on the entire foot and commonly leaving a track showing both the large palm and sole pads and, in some, the claws. In the skeleton the bones of the forearm (radius, ulna) and of the lower leg (tibia, fibula) are separate—again a human parallel. In consequence, bears have considerable ability to rotate the forearm, which makes for skill in hunting, digging, or manipulating food, and facilitates climbing by some. The teeth of bears are sturdy. The molars, or "cheek teeth," in the hinder part of both upper and lower jaws have broad crowns and are nearly flat, with low tubercles to facilitate crushing; they lack the "carnassial" shearing mechanism of other carnivores in that the last upper premolar and first lower molar do not have cutting edges.

The bears are usually classified as comprising the family Ursidae, which separates them from other groups within the order Carnivora, such as the dogs (Canidae), the raccoons and pandas (Procyonidae), and the cats (Felidae). The bears are considered to have differentiated from the dogs late in geological time, in the mid-Miocene or early Pliocene. Indeed, some zoölogists believe that the bears and dogs should be included in the same family (Simpson, 1945 : 224–225).

In the opinions of some mammalogists, the living bears rep-

1

resent five to seven genera, although one recent student (Erdbrink, 1953) puts all existing bears in the one genus, *Ursus*. We
here employ the usual generic designations for the several kinds
of bears, except the Asiatic Black Bear.

The Polar Bear, *Thalarctos maritimus* (Phipps), of the Arctic regions has a small, slender head, small ears, a long neck, and
an elongate body. The very dense fur is uniformly yellowish
white and the black soles of the feet are partly haired. The bear
lives in the northernmost parts of Eurasia and North America, in
some places on the mainland but principally oñ the ice, whence
it occasionally is carried south on floating cakes of ice or icebergs
to Hudson Bay, Ungava Bay, and the coasts of Greenland. The
"ice bear," or "water bear," dives readily and swims long distances
in the chilly polar waters.

The Sloth Bear, *Melursus ursinus* (Shaw), of southern Asia
is small as bears go, weighing only about three hundred pounds.
It has a shaggy coarse black coat, gray tip on the snout, an inverted white horseshoe mark on the chest, and white claws. It is
native from Ceylon and southern India north toward the Central
Provinces, Bihar, Bengal, Assam, and possibly Darjeeling. The
sloth bear hunts partly by day and does not hibernate. Its food
consists almost entirely of wild fruits and insects.

A second native of southeastern Asia is the even smaller Ma-
layan Sun Bear, *Helarctos malayanus* (Raffles). This bear inhabits
the Malay Peninsula, Thailand, Indochina, Burma, possibly parts
of southern China, and Borneo and Sumatra. The animal is a forest dweller, climbs well, and is primarily a fruit-eater.

The Spectacled Bear, *Tremarctos ornatus* (F. Cuvier), inhabits forests from the base of the Andes to altitudes of about
9,800 feet. It occurs from western Venezuela across Colombia and
Ecuador to Peru and Bolivia, lives in remote places, and is rather
rare. The pelage is long, dense, and coal black, and each eye is
more or less encircled by a yellow or white line. The animal is
strictly vegetarian, eating palm fruits and young leaves. It makes
beds of sticks and leaves in trees. (Cabrera y Yepes, 1940 : 141–
143.)

The Asiatic (or Himalayan) Black Bear is commonly designated as *Selenarctos thibetanus* (G. Cuvier) or as *S. torquatus*

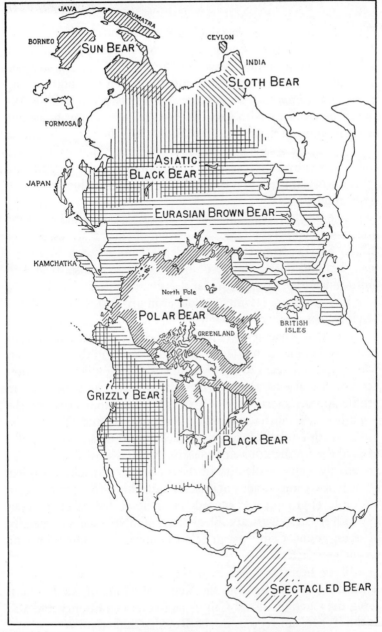

Approximate original distribution of living bears.

Grizzly and brown bears, ≡ ; black bears, |||.

(Wagner) and thus is placed in a different genus from the North American Black Bear. G. M. Allen (1938a : 330–332), however, in his monograph on mammals of China and Mongolia, weighed the evidence for this separation and concluded that both belong in the genus *Euarctos*. We accept Allen's conclusions. The Asiatic species is black with a narrow inverted crescent or horseshoe mark of white on the chest. The chin is white and the nose reddish brown. This bear ranges from eastern Siberia, Manchuria, Japan, and Formosa over most of China to Indochina and Thailand; and from Burma and Assam west to Nepal, Kashmir, Baluchistan, and Afghanistan.

The Black Bear of North America is distinguished from the grizzly–brown bear complex by certain obvious features, such as shorter muzzle; front claws of moderate length, not much longer than the hind ones; hind feet relatively short; fur short and of uniform length; last upper molar decisively shorter (less than 1¼ inch in black bear; 1¼ inch or longer in grizzly). In general, the black bear is smaller, although occasional old males are as large or larger than small adult female grizzlies. The grizzly has a conspicuous shoulder hump (figs. 1, 2) not present in the black bear, which makes it seem to stand higher at the shoulder. The forehead of the black bear is slightly more raised, and the head in profile appears shorter and less slender. The front claws of the black bear are shorter, narrower, more curved, and darker. The canine teeth (tusks) are larger at the base and taper more rapidly toward the tip than those of the grizzly. Tracks of bears are not necessarily a clue to identity of the species. The black bear is less likely to leave impressions of the claws; any hind-foot print in firm soil on level ground that exceeds 11 inches in length or 6 inches in width is that of a grizzly. Black bears climb regularly and easily in trees; young California grizzlies reportedly climbed, but the adults were strictly terrestrial.

Black bears are native to the forested regions of North America. They occur from the Kenai Peninsula of Alaska south along the Pacific Coast in California to Sonoma County and also in the Tehachapi region; in the Rockies they were represented as far south as the Sierra Madre in Chihuahua, Mexico; and in eastern North America from Labrador and Nova Scotia south to Florida and Louisiana. In recent years, black bears have been introduced into some regions of California where they were not native.

Most mammalogists consider that the North American population of black bears represents one wide-ranging species, *Euarctos americanus* (Pallas), divided into a number of subspecies. Others characterize certain of the regional forms as distinct species: *floridanus* (Florida), *luteolus* (Louisiana), *machetes* (Chihuahua). The "black" bear (*Euarctos americanus* and its subspecies or closely related species) has various color phases, the most common being black and cinnamon although some are literally yellow.[1]

Last but most outstanding among all bears are the Grizzly Bear and the Brown Bear that originally were native in much of the Northern Hemisphere. They are distinguished for their size, their shaggy coats, the long, curved front claws, and their terrestrial habits. Largest and most powerful of the bears, they are the peerless "big game" of rugged hunters who have contested with them for centuries. The largest are found on Kodiak and Afognak islands and the Alaska Peninsula. Of these Rausch (1953 : 97) says: ". . . the greatest reliable weight of which I know is 1,200 lbs. Weights exceeding 1,500 lbs. are frequently reported, but these are estimates . . ."

The Old World Brown Bear is considered by Ellerman and Morrison-Scott (1951 : 236–238) to be a single wide-ranging species, *Ursus arctos,* named formally by Linnaeus in 1758, and having six subspecies in Eurasia. Its original distribution reached from Spain, Greece, and northern Africa to Sweden and Norway, over most of the U.S.S.R. (in summer far into the tundra), and across Asia to Japan.

There is general agreement that the brown and the grizzly bears all belong to one genus, *Ursus;* but there are differences of opinion concerning the relationships between the two kinds, and the number of species or varieties represented. Earlier it was thought that two distinct kinds existed, brown and grizzly, and

---

[1] These differences are genetic, probably simple Mendelian characters, since litters are reported containing both black and cinnamon cubs; and in Alaska both "glacier" (blue) and typical black bear cubs have been seen in the same litter (Nelson, 1918 : 437).

that the brown bears of southeastern Alaska were more closely related to the Eurasian brown bear, *Ursus arctos*, than the grizzly. Even Merriam (1918 : 12–13) stated that "The typical brown bears . . . [have more uniform color] with less of the surface grizzling due to admixture of pale-tipped hairs; the claws are shorter, more curved, darker, and scurfy instead of smooth; the skull is more massive . . . But these are average differences, not one of which holds true throughout the group."

Anthony (1928 : 83–84) recognized seven kinds of brown bears ranging from Unimak Island in the Alaska Peninsula northeasterly to Baranof Island, and one from north of Great Slave Lake. Another competent authority, G. M. Allen (1938*b* : 266), stated: "There can be no doubt that [the bears from western China and eastern Tibet, of the *Ursus arctos* group], as Miller suggests, are closely related to the North American grizzly bears, which doubtless represent the Brown Bears in the New World." Indeed, Allen (1938*a* : 328) uses the term "Black Grizzly" for *Ursus arctos lasiotus* of northern China and Mongolia.

A European biologist, D. P. Erdbrink (1953 : 339), who studied fossil and recent bears of the Old World, working solely from the literature and museum material in The Netherlands, thought the case for the grizzly as a distinct species or subspecies ill-founded and made the novel suggestion "to recognize a 'grizzly mutation' occurring now and then among the members of the species *U. arctos*, and more frequent in some geographical regions than others." In his opinion, this "mutation" is characterized by curved, nontapering, bone-colored claws; fur often containing whitish or yellowish hairs among the darker ones and producing the grizzled appearance; the skull usually broad in front; and a tendency to develop accessory cusps on the last premolars and molars. His map (p. 536) shows approximate "areas where the grizzly mutation chiefly occurs" extending in North America from the Canadian prairies into Mexico (but excluding much of the Pacific Coast States) and in the Old World from central China to eastern Turkey!

Even in the eighteenth century the grizzly was noted as being different from other kinds of bears and was commonly called the "red bear," or "grizzle bear," to distinguish it from the black bear; but it was not characterized scientifically until early in the nineteenth century. The second edition of William Guthrie's "Geography"[2] published at Philadelphia in 1815 contains a section on "Zoology of North America," by George Ord, in which scientific names were first applied to several animals of this continent. The grizzly bear was designated as *Ursus horribilis* (p. 291). Curiously, Ord provided no description of the animal; instead he quoted (pp. 299–300) from H. M. Brackenridge (1814 : 55–56), *Views of Louisiana* (that is, the then vast Louisiana Territory), which had borrowed from the first account of the Lewis and Clark expedition, published in 1814 (see Coues, 1893 : 297, 298, Lewis and Clark Journals). The only precise information included by Ord is of a bear killed "near the Porcupine river." This bear, the largest seen until that time by Lewis and Clark, was shot on May 5, 1805, near old Fort Charles at the mouth of Little Dry or Lackwater Creek, in what became Dawson County in northeastern Montana (Rhoads, 1894 : 28). This place, in zoölogical parlance, is the type locality for the species; it is restated by G. S. Miller (1924 : 92) as being "Missouri River, a little above the mouth of Poplar River." The grizzly measured as follows: from nose to extremity of hind feet, 8 ft. 7½ in.; girth of neck, 3 ft. 11 in.; girth of body near forelegs, 5 ft. 10½ in.; girth of forelegs at middle, 1 ft. 11 in.; length of [front] claws, 4⅜ in. Its weight was estimated to be between five and six hundred pounds (Ord, in Guthrie, 1815 : 300).

Another name, *Ursus ferox* Rafinesque 1817, was used by some early writers, but is a *nomen nudum*, without description; being later in date, it is a synonym for Ord's name.

The earliest scientific description of the grizzly in California is that by Spencer F. Baird, first Secretary of the Smithsonian Institution, in his monograph (1857) on mammals collected by the Pacific Railroad surveys.[3] He had five skins and seven skulls from

---

[2] This exceedingly rare book is of interest to zoölogists because of the scientific names applied by Ord. The zoölogical section was reprinted by Samuel N. Rhoads (1894).

[3] In March, 1853, Congress appropriated $150,000 for the survey of possible routes for a railroad from the Mississippi River to the Pacific Ocean. Eight or more parties were organized to explore routes between the 32d and 47th parallels. Specimens collected by members of these parties were forwarded to the Smithsonian Institution in Washington.

California, together with eight skins and five skulls from other parts of the western United States. His characterization of the grizzly as a single species, *Ursus horribilis* Ord, was as follows (p. 219):

Size very large. Tail shorter than ears. Hair coarse, darkest near the base, with light tips. An erect mane between the shoulders. Feet very large; fore claws twice as long as the hinder ones. A dark dorsal stripe from occiput to tail, and another lateral one on each side along the flanks, obscured and nearly concealed by the light tips; intervals between the stripes lighter. All the hairs on the body brownish yellow or hoary at tips. Region around ears dusky; legs nearly black. Muzzle pale, without a darker dorsal stripe.

A specimen from New Mexico was named by Baird *Ursus horribilis* var. *horriaeus*. In 1838 Swainson described *Ursus richardsoni* from the shores of the Arctic Ocean, and in 1903 D. G. Elliot characterized *Ursus hylodromus* from British Columbia. With these exceptions, all other names proposed for grizzlies were the work of one man, C. Hart Merriam (1855–1942).

The classification of grizzlies and brown bears in North America was a major interest of Dr. Merriam, who was Chief of the Bureau of Biological Survey, United States Department of Agriculture, from its establishment in 1885 until 1910, and thereafter continued his research, under a private endowment, for the remainder of his life. He collected large series of specimens of the small native mammals of North America and classified and named many new species and outlined their geographic distribution. His methods and points of view are the basis of much current mammalogical research in taxonomy and distribution.

About 1891 he turned to study the large bears, and shortly recognized eight species, five of which he described as new (Merriam, 1896 : 65–83). In 1914 he characterized thirty apparently new grizzly and brown bears in North America (Merriam, 1914 : 173–196), and in 1918 he published a "Review of the Grizzly and Brown Bears of North America" (Merriam, 1918), which included eighty-six kinds.[4] In this, most of the named forms were

---

[4] The entire series was subdivided into fourteen "groups," but these were not defined. The characterization of individual species and subspecies is not of a sort readily understood or interpreted by other persons. For his 1918 "Review," Merriam had assembled practically all skulls of grizzlies and brown bears in museums within the United States; in his 1914 paper (p. 173) he mentions "more than 500"

# Introduction

considered as full species, but a few were listed as subspecies. Dr. Merriam was long engaged in writing a book on bears, but the task was incomplete at his death and the volume has not been printed.

Merriam ascertained that in most species of bears the males are much larger than the females (*magister* of southern California being conspicuous in this respect), but in a few the difference is slight. It was his opinion that cranial (bony) characters are more permanent and of greater significance for classification than minor tooth characters. "The teeth are strongly modified by food and consequently in some cases present marked variations in the same group" (p. 13).

Merriam held to the belief that hybridization among animals in nature is rare. Because of this he wrote: "One of the unlooked-for results of the critical study of American bears is the discovery that the big bears, like mice and other small mammals, split up into a large number of forms whose ranges in some cases overlap so that three or more species may be found in the same region" (*ibid.*, p. 9).

Because of Dr. Merriam's long preoccupation with the bears, other mammalogists did little with them. When necessary to cite scientific names and geographic ranges they usually followed his 1918 paper, but most of them were doubtful whether the large bears were so finely and intricately subdivided. G. G. Simpson (1945 : 225) wrote that "C. H. Merriam distinguished about ninety species of North American bears alone, but he had a . . . unique conception of the character of a species, giving it less scope than most authors give a minor geographic race, not much more than an individual genetic family group. In such a system twin bear cubs could be of different species." H. E. Anthony in preparing his *Field Book of North American Mammals* made a synopsis for "every one of the 84 forms," but "in the final analysis" discarded it in favor of "a much briefer, more comprehensive treatment" (Anthony, 1928 : 80). He finally recognized eleven species of grizzlies and seven of brown bears.

---

then brought together. But there is no indication of the number of specimens of each form or of the actual total. Museums of the United States contain only 34 skulls ascribed to California, according to a survey by the present authors in 1952 (see App. A).

The limited number of specimens of grizzlies in museums, many with fragmentary or dubious data for locality and sex, precludes a study of these bears comparable to the monographic analysis possible for many other genera of North American mammals.

A recent investigation of big bears in Alaska by Rausch (1953 : 95–107) tends to clarify the situation by emphasizing the high degree of individual variation among these animals. In any one place "there is rarely size-uniformity in a grizzly population. The majority of animals may be of about the same size, but unusually large and unusually small individuals are seen" (*ibid.*, p. 97). "Maximum size in *Ursus arctos* is attained where an unlimited supply of high protein food is available" (*ibid.*, p. 98),[5] as in the salmon runs in southeastern Alaska. "The colour of grizzlies is highly variable and . . . has little or no taxonomic value. . . . Grizzly claws are very variable in shape, size, and colour"—white to nearly black (*ibid.*). The skulls are nonuniform in gross size, length–width ratio, profile, length of rostrum (nose), dimensions of the palate, form and length of mandible, size of teeth, and other features. A series of twenty-two from one region (Brooks Range) when arranged in presumed sequence of increasing age, as judged by over-all length and degree of tooth wear, shows no correlation in the change of parts; they tend to grow independently. Rausch has photographs and measurements of skulls of four old or aged males from among thirty-five bears killed in one region in the course of five years (Anaktuvuk Pass, central Brooks Range). "According to Merriam's system of bear classification, it would be acceptable to consider each of these four animals representative of a different species" (*ibid.*, p. 101), but "they probably comprised interbreeding members of a single population" ( *ibid.*, p. 99).

Because of the wide variation in skulls, bodily size, and coloration, Rausch concluded that all big bears in North America, "grizzly" or "brown," belong to one highly variable Palearctic species, *Ursus arctos* Linnaeus.[6] The stocks in all of continental

---

[5] Abundance of range livestock in Spanish and Mexican days in California may have had a similar effect on the size of bears.

[6] Couturier (1954), in a book issued after completion of our manuscript, also

North America, past or present, he ascribed to a single subspecies, *U. a. horribilis* Ord; those of the Alaska Peninsula to another, *U. a. gyas* Merriam; and those on Kodiak, Afognak, and Shuyak islands to a third, *U. a. middendorffi* Merriam.

Merriam described seven kinds of grizzlies in California: he assigned *californicus* (Monterey), *tularensis* (Fort Tejon), and *colusus* (Sacramento Valley) to the "horribilis" group that extended to Colorado, Montana, and Yakutat Bay, Alaska; *klamathensis* (Klamath River), *mendocinensis* (Mendocino County), and *magister* (southern California) to the "arizonae" group that had representatives from northern Mexico and New Mexico to Yukon Territory and southeastern Alaska; and *henshawi* (southern Sierra Nevada) to the small "horriaeus" group having two other species in Arizona and New Mexico. Grinnell (1933), in summarizing the distribution of California mammals, accepted all seven of Merriam's varieties, and these were also included in the work on California fur bearers (Grinnell, Dixon, and Linsdale, 1937).

The diversity of physical, climatic, and biological environments in California, such as the northwestern humid coast, the drier central and southern coastal regions, the Great Central Valley, and the Sierran foothill and mountain areas, are commonly reflected in subspecific differences among the native mammals, both large and small. The lines or areas of demarcation vary from one species or genus to another. It is possible that the grizzlies of California in the several regions indicated showed some small differences in average adult size, coat color, and skull characters; but when grizzlies were abundant and available there were no local mammalogists, and now there is not enough museum material. Whether California had one, seven, or some other number of varieties of grizzly, it is impossible to determine.

It is reasonable to think that the grizzly stock in California was different in some degree from that in the Rockies and to the north in Canada and Alaska. In 1896 Merriam characterized a bear from this state by the name *californicus*, based on a speci-

concludes that all the brown and grizzly bears constitute one species, *Ursus arctos*. He would abandon all attempts to recognize subspecies and merely refer to local groups as "populations."

men from Monterey. We may therefore use the name *Ursus arctos californicus* Merriam as appropriately designating the once common and widespread native of this western state and linking to it the name of the most distinguished student of grizzlies.

The grizzly bear originally ranged over an enormous territory in North America (p. 3). A map by the United States Fish and Wildlife Service based on specimens in the United States National Museum and records in the literature show the grizzly as ranging north to the Arctic Coast of Alaska and Canada (east to Longitude 100° W.) and thence down to southern Durango, Mexico.

It once occupied three limited areas in the north-central, northeastern, and southeastern parts of Washington (Dalquest, 1948 : 177); and in Oregon was in the Coast Range (but not on the coast itself), the Willamette Valley, the Cascade Range, and some eastern parts of the state (Bailey, 1936 : 324). Some early explorers traveling southward in Oregon, such as the Wilkes party of 1841, met none until they reached the Umpqua River (Cassin, 1858 : 13). Grizzlies reportedly were abundant in the Rogue River Valley and up into the region that is now Crater Lake National Park (Wright *et al.*, 1933 : 121–122). In California there were grizzlies from about Siskiyou and Humboldt counties to San Diego County. None lived in the Great Basin. The easternmost extent of their range is uncertain. In Texas one was recorded in the Davis Mountains, Jeff Davis County (Bailey, 1905 : 192). For Kansas there are three records: Castle Rock, Gove County, and Smoky Hill River, Logan County, in the west, and Council Grove, Morris County, in the central part of the state (Cockrum, 1952 : 238–239). We have no record of grizzly bears in Nebraska. Only two are recorded in South Dakota—in the extreme western part—but there are records of several in North Dakota from the eastern border westward. In Canada the easternmost records are from near Calgary, Alberta, northeasterly to the Arctic coast near Simpson Strait, Mackenzie District (Anderson, 1946).

The steady pressure of man on grizzlies has contracted their territory and numbers to a minute fragment of the original. A summary of information about grizzlies in 1922 led Merriam to conclude that a few remained in Washington, although the last two records for Oregon were in 1894. The last grizzly in Texas was killed in 1890. A few were believed to exist in Arizona and New Mexico and in Utah in 1922, and Colorado still had a few; Idaho, Wyoming, and Montana had the most, many in the national parks and lesser numbers outside (Merriam, 1922). By 1941 their range in the United States had shrunk to small areas in Idaho, Montana, and Wyoming adjacent to and including national parks; their numbers in that year were roughly estimated as follows: national forests, 775; Indian reservations, 50; national parks and monuments, 420—a total of 1,345 (Jackson, 1944*a*, 1944*b*). In 1951 it was estimated that three national parks in the northern Rockies had 310 grizzlies: Grand Teton, 10; Yellowstone, 180; Glacier, 120 (Cahalane, 1952). In 1952 the estimated total was 908, of which 737 were in fourteen Montana counties (Hickie, 1952; letter, Montana Dept. of Fish and Game, May 3, 1954). These figures are not an accurate census but are based on collected opinions of forest, game, and park officials.

Outside the United States grizzlies are still present in British Columbia and other parts of western Canada and are common in much of Alaska. Some persist in parts of northern Mexico, but it is not certain that they occupy all the area indicated in a recent map by Couturier (1954 : 211–212).

Throughout its range, the grizzly, as an animal type, experienced a wide diversity of climatic and biological environments. To the north, grizzlies inhabitating the Barren Grounds and interior Alaska had short summers of continuous daylight alternating with long, severe, dark winters and extremely low temperatures. By contrast, the bears in the Great Central Valley of California experienced mild, rainy winters, only occasional frosts, and long, hot, dry summers. Between these extremes still others lived in the intermediate climatic regimens—in the Rocky Mountains, on the western edge of the Great Plains, in the moderate altitudes of southern Arizona, and in the northwestern coastal region of California.

Coronado was probably the first among the early explorers to see grizzly bears. In 1540 he marched from the City of Mexico to the Seven Cities of Cibola (the Zuni region of west-central

New Mexico) and on to the buffalo plains of Texas and Kansas. It is the opinion of Seton (1929 : 13) that Coronado "certainly saw many grizzly-bears," but the Spaniard's account merely states that the natives had "many animals—bears, tigers [jaguars?], lions," and so forth, and "the paws of bears." His lieutenant, Pedro de Castañeda, wrote, "There are many bears in this province . . ." (Winship, 1896 : 569–570, 518).

Baron Lahontan, who traveled in Canada from 1683 to 1691, listed "reddish bears" from the southern part of that country. "The reddish bears are mischievious creatures," he wrote, "for they fall fiercely on the huntsmen, whereas the black bears fly from them. The former sort are less [abundant?], but more nimble than the latter" (Lahontan, 1703, 1 : 234; in Pinkerton, 1812, 13 : 350, 351). It is not clear whether he actually saw grizzlies.

Edward Umfreville, who wintered on the Saskatchewan and at Cumberland House of the Hudson's Bay Company in Canada from 1784 to 1787, wrote: "Bears are of three kinds: the Black, the red, and the Grizzle bear" (1790 : 167–168). He mentions the savage nature of the "grizzle" and red bears and the number of maimed Indians who had been attacked by them. Samuel Hearne was possibly the first white explorer actually to see grizzlies in Canada. In July, 1771, he "saw the skin of an enormous grizzled bear at the tents of the Esquimaux at the Copper[mine] River" and said that "many of them are said to breed not very remote from that part." Hearne's account, however, was not published until twenty-four years later (1795 : 371–372).

Unnoticed by zoölogical writers, however, is another early record, and the first for California, which antedates all except that of Coronado. On a voyage of exploration, Sebastian Vizcaíno stopped at the site of Monterey from December 16, 1602, until January 3, 1603. While he was there, bears came down at night to feed on a whale carcass stranded on the beach (Ascensión, 1611; see Bancroft, 1884 : xxix, 102; Wagner, 1929 : 247). (See our chap. 5.) These animals could only have been grizzlies, since black bears were not native there. The reports of the Spanish author had no general circulation, however; those of the Canadian explorers in the eighteenth century were the first to announce grizzlies to the world at large.

# I  *The Record of Grizzlies in California*

## PREHISTORIC TIMES

In the eons of prehistoric time the form and character of the area now known as California changed again and again. As the land was successively elevated and eroded in diverse patterns the sea sometimes was pushed out to the west and at other times penetrated the coastal region. Not until late in Tertiary time—yesterday in the long calendar of earth history—did the central Pacific Coast of North America begin to assume its modern conformation. The Pleistocene, just before the Recent period in which we live, was a time of repeated glaciations in the highlands; the last glacial scars still burnish granite masses in the Sierra Nevada, and moraines in mountain valleys mark the extent of the ice. The fossil record of California's Pleistocene testifies that many of the present-day mammals and birds then inhabited the lowlands, along with others now extinct, such as the giant ground sloth, camel, horse, elephant, mastodon, and the powerful saber-toothed cat.

A cavalcade of monster beasts sought out this land, occupied it for a time, then faded away or was rapidly extinguished by man. Among the northern creatures were the bears, well adapted to survive the cold climate. Even these retreated before the ice that finally covered the center of the continent. An early California bear was a short-faced brute (*Tremarctotherium*) like the spectacled bear still living in the northern Andes. His remains were found in the caves of Shasta County . . . Bones of a large short-faced bear were [present but] scarce in the tar pits of Rancho La Brea, and remains of a huge

15

one were found in Alameda County in the Irvington early Pleistocene. (Camp, 1952 : 64.)

None of the fossil bears is definitely known to be ancestral
to any of the modern species in North America; they are mentioned here merely to show that bears of one kind or another
have been part of the California fauna for up to a million years—
the estimated duration of the Pleistocene.

Somewhere during the counterplay of natural forces and
climates that pushed forests and other vegetation first south and
then north with the alternate ice ages and interglacial periods,
the climbing black bear and the essentially terrestrial grizzly came
to inhabit the California landscape. Paleontology gives no clear
indication of how these two kinds arrived here. In all probability
grizzlies were present in western North America before races of
men migrated across the North Pacific land bridge in the vicinity
of Bering Sea and invaded the Americas.

## MODERN TIMES

In the Recent period, there were grizzlies from northern
Alaska and the Barren Grounds of Canada south to the western
Great Plains, down the Rockies to beyond the Mexican bound-
ary, and south to southern California or slightly beyond (p. 3).

The role of grizzlies in primeval California is unknown.
None of the early naturalists attempted to describe the "balance
of nature" that had existed in this region before the advent of
white men with firearms. The grizzlies were big and, from all
evidence, exceptionally abundant. In the scheme of animal classification the grizzly and other bears are technically "carnivores."
Actually they are omnivorous, eating any sort of animal food
from ants and rodents to whale carrion and also a wide diversity
of plant materials.

The undisturbed California landscape of the lowlands was
tenanted by elk, antelope, and deer; by ground squirrels, pocket
gophers, hares, and rabbits; by many kinds of ground-dwelling
birds, including an abundance of waterfowl in the extensive
marshy areas; and by salmon and other fishes in streams and
ponds. The ocean beaches, at times, had carcasses of whales,
seals, and sea lions. The grizzlies probably did not kill any of

the larger hoofed mammals, as a rule, although they very likely ate maimed or diseased individuals and may have expropriated some slain by the more agile mountain lions. Injured and sick waterfowl could be had in the marshlands.

The flora included many bulbous plants, clovers, and grasses. Several of the foothill shrubs produced berries in season—the elderberry, manzanita, blackberry, and others. The many species of oaks yielded acorns, sometimes abundantly and over a period of several months. All these plants and animals and others unmentioned were food for the grizzlies.

We think that the grizzly must have been a dominant element in the original native biota of California—it was usually avoided by the Indians, and because of its size, prowess, and temperament it could preëmpt any available food before other large mammals. Its adaptability in diet was not exceeded by that of any other local mammal and was paralleled by few. It was a big, resourceful beast. The energy needs of this active mammal weighing up to a thousand pounds or more were obviously great. Its numbers multiplied by its average daily metabolic requirement must have made the grizzly an outstanding factor in the total food consumption by mammals.

Before the invasion of California by white men there was space for plenty of grizzlies along with the rather sparse population of Indians (133,000 estimated by Kroeber, 1925 : 882). Both bears and Indians occupied the land, though they killed one another. With establishment of the mission herds, some of the bears found it easier to kill the cattle and other livestock than to hunt wild food. Early American travelers had some contact with the big animals, but with the gold rush the battle between man and bear became intense. Bear meat was food, and bear rugs served for beds. What was more important, however, miners and early farmers could not tolerate so large a carnivore, peaceful though it apparently was when not surprised or injured. Because of the size, temperament, and numbers of the grizzly, the settlers found the bear a prime element of concern in their use of the California valleys and hills for farms, homes, and communities. Bears wandered into the outskirts of the new settlements, even to Mission Dolores in San Francisco in 1850. They were a hazard

to the travels and daily activities of the miners; they killed livestock and disturbed peaceful farmers; and for big game hunters
they provided plenty of surprise and excitement—and also injury
and death. The improving quality of rifles and other weapons
gave the whites an ever-increasing advantage.

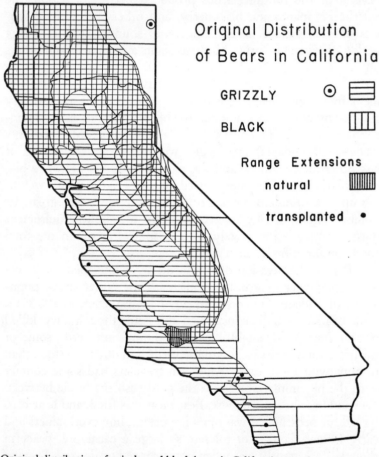

Original distribution of grizzly and black bears in California, together with range
extension and transplants of black bear.

## ORIGINAL DISTRIBUTION

Most of California was formerly the domain of grizzlies.
They probably ranged continuously from Oregon southward;

but records of any in the vicinity of the Oregon-California boundary are lacking, and none has been found for Del Norte County. There were "great numbers" in the Coast Range between the Russian and Eel rivers, on the "Bald Hills" between Humboldt Bay and Klamath River, and in the mountains between that river and the Trinity (Suckley and Gibbs, 1860 : 119–120). In Humboldt County, besides those in the Bald Hills (Wistar, 1937 : 83) there were some in the Mattole Valley (Grinnell *et al.*, 1937 : 70) and in other places between those localities. For Siskiyou County there are records at Fort Jones in Scott Valley (Anon., 1949 : 211; Wistar, 1937 : 262) and on Goose Nest Mountain about twenty miles north of Mount Shasta (Merriam, 1899 : 107). Some early travelers made no mention of grizzlies until they were near the Sacramento River: John Work noted them at Cow Creek (near what is now Anderson, Shasta County) in 1832 (Maloney, 1943 : 209), and P. B. Reading saw one at the junction of Pit River and the Sacramento in 1843 (O'Brien, 1951 : 135).

Newberry (1857 : 48) encountered some grizzly sign daily while crossing in 1855 from the Pit River drainage to Klamath Lake. The single mention of the grizzly in Modoc County is that of Symons (1878 : 1538); he wrote that at the head of Pine Creek in the Warner Mountains "grizzly bears and bear signs were seen" in mid-August, 1877. About Shingletown and McCumber's Flat, northeast of Fort Reading (which was five miles northeast of Anderson, Shasta County), and around the base of Lassen's Butte (now Lassen Peak) they were very numerous in 1855 (Newberry, 1857 : 47–48). There are records of an encounter with one near Susanville (Fairfield, 1916 : 237); of a man killed by a grizzly near the head of Sierra Valley in December, 1874 (N 85); and another killed at Independence Lake, Nevada County, in the same month (N 84). A large grizzly was killed just west of Donner Summit, Nevada County, on August 23, 1849 (Wistar, 1937 : 115), and another by a member of the Donner party in 1846 in the same region (Thornton, 1855, 2 : 124–125). Still another author (McCauley, 1910 : 5) mentions grizzlies as numerous "in a little valley near the summit of the Sierras in 1850." Grizzly Adams (Hittell, 1860 : 295) encountered one in 1854 on the east slope "as soon as we camped under the Sierra" (presumably near the eastern end of the Sonora Pass road).

No records have been found for Owens Valley, and none for the highest eastern parts of the southern Sierra Nevada. For the Yosemite region the easternmost records are those of Crescent Lake, Madera County, some ten miles east of Wawona "about 1895" at 8,500 feet elevation (Grinnell and Storer, 1924 : 70), and the capture of "Ben Franklin" in 1854 by Grizzly Adams in an area vaguely southeast of Yosemite Valley (Hittell, 1860 : 207 ff.). On the South Fork of the Kings River, about 7,500 feet above sea level, a female with two cubs was seen on June 28, 1864, by William H. Brewer (Farquhar, 1930 : 523). The last specimen taken and the last substantiated sight records were from Sequoia National Park and vicinity. Grizzlies also were present in the lower altitudes at the southern end of the Sierra Nevada—at Havilah, Kern County (USNM 15671), for instance, and at Fort Tejon (USNM 3536–3538).

In southern California, grizzlies ranged seemingly to the edge of the deserts. They were recorded on the desert side of Cajon Pass (McAllister, 1919 : 172), in Bear Valley in the San Bernardino Mountains (Wilson, in Cleland, 1929 : 383), in the San Jacinto range (Grinnell and Swarth, 1913 : 375), at Palomar Mountain (Bell, 1930 : 108), and in Pine and Cuyamaca valleys, San Diego County (Abbott, 1935 : 151).

For Lower California there is a single report, in 1828, about seventy miles south of the Mexican Boundary near Santa Catarina Mission in the Sierra Juarez; this was in Pattie's narrative (Thwaites, 1905 : 221).

West of the limits indicated, grizzlies inhabited all the coast ranges and valleys but with decreasing numbers in the denser forest areas of the extreme northwest; they lived on the plains and in the hills and mountains of southern California; they were in most parts of the Sacramento and San Joaquin valleys and on the western slopes of the Sierra Nevada. They were numerous at the lower altitudes but some ranged up to the crest and slightly beyond.

Both black bears (*Euarctos americanus*) and grizzlies were

native to California, and under original conditions their ranges overlapped in some degree. There are a few early remarks about the scarcity or absence of black bears where grizzlies held sway. Newberry (1857 : 49) wrote that the black bear extended its range from Oregon "into California only near the coast. Near Fort Jones [Siskiyou County] it has occasionally been killed, but south of that point it is replaced by the grizzly." Brewer (Farquhar, 1930 : 523), however, met the two species together high on the Kings River in 1864. It may be that the black bear has spread—certainly it has become abundant—since its larger relative disappeared.

The original range of the black bear in California, so far as known (p. 18), was from the Oregon line south in the Coast Range to Bodega, Sonoma County, and in the Cascade–Sierra Nevada from Siskiyou County southward through the Tehachapi Mountains to Tejon Ranch (Grinnell, 1933 : 96). There were and are no records of naturally occurring black bears for the coast ranges and counties south of the Golden Gate or for southern California. This situation means that during early days any mention of "bear" south of the limits indicated pertained to grizzlies, whether so specified or not.

Originally the region west of Tejon Ranch, Kern County, had only grizzlies; but as these were exterminated black bears spread into parts of Santa Barbara County and even into San Luis Obispo County (Grinnell *et al.*, 1937 : 104–105, fig. 24).[1]

## ECOLOGICAL DISTRIBUTION

Grizzlies in early California were highly adaptable to their various environments, and this adaptability may have been a major factor in their attaining a high level of population. Over the relatively large area they then occupied, the climatic cycle, the physical terrain, the places available for shelter, the plant cover

---

[1] More recently, black bears have been planted in places where the species was not native. In 1933, some trapped on the floor of Yosemite Valley were released in southern California, as follows: Crystal Lake, Los Angeles County, 11; Bear Lake, 6, and Santa Ana Canyon, 10, both in San Bernardino County (Burghduff, 1935 : 83–84). One or two also were released in San Diego County (Abbott, 1935 : 149–151) about 1917-1919, and one was killed there in 1934. A few were released in Monterey County near the Big Sur some years ago but subsequently were killed out.

(fig. 5), and the various animals usable as food were of widely differing kinds.

In the northwestern part of the state where winters were mild but rainy and there was much dense forest, grizzlies were commonest along the river bottoms and in the sparsely forested oak-prairie regions. Those living to the northeast—on the "Modoc Plateau" beyond the summits of the Cascade Range and the Sierra Nevada—were exposed to a rigorous continental type of climate during the winter; but there were lava caves in which they could take shelter, and they ate berries and fruits of wild shrubs and trees.

Grizzlies in the Great Central Valley inhabited the dense growths of trees, vines, and cattails that bordered lowland rivers and creeks before white men cleared and reclaimed such areas for agriculture. These thickets were nurtured by the silt of overflow waters, and there were extensive marshes and many ponds resulting from spring freshets and from the work of beavers. The bears pastured on wild grasses and clovers during the spring and ate acorns from valley oaks in the autumn. Fish and other aquatic animals were available. Some bears dug their own dens in the valley flats, though some may have deserted the flatlands in summer to live in adjacent hills.

Throughout the foothill region the "elfin forest" or chaparral of many shrubby plant species clothed the slopes with a dense cover sometimes up to twelve feet in height (fig. 6). Beneath the interlocking branches, the grizzlies had trails and escape or resting places. The manzanita and some other chaparral plants afforded seasonal crops of berries, and on adjacent slopes several species of oaks provided acorns.

For the higher mountains there is little or nothing recorded in regard to the habitats used. We know that grizzlies lived there, because some were shot in summer and some when the mountains were blanketed with snow. We infer that dens among rocks or under large trees were more commonly available—and some were dug—to serve for sleeping, hibernation, and the rearing of cubs. Food supplies, on the whole, were less prolific. Meadows afforded grasses, clovers, bulbs, and roots; some shrubs of montane chaparral yielded berries; there were oaks at middle altitudes; and there was a fair supply of small rodents for the bears to eat.

In southern California, chaparral was widespread, but river-bottom environments were more restricted. Some grassy pasturage was to be had in both lowland and upland valleys. The absence of grizzlies on the southeastern deserts was obviously due to the lack of food for such large animals. No mammal bigger than a coyote could make a living there. The desert, for the most part, has no hoofed mammals and has no supplies of grass, clover, roots, or acorns such as grizzlies ate elsewhere.

Along the entire coast, grizzlies within forage distance of the ocean could rely on the more or less continuous supply of marine animals washed ashore, and could supplement or vary their diet with acorns or berries from the adjacent hills.

The black bear, although omnivorous in diet, seems less catholic in other environmental needs than its bigger relative. At present it is mainly an inhabitant of the coniferous forest region at middle elevations in the Sierra Nevada; it has been recorded at 1,200 to 8,500 feet. In the northern Coast Ranges it similarly lives amid evergreen trees but at lower altitudes. As Newberry (1857 : 47) said, "he is the bear of the forest, while the grizzly is the bear of the 'chaparral.' "

## POPULATION

There is no means of knowing the grizzly population in primeval days, since no California Indian diary or numerical record of animals has been discovered. The first hints come in diaries of the Spanish explorers. In September, 1769, near the site of San Luis Obispo, the Portolá expedition saw "troops of bears" and found the land plowed up where the animals had been grubbing for roots (Teggart, 1911 : 59–61). In subsequent years the Spaniards met bears singly or in groups and, as we have indicated elsewhere (see our chap. 5), there is evidence to show that grizzly numbers were substantially augmented by the food supply afforded in the livestock of the mission herds. During the second quarter of the nineteenth century when several exploring parties visited the California coast and as Americans began to enter and settle, the bear population was large.

Ludovik Choris, who was with the 1816 Kotzebue party at

San Francisco, said "bears are very plentiful" (Garnett, 1913 : 15). In January, 1827, Duhaut-Cilly wrote that "bears are very common in the environs; and without going farther than five or six leagues from San Francisco, they are often seen in herds" (Carter, 1929 : 145). George C. Yount was among the first American pioneers in California, arriving in February, 1831. Of grizzlies in the Napa Valley (where the town Yountville carries his name) he said: "they were every where—upon the plains, in the valleys, and on the mountains . . . so that I have often killed as many as five or six in one day, and it was not unusual to see fifty or sixty within the twenty-four hours" (Day, 1859 : 1). When Don Agustín Janssens rode between San Marcos and Santa Ynez in 1834 he said, "All the way we saw bears, for it was winter and . . . the acorns were dropping" (Ellison and Price, 1953 : 25). John Bidwell, in the Sacramento Valley in 1841, saw sixteen in one drove and said that "grizzly bear were almost an hourly sight, in the vicinity of the streams, and it was not uncommon to see thirty to forty a day" (Bidwell, 1897 : 75–76, 73). Even in Humboldt County, where much land is forested and unfavorable to the species, there is early mention of nine seen in one place, and again of "40 bears in sight at once from a high point in the Mattole country," where a great extent of open land could be seen; all or most of these presumably were grizzlies, since black bears then were uncommon (Grinnell *et al.*, 1937 : 69, 70).

Two men counted "eighteen grizzlies in one afternoon in the fall . . . under the oaks eating acorns" in Cholame Valley, San Luis Obispo County, about 1840 (*ibid.*, p. 88). Several writers commented on their numbers at Fort Tejon in Kern County. John Xántus wrote from there on June 5, 1857, to Spencer F. Baird, Secretary of the Smithsonian Institution: "We have here grizlys in great abundance, they are really a nuisance, you cannot walk out half a mile, without meeting some of them, and as they just now have their clubs [cubs], they are extremely ferocious so, I was already twice driven on a tree, and close by to the fort" (Madden, 1949 : 83).

Benjamin D. Wilson, later the first mayor of Los Angeles, led a punitive expedition against the Mohave Indians in July or August, 1845. He took twenty-two vaqueros into the San Bernardino Mountains and found a valley where many bears were eating clover. In pairs his men roped eleven grizzlies, and a few days later they repeated the performance. "The whole Lake and swamp seemed alive with bear." He therefore named the area Bear Valley. (Cleland, 1929 : 383, 387.) Ramón Ortega, *mayor-domo* of Rancho Sespe, asserted that he once counted a hundred of these huge bears between Mission San Buenaventura and the ranch! (Cleland, 1940 : 105).

In some favorable places grizzlies persisted in numbers and kept their gregarious habits long after California became a state. Groups of ten or twelve were seen on the beach at the Little Sur River in Monterey County in 1870 (Isobel Meadows, in Anne B. Fisher notes). Nine were seen in the head of Matilija Valley, Ventura County, on a day in September, 1882 (N 93); and during June, 1885, in the foothills southwest of Bakersfield fourteen were in sight at one place (Seton, 1929 : 20).

Further attest to the numbers of grizzlies is found in the reports of some early hunters. On June 30, 1823, a Mexican officer and ten soldiers killed 10 bears in Suisun Valley (Anon., 1860*a* : 115; Hittell, 1885, 1 : 497). John Work and a hunting party in the Sacramento Valley killed 45 between November, 1832, and May, 1833 (Maloney, 1943). In the area now San Luis Obispo County, George Nidever (1937 : 49–51, 53) killed 45 and injured others during 1837; he estimated having slain upwards of 200 in his early years in the state. William Gordon of Yolo County killed nearly 50 in one year in the 1840's (Gilbert, 1879 : 31); and Colin Preston, in the central coast region during the same period is said to have killed about 200 in a single year (Anon., 1857 : 819). When Frémont (1887 : 571) was in the Salinas Valley in September, 1846, he and his thirty-five men came on a large number of grizzlies in the oaks and on the ground; the party killed 12 and others escaped. Two years later, in the San Luis Obispo Valley, he reported killing 12 in one thicket (Frémont and Emory, 1849 : 27). Three hunters in the Tejon Pass region in 1854 are said to have killed 150 bears in less than a year (N 28).

A numerical estimate of the California grizzly population was once attempted by Joseph Grinnell (1938 : 75). Assuming one bear per "20 square miles of suitable territory" and one-third of the state occupied in that density, he offered a figure of 2,595 adults for the period preceding 1830; but he had examined very few of the sources we have consulted. The weight of evidence, from the examples cited above and others not mentioned, leads us to believe that the grizzly population was perhaps closer to 10,000. Many people met grizzlies in groups, large numbers of the bears were killed, and there were conspicuous well-worn grizzly trails in both streamside thickets and foothill chaparral—the grizzlies *were* abundant.

We infer that a graph of California grizzly population would have shown a long plateau through the centuries with minor ups and downs, then a rise in numbers—particularly in the coastal regions adjacent to the missions where cattle were abundant—that reached a peak about the time of the American occupation. This was followed by a quick descent—a half century or less of small numbers and spotty occurrence—to the base line, extinction.

## LAST RECORDS

The decline in numbers and final extinction of the California grizzly have left a faded trail in the literature, most of it too dim to be read with accuracy. In a few localities, the year in which the last bear was killed—and sometimes the precise date—is known (see Appendix B); but in many other places the last reported killing may have been some years before the grizzly actually became extinct. Isolated males and lone females too old or without mates may have lived on to die from natural causes.

There was active hunting of grizzlies in California before the gold rush, but the greatest reduction in numbers was probably between 1849 and 1870. Hunting then was intensive, both near cities and towns and farther afield; the bears were killed not only for their meat but also to prevent them from injuring the settlers and killing livestock. Besides the many accounts of grizzly hunting in early diaries and other books, we have found reports of forty to fifty encounters between men and bears in newspapers of the years 1851–1860. For the next decade there are fifteen, and for the years 1871–1879 there are sixteen. After 1860, reports are fewer near communities; some pertain to ranchers and others are of hunting experiences. Exhaustive search of rural newspapers would probably reveal additional articles; but since the Sacramento and San Francisco dailies often reprinted items from country papers, the residue of undiscovered news notes may not be large.

By about 1880, grizzlies in California no longer were in the lowlands they had dominated for centuries; those that had survived inhabited hilly and mountainous areas. In the opinion of Grinnell (1938 : 78),

the parts of the State in which grizzlies were able to persist longest were thus those where heavy and continuous chaparral, therefore lack of any grassland, kept out the sheep-herder. These were not, however, necessarily the parts of the State in which the bears were originally most numerous. In general, their last strongholds were in the Santa Ana Mountains, Orange County, the San Gabriel Mountains of northern Los Angeles County, the mountains of Santa Barbara County, and the western flank of the southern Sierra Nevada in Tulare County . . .

The last grizzly in the northern half of California was killed near Hornbrook, Siskiyou County, in 1902 by Gordon Jacobs (Schrader, 1946 : 15). In the Yosemite region one was taken on October 21, 1887 (MVZ 27928); Allen Kelley (1903 : 120, 123–124) mentions two wounded by members of the Fourth Cavalry serving as guards in the Park about 1891; and the last known specimen there was taken about 1895 (MVZ 31826). In the winters of 1908 to 1911 a big bear was reported living on Bullion Mountain, Mariposa County. It had long claws—the track was said to be "9 by 17 inches (or a little more)"—and it had trails from chaparral to oak groves. Chased by men and dogs, it "left the country" and was not seen again. (Grinnell and Storer, 1924 : 70.) According to A. Vela of Jackson, Amador County, a grizzly was killed "on the summit" (toward Carson Pass) about 1902 or 1903, elevation 10,000 feet; the hind-foot track was 13 inches long (reported July 10, 1925, to T. I. Storer).

No grizzly was killed in the Tejon–San Emigdio region of Kern County after 1898, according to W. S. Tevis; the last was shot in May of that year in Salt Creek Canyon near Mount Pinos. In the Cuyama Mountains of northern Santa Barbara County, however, some grizzlies were present—specifically a female and two cubs—as late as 1912, according to local informants (Grinnell *et al.*, 1937 : 78). In Los Angeles County, "Monarch" (see our chap. 10) was taken in 1889. The San Gabriel Mountains north of Pasadena had grizzlies in the 1890's. In July, 1895, and July, 1897, Joseph Grinnell saw fresh bear tracks daily near Waterman Mountain and Mount Islip. The bears roamed open-forested country at night, approached camps, and stampeded campers' burros; but their trails to daytime retreats led into heavy chaparral of Bear Creek in the San Gabriel drainage. (Grinnell *et al.*, 1937 : 83.) The last specimen, a nearly full-grown male was shot by Walter L. Richardson on May 16, 1894, in Big Tujunga Canyon. Both skull and pelt were saved and are now the best-preserved museum example of a California Grizzly (MVZ 46918).

In Riverside County, the last grizzly was killed about 1895 (Grinnell, 1938 : 78); but farther south one was shot during August, 1900 or 1901, in the northwestern tip of San Diego County at the head of San Onofre Canyon in the Santa Ana Mountains (skull, USNM 160155). The last taken in southern California was shot in January, 1908, in Trabuco Canyon in the Santa Ana Mountains, where Orange and San Diego counties meet (skin and skull, USNM 156594). These two were reputedly mates, according to a note on the Trabuco Canyon specimen.

A grizzly was trapped and killed by Cornelius Johnson near Sunland in Tujunga Canyon, Los Angeles County, on October 28, 1916 (MVZ 24408; Grinnell *et al.*, 1937 : 90–92, fig. 22). This was considered the "last grizzly" in southern California; but inquiry by Dr. Elmer Belt of Los Angeles, who was well acquainted with Mr. Johnson, revealed that shortly before this animal was shot a grizzly had escaped from the Griffith Park zoo. The keeper did not publicize the escape, for obvious reasons; but after public excitement over the killing had subsided, he told Mr. Johnson, who in turn informed Dr. Belt (letter, MVZ files, Jan. 10, 1939). The skull does not resemble that of other southern California grizzlies now in museum collections.

The final chapter of the grizzly in California centers in and about Sequoia National Park. In 1920, Raymond J. Palmer, when in the Mount Whitney country near Bench Lake, not far from

Kearsarge Pass, watched a female bear (with a cub) that "seemed to be three times the size of an ordinary black or cinnamon bear" (verbal report, April 24, 1953). On August 7, 1921, eleven visitors at the Giant Forest bear pit reported a large gray bear, with a distinct hump above its shoulders, that was twice the size of adult black bears. When the big one appeared, the black bears ran away (Fry, 1924 : 1). In August, 1922, after several calves had been lost, Jesse B. Agnew shot a bear near his cattle ranch at Horse Corral Meadow, Fresno County, at 7,500 ft. elevation. A tooth was sent to C. H. Merriam, who replied: "This tooth in itself is sufficient to prove beyond doubt that the bear was a Grizzly. It is the lower canine of what appears to be an adult female Grizzly . . ." (Merriam, 1925 : 3). The tooth could not be found at the U. S. National Museum in May, 1953, when one of us inquired about it. The disposition of the skin is in doubt. J. B. Agnew wrote to Joseph Grinnell (letter, MVZ files, Aug. 14, 1928): "I saw the skin of the grizzly I killed over in Korea (Chosen) in 1925 where my nephew has it and there is no question but that it was a grizzly." But Sumner and Dixon (1953 : 463) reported: "The skin of this bear, somewhat weathered, in 1950 was still nailed to the barn in Horse Corral Meadow." Mr. Halstead G. White of Berkeley examined the pelt on the barn on June 3, 1954, and obtained for us two claws: one from the right forefoot, the other from the right hind foot. These are clearly claws of a black bear. A cloud thus hangs over the last reported specimen of *Ursus arctos californicus*.

There are, however, a few hints that Agnew's bear may not have been the last *living* individual, because

During the month of April, 1924, Mr. James B. Small and his road working crew . . . reported on several occasions having seen a large grizzly colored bear in the vicinity of their camp near Moro Rock [Sequoia National Park]. Mr. Small, and some of his men, had previously worked in Yellowstone National Park, where grizzly bear are numerous, and all these men pronounced the bear they saw as a grizzly. They all made mention of the hump the bear had on the top of its shoulders. On October 13, 1924, Mr. Alfred Hengst, a cattleman of Three Rivers, came into very close contact with a huge bear near the head waters of Cliff Creek. Undoubtedly this bear was the same animal . . . "It was the biggest bear I [Hengst] ever saw—bigger than any cow, and looked as though sprinkled over with snow. I had a close view of the beast which was undoubtedly a grizzly" (Fry, 1924 : 1).

In 1949 there were two newspaper stories of persons who had seen bears or tracks of bears thought to be grizzlies. These were in Butte and Siskiyou counties, respectively (Sacramento *Bee*, Aug. 2 and 17, 1949); but no satisfactory substantiating evidence was presented and no specimens were produced.

In summary, the last California grizzly of record to become a museum specimen (USNM 156594) was taken in 1908; but there is evidence that a grizzly was shot in 1922 and that another was seen two years later. The last living captive ("Monarch"; see chap. 10) survived until 1911. Since the Sequoia National Park reports of 1924 none of any presumptive validity have appeared. The native grizzly, once numerous and dominant, is no more.

# 2 *Physical Features of the Grizzly*

## GENERAL APPEARANCE

Among "carnivorous" mammals, the bears as a group are characterized by their stout form and large size. The head is proportionately small and rather acutely tapered, with less length of snout and greater over-all bulk toward the base of the skull, because of the powerful jaw muscles. The eyes seem rather diminutive for the general size of the animal, and the ears are short. The neck is of moderate length but is large in diameter, because of its thick musculature. The body is heavy, and is wider than deep. The four limbs are of about equal length but are heavy in build and conspicuously tapered, and the feet are large. The tail is so short as scarcely to be visible in ordinary view.

Features that set the grizzly apart from other bears—except the brown bears—are the shoulder hump, the long front claws, the color of pelage, and the structure of the skull and teeth. As compared with the black bear, it has higher shoulders, a longer body, a straighter back, and lesser elevation of the haunches. Its head is narrower, and the snout and jaws are longer and less blunt. (Mills, 1919 : 251.)

## THE HUMP

Over the shoulders there is a characteristic hump, evident in both young and old grizzlies (figs. 2, 3). The hump results from the size and placement of the muscle mass above the shoulder blades, according to Dr. Robert K. Enders, who has anatomized several Yellowstone grizzlies in recent years (letter, Jan. 11, 1953). There is no pad of gristle in that region as might be

31

supposed, and the dorsal spines of the chest vertebrae are not longer, proportionately, than in a black bear.

## FEET AND CLAWS

The feet of bears are large because the animals are plantigrade, the palm or sole as well as the toes being provided with durable pads that regularly make contact with the ground (fig. 7). In this respect they differ from the speedier dogs, cats, and

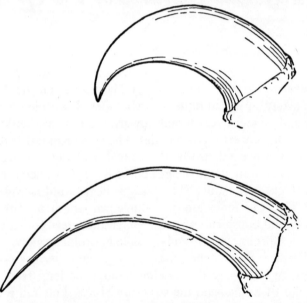

Front claws of California bears, natural size. Black bear above, grizzly below. (After Grinnell *et al.*, 1937.)

other lithe carnivores that habitually run on their toes. The forefoot of grizzlies—and that of black bears—has a large pad on the palm that is wider than long and somewhat rectangular. Each of the five toes has a small oval pad. The hind foot of the California grizzly was a huge structure in fully adult males. For example, that of the big individual which served as the type of *Ursus magister* (USNM 160155) measured 12 inches in length without the claws, and 8 inches in greatest breadth. The sole pad is crudely triangular, with the base foremost against the pads of the five toes. All the pads are surfaced with tough, cornified epidermis over a substantial mass of resistant connective tissue. This covering of the foot is the sturdy, self-renewing "shoe."

The front claws of grizzlies, as compared with those of black bears, are heavier, longer, broader, and only slightly curved (fig. 7, and p. 32). They serve the grizzlies in their extensive digging operations for bulbs, roots, and rodents and in fighting. Many human beings were deeply gashed by these great hooks that are powered by heavy arm and shoulder muscles. The length and size of the claws vary with age, and possibly bears of different regions had some differences in relative proportions and dimensions of these members. The activities of the animals also affected the claws. According to Xántus, grizzlies near Fort Tejon had claws much worn from digging. Of the grizzlies in the Rockies, Mills (1919:91) wrote that when the animals entered hibernation the claws were worn, blunt, and broken, but when they emerged from the winter rest period the claws were long and moderately pointed. Replacement growth would not have been conspicious in lowland California grizzlies that did not hibernate.

Measurement of the claws on four California skins (MVZ 46918, 16615, 16616, 27928) taken round the external curve, in millimeters, were as follows (averages and extremes; 8 of 80 claws missing):

| | FRONT | | | REAR | |
|---|---|---|---|---|---|
| | *Left* | *Right* | | *Left* | *Right* |
| (1) | 82 (76–89) | 79 (78–82) | (1) | 49 (44–59) | 42 (39–47) |
| (2) | 83 (77–93) | 81 (77–89) | (2) | 38 (33–49) | 34 (25–42) |
| (3) | 77 (71–87) | 76 (74–77) | (3) | 37 (30–45) | 35 (31–36) |
| (4) | 73 (68–87) | 72 (67–79) | (4) | 35 (31–42) | 31 (26–34) |
| (5) | 68 (62–77) | 65 (63–68) | (5) | 41 (34–49) | 37 (33–43) |

The front claws thus are 2.4 to 3.7 inches and the hind ones 1.0 to 2.3 inches in extreme lengths, any hind claw being about half the length of the foreclaw on the corresponding toe. Elsewhere there are reports of front claws considerably longer than these on California specimens. Front claws of black bears rarely grow to be 2 inches in length. Claws of grizzlies are pale brownish to yellowish; those of black bears are black or dark brown on individuals of the cinnamon phase.

## PELAGE

The coat of bears consists of an underfur of relatively fine and shorter fibers and the main mass of coarser and longer guard hair that provides the scheme of coloration and external contouring. The hair of a grizzly is long and relatively coarse, and the coat seems somewhat shaggy compared with the sleeker covering of a black bear, especially when the latter is in fresh new pelage; but this may be partly an illusion deriving from the uniform coloration of the black bear. Either species may appear unkempt—mangy, older writers called them—when the coat is far worn or the animal is thin. One of these early writers, S. F. Baird (1857 : 220), described a mane along the neck from the occiput to the shoulders, where the hair is stiffer, even "wiry," and longer than elsewhere on the body. On the largest California skin studied by Baird the hairs of the mane were five inches long. On the lesser pelts these hairs measured only three inches.

The pelage of grizzlies in California and elsewhere is more varied in coloration than that of many other kinds of mammals. The black bears (including the cinnamon phase) have a rather uniform color scheme as compared with that of grizzlies. Basically the grizzly coloration was brown, but the shade differed from one individual to another and on the parts of a single bear. The feet and legs commonly were the darkest, being blackish brown or even nearly black.

An old female from the Santa Ana Mountains was described by Merriam (1918 : 75) as: "Dusky or sooty all over except head and grizzling of back; top of head and neck very dark brown, sparsely grizzled with pale tipped hairs; back dusky, grizzled with grayish; muzzle gray or mouse brown, palest above; legs and underparts wholly blackish" (presumably USNM 156594). This animal was reportedly 6 feet 3 inches long.

Five California skins taken during the 1850's and hence relatively freshly prepared, led Baird (1857 : 219) to describe the coloration thus: "A dark dorsal stripe from occiput to tail and another lateral one on each side along the flanks, obscured and nearly concealed by the light tips; intervals between the stripes lighter. All the hairs on the body brownish yellow or hoary at the tips. Region around ears dusky; legs nearly black. Muzzle pale, without a darker dorsal stripe."

Of the five California skins in the Museum of Vertebrate Zoölogy at Berkeley[1] the body pelage is predominately dark brown with some paler grizzling on hair tips; a warm brown, slightly reddish (skin B); and predominantly pale, almost yellowish white (A). The amount of pale or "silver" tipping varies. In one skin (A), the hairs for nearly half their length are pale; in another (D), far less than half.

Only one specimen (E) gives a hint of the three dark longitudinal stripes mentioned by Baird (1857 : 220). This pelt also has slight evidence of a pale arc on each side from the back of the head to beyond the scapular region, convex toward the vertebral line. These two arcs, on the right and left sides, as seen from above, give a pattern thus ⌒⌒, with a dark area between, and may be the basis for early remarks about the "X-bear" (such as killed Peter Lebec; see chap. 8). The other four skins show no hint of this design.

These five specimens are of diverse origin, three having been used as rugs for varying periods. The exact date of capture is known for only one (MVZ 46918: May 16, 1894). Although the degree of seasonal wear between molts and the amount of wear and fading in those serving as rugs are impossible to determine, these pelts no doubt display some of the color differences among individual California grizzlies.

Writing of grizzlies in Colorado, Enos Mills (1919 : 250–251) said: "The color of the species runs through many shades of brown: among them are cream, tan, mouse-color, cinnamon, and golden yellow. Black or almost white may be the fur of the grizzly, but shades of gray and brown predominate. . . . Cinnamon and brown are common colors of both grizzlies and black bears."

In retrospect we visualize the grizzly as a beast with a rather long dense pelage when in good coat. The annual molt, which came in late summer or autumn, gave the animal a new coat with more insulation for the winter and hibernation. (Body fat,

---

[1] (A) MVZ 31826; (B) MVZ 16616; (C) MVZ 27928; (D) MVZ 46918; (E) MVZ 16615; see App. A for list of specimens.

however, probably was of greater heat-conserving value.) The differences noticed in general tone of coloration may have been variations of several categories: individual, sex, age, or regional. Occasional bears were described as being whitish on the face or all over; "red" bears are mentioned; and a considerable range of other terms was applied. Some early writers mentioned grizzlies and "brown" bears separately.

## SKULL AND TEETH

The skulls of bears are stout and heavy, with jaws of moderate length as compared with those of many other carnivores. The greatest single difference between black bear and grizzly skulls is seen in profile: that part immediately forward of the brain case is more or less convex in the black bear but is flattish or concave in the grizzly (fig. 8). Because of this, the grizzly has been described as "dish-faced." In two skulls of approximately the same total length, the grizzly palate is longer, mainly to accommodate the greater size of the last upper molar. In old grizzlies, massiveness characterizes the glenoid fossa, the concave surface on which the articular process of the mandible moves, and also the mastoid process at the hinder outer corner of the cranium.[2] According to Merriam (1918 : 10), "Bear skulls undergo a series of changes from early life to old age, and in most species do not attain their mature form until seven or more years of age."

The teeth number up to 42; on each side there are 3 pairs of rather small incisors at the front of the jaws, 1 pair of heavy conical canines or tusks, 4 pairs of premolars or "milk molars" (some of which are lost early), and the molars, 3 above and 2 below. Most distinctive and positively diagnostic of the grizzly is the last, or third, upper molar, a huge crushing tooth, never less than 1¼ inches (31 mm.) in length and seldom less than 1½ inches (38 mm.). In the black bear this tooth does not attain a length of 1¼ inches (fig. 9).

---

[2] In a huge early grizzly skull from California (USNM 1219; condylobasilar length 367 mm.) the mastoid process is about 30 mm. wide and 15 mm. thick, extending laterally to the outer (distal) fifth of the mandibular condyle. This far exceeds, in comparable development, that on an African lion (USNM 216604) having nearly the same total length of skull.

The limited series of California grizzly skulls extant (not all of which have been seen by the authors) affords scant information on tooth wear. In one big Coast Range specimen (USNM 1219) the inner incisors are down to stumps, though the big upper molars show slight wear. A skull (USNM 156594) from the Santa Ana Mountains has all teeth far worn; the upper molars are ground to a — shape in cross section, and the incisors are mere blunt stubs. The rugose deposits of bone (exostoses) on the underside of the cranium and the rear of the mandibles denote extreme age—this is the last grizzly killed in California of which the skull is in a museum. A skull from the Siskiyou Mountains (USNM 178735), not particularly old, as judged by slight wear on the teeth, lacks the first upper molars on both sides and the sockets for these teeth.

## SKELETON

The bodily framework of the grizzly is substantial, to support the weight of the animal; yet the bear has a greater degree of flexibility in its movements than is possible in many other sturdily built mammals. This freedom of motion is a correlated function of the bones, ligaments, and muscles. The skeleton of a bear—grizzly or other kind—is much like that of related carnivores, but there are many small differences.

A noticeable massiveness is evident in all the bones. The neck vertebrae are large but are capable of much rotational movement, the spinous processes along the back on the dorsal vertebrae are heavy, and the shoulder blade is ample. The limbs are of nearly equal length. Both fore and hind feet are fully plantigrade: the entire surface of each foot comes in contact with the ground as the bear walks. The bones used in lifting or extending the feet (the pisiform on the fore foot, the calcaneum on the hind) are larger than in some other carnivores. All bones of the legs, both front and rear, are separate. In the front leg, the radius and ulna are of nearly equal size for easy and powerful rotation of that member; and in the hind leg, the fibula, which is involved in twisting movements, is free and larger in relation to the tibia than in mammals unable to make such movements. These skeletal features, together with the muscles attached to them, give the bears dexterity in using their limbs—more or less in the manner of human beings (fig. 11).

Our first inventory of specimens of the California grizzly (see App. A) included record of only one partial skeleton—that of a long-time captive ("Monarch," see chap. 10; MVZ 24537). Then we found an obscure bibliographic hint about a skeleton from Monterey taken by a French exploring expedition. Following this slender lead, we obtained a photostat from the zoölogical report of the voyage of the frigate *Venus* (Geoffroy Saint-Hilaire, 1846 : 125, atlas pl. 5); this showed an entire skeleton (fig. 10), mounted, 1/5 natural size; the expedition was at Monterey from October 18 to November 14, 1837. An inquiry to the Museum National d'Histoire Naturelle in Paris brought a quick reply from Professor J. Millot that the skeleton (A2754) of an aged male was still extant, minus a few parts, and he kindly provided detailed measurements, as follows:

Total length (straight line), base of incisor teeth to end of pelvis, 69.5 in. (1,766 mm.)

Vertebral column, atlas to last tail vertebra, 62.87 in. (1,597 mm.)

Height, 39.05 in. (992 mm.)

Skull, base of incisor teeth to foramen magnum, 15.8 in. (401 mm.)

Skull, breadth across zygomatic arches, 9.0 in. (230 mm.)

Mandible, base of incisor teeth to end of angular process, 10.6 in. (270 mm.)

Humerus, 15.0 in. (381 mm.); radius, 12.7 (323); forefoot without claws, 9.2 (238)

Femur, 18.2 in. (460 mm.); tibia, 12.6 (320); hind foot without claws, 11.2 (285)

Spinal processes of thoracic vertebrae, height: 1st, 4.0 in. (103 mm.); 2d, 4.5 (115); 3d, 4.7 (119); 4th, 4.85 (123); 5th, 4.7 (119); 6th, 4.8 (122)

## DIMENSIONS

Seemingly, no one ever measured and recorded in print all the usual dimensions of a wild California grizzly. The first notations of size are in the Portolá expedition diary of Fray Juan Crespi (Bolton, 1926 : 168–169). In what is now San Luis Obispo County, on September 2, 1769, the soldiers killed a bear which "measured . . . fourteen spans [10 ft. 6 in., if the span was 9 in.] from the soles of its feet to its head. The paws were one span long, and it must have weighed three hundred and seventy-five pounds."

Ten other and later reports give the length—presumably from tip of nose to end of body (without or with the tail)—as 6 to 10 feet. For three of these, implying measurements actually taken, the figures are 6 feet 3 inches for an old female (Merriam, 1918 : 76), 7 feet 10 inches "tip to tip" (Reid, 1895 : 129), and 8 feet 10 inches (Bachman, 1943 : 79), the last being of a dressed carcass hung for market. In addition, the skin of the "Wellman" specimen from Yosemite National Park (MVZ 27928), killed in 1887, measured nearly 10 feet, nose to tail, when freshly pegged out; after tanning it was 7 feet 6 inches long and 5 feet wide (Grinnell and Storer, 1924 : 69).

The height at shoulder is given as 3 feet (Engelhardt, 1924 : 40) or 4 feet (J. S. Hittell, 1863 : 108; Merriam, 1918 : 76). A 7-foot specimen mentioned by Revere (1849 : 143) measured 9 feet "across the hams"—presumably between the tips of the outspread hind feet. The girth of the chest as given by Colton (1850 : 119) was 4 to 5 feet. The skinned animal mentioned by Bachman was 3 feet in minimum circumference of the neck.

## WEIGHT

The weight of California grizzlies is a topic on which there are many statements and some estimates but few facts. We have found fully fifty references on the subject, including a few precise figures. Some state that the animal was actually weighed, but other "weights" are sheer guesses. We know that the newborn grizzly was a relatively tiny creature, weighing less than two pounds; and we can be certain that some individuals attained to huge size—excluding exaggerations, there is adequate testimony on this point. The weight of any individual would depend on its age, sex, state of health, and nutrition, and possibly on the season of capture. The grizzly evidently had a growing period that lasted for several years. Data on grizzlies elsewhere indicate that males attain a larger size than females. It is possible that some grizzlies in California lived in places where a greater food supply was available than in other localities; and seasonal food supplies may have caused grizzlies to be fatter at certain times of year, such as after the acorn harvest. Data are lacking, however, on all these variables.[3]

The two extreme statements we have found in regard to weights of California bears are these: "a young grizzly, weighing some eighty pounds" (Oct. 4, 1866; N 67) and "the bear tipped the beam—forbid it that anyone should question the reading of the scales!—at two thousand, three hundred and fifty pounds" (Newmark, 1926 : 447).

The last captive, "Monarch" (fig. 33), when killed after a long life in a public zoo where he was underexercised and probably overfed, weighed 1,127 pounds (Grinnell *et al.*, 1937 : 89). Adams' big captive, "Samson," was several times reported to weigh more than 1,500 pounds (Hittell, 1860 : 295). One report of 1856 (Herrick, 1946 : 179) states that a "mammoth grizzly," taken in what is now El Dorado County, afforded no less than 1,100 pounds of meat (which yielded the hunter $1,375). Of two killed in the hills near Matilija Canyon, Ventura County, in September, 1882, it was stated: "The largest . . . would weigh about 1500 pounds; it was all two strong horses could do to drag it . . ." (N 93).

Our records of animals with weights below 1,000 pounds, mainly from early newspapers, are as follows: 250 pounds, one; 300 pounds, two; 500–525 pounds, four; 630–642 pounds, three;

---

[3] The only record found of weight increments in a big bear is that of an Alaskan brown bear (*Ursus a. gyas*) taken May 24, 1901, near Douglas settlement, at the western entrance to Cook Inlet, when probably about 3½ months old. It was kept in the National Zoölogical Park, Washington, D.C., where its successive weights in pounds, were as follows (Baker, 1912):

| | |
|---|---|
| May 24, 1901, 18 | March 11, 1907, 970 |
| Jan. 4, 1902, 180 | March 21, 1908, 1,050 |
| Jan. 15, 1903, 450 | March 5, 1909, 960 |
| Jan. 18, 1904, 625 | Jan. 20, 1911, 1,160 |
| Jan. 28, 1905, 770 | Dec. 13, 1911, 1,090 |
| Feb. 28, 1906, 890 | |

It seemed heaviest about Dec. 1, 1910, but could not then be weighed. The decrease in weight during 1909 resulted from removal of extensive "corns" from all the feet on June 15, 1908, which crippled the animal for some time. Since bears confined in public parks usually receive much supplemental food from visitors and have limited opportunity for exercise, the gains in weight of bears in their native environments may not be comparable.

700–800 pounds, four; 900–932 pounds, four. The few weights not given in round numbers may indicate that they were of bears actually weighed. There are fully fifteen statements in early newspapers and a dozen or more in books, of weights of "1,000 pounds" and upward, practically all in round numbers.

The maximum weight of male California grizzlies was estimated at 1,200 pounds by Grinnell (1938 : 72) and by Hall (1939 : 238), neither of whom had access to the numerous reports we have found on the subject. We are inclined to believe that the maximum was somewhat higher. Seton (1909 : 1032) was of the opinion that no true grizzly ever weighed 1,500 pounds or that any but the California grizzly reached 1,000 pounds; he gave 600 pounds as the average weight for males, and 500 for females.

Writing from Colorado of the bears there, Mills (1919 : 251–252) said:

The grizzly always appears larger than he really is. The average weight is between three hundred and fifty and six hundred pounds; males weigh a fourth more than females. Few grizzlies weigh more than seven hundred pounds, though exceptional specimens are known to have weighed more than one thousand. . . . It may be that years ago, when not so closely hunted, the grizzly lived longer and grew to a larger size . . .

# 3 Habits of California Grizzlies

How did the grizzlies live? What did they do at various times of the day or night and throughout the year? When did they produce and rear young? What were the food habits of these bears and what were their relationships with other animals? These questions and many others come to mind as we attempt to visualize the role of the big bears in California before its invasion by white men. Only fragments of the natural history are found in the literature, and some of these are of secondhand information or hearsay. The grizzlies of California probably resembled their counterparts elsewhere in North America, although the local peculiarities of topography, climate, vegetation, and other environmental factors led to some differences in behavior and activities.

The grizzly was a paradox. Attacked in the wild by man it was a dangerous and deadly adversary, yet some individuals taken into captivity (see chap. 10) became docile and trustworthy. When in search of animal food to satisfy its hunger, the bear would strike down any large beast; and in the fighting ring it was a savage contender against a wild bull. But when grizzlies congregated with their kind to relish some desirable food—clover, acorns, or whale—they ate together peaceably. If approached slowly and quietly by man, the bear would retire, unless it was a female with cubs. The grizzly had the power to be dominant but only exercised it when alarmed, disturbed, or injured; otherwise it was a well-behaved member of the animal community.

## DAILY AND SEASONAL ACTIVITY; DORMANCY

Under primeval conditions the grizzlies were active both day and night—the earliest travelers in California met them at various hours. The bears had nothing to fear from other animals, and the time of foraging probably reflected their state of hunger. Those living in the hotter lowlands may have retired to shelter during the peak of midday heat. The first mention of grizzlies in California, by Vizcaíno's party at Monterey in the winter of 1602–1603, is of bears going to the beach at night to feed on a whale carcass (Wagner, 1929 : 247). Some early observers state that the grizzlies were active mainly at night, but others mention seeing or encountering bears either singly or in aggregations in daylight hours (see "Population" in chap. 1). Many people met females with their cubs in the daytime—the fast-growing young had good appetites!

As settlement of the state proceeded and the bears were hunted intensively, they became more wary and more nocturnal. Mackie (1915 : 285), who was in the Bakersfield–Fort Tejon region in 1859, wrote that "it was not safe to travel certain trails after dark, as the bears would sometimes attack men even on horseback." William H. Brewer (Farquhar, 1930), who traveled in many parts of California in the early 1860's, often saw bear trails but encountered few grizzlies in the daytime; he avoided journeys after dark for fear of meeting them. J. S. Hittell, writing in 1863 (p. 110), said that "the grizzly, though he often moves about and feeds in the day, prefers the night, and almost invariably selects it as the time for approaching houses, as he often does in search of food." Yet as late as 1885 two hunters southwest of Bakersfield saw fourteen bears rooting at once in the daytime (Seton, 1929 : 20).

When not abroad to seek food or mates, or traveling, the bears rested or slept in shelter of various kinds—under dense chaparral or similar cover, in the open bases of large trees, in caves, or in dens they themselves had excavated.

Grizzlies in more northerly parts of North America and those in the Rockies denned up and were dormant for varying periods during the winter, but those in the southern part of the original range, including California, were abroad throughout the year. According to Seton (1909 : 1046), the males were active through a longer season than the females. Early newspapers contain reports of four encounters with grizzlies in each of the three winter months. Most of these were in the lowlands or warmer foothills, but two were in Shasta County during December.

There are several reports of grizzlies active at higher elevations of the Sierra Nevada in winter. A bear was abroad at Leek Springs, El Dorado County, during the winter of 1851 when the snow was 8 feet deep and frozen on the surface (N 19). Grizzly Adams (see chap. 9), caught or killed bears in the snow season in the central Sierra Nevada (probably at 4,000 to 5,000 feet elevation)—some when they were abroad, others in their dens. A man was killed by a grizzly during December, 1874, in Sierra Valley, four miles west of Sierraville, while snow lay on the ground (N 85).

We think that, of the grizzlies living at lower elevations, the females with their young cubs were sequestered for a time but the others were active through much or all the year. In the higher mountains, dormancy in limited degree may have been necessary because food in the quantity needed by such large animals would be scarce in mid- and late winter; the deer then would have moved to middle or low altitudes, and most of the plant and animal food on or near the ground surface would be covered with snow.

## THE GRIZZLY'S "HANDS"

The grizzly was a heavy, rugged beast and might be thought to have been awkward or cumbersome; actually it was relatively agile and skillful, capable of various postures and movements (fig. 11). This resulted from the flexibility of its body (vertebral column) and particularly that of its limbs and feet. It could sit upright with the hind legs extended on the ground, keeping the forelegs free for any use. The bear usually walked on all fours when traveling and also when feeding or digging, and probably spent much of its time in that position. But it could rise on the hind feet to an erect posture with good balance—it did this when reconnoitering, reaching for food, and offering combat. In the upright position a large grizzly could reach to a height of nine to ten feet, and thus was able to obtain fruit, nuts, or vegetation beyond the reach of most terrestrial mammals. The bear would rear up to smite a large animal, such as a cow or bull, at close range; and many California grizzlies did this in the bear-and-bull fights and also in encounters with men. Some hunters that took shelter in small trees were injured by grizzlies that stood up and chewed or clawed their legs and feet.

Mills (1919 : 253), writing of the animal in the Rockies, said: "The grizzly is exceptionally expert and agile with his paws. With either fore paw he can strike like a sledge-hammer or lift a heavy weight. He boxes or strikes with lightning-like rapidity. Most grizzlies are right-handed . . . If a small object is to be touched or moved, he will daintily use but one claw. The black bear would use the entire paw."

## WALKING, RUNNING, AND CLIMBING

The gait most often described was an ambling walk with more or less side-to-side rolling of the body. Of this, Palliser (1859 : 181) wrote, "as he walks and trots, he moves the hind and fore foot together on the same side [like a pacing horse], and rolls his head at every step." Among Charles Nahl's illustrations, the original painting at Colton Hall (fig. 2) and the wood-cut in *Hutching's Magazine* (Anon., 1856 : 106) indicate this gait, but the sketch of "Samson" by the same artist (Hittell, 1860 : 300) does not.

In general, a grizzly would take the path of least resistance, going around rather than over obstacles, usually pursuing a less steep course than deer or other more agile creatures. Adams (Hittell, 1860 : 58) told of a female grizzly descending a steep hill rear end foremost, "allowing her weight to carry her, while she retarded what otherwise would have been too rapid a descent by holding on to the rocks and bushes with her claws."

For rapid travel the grizzly would break into a gallop, some-what like the gait of a "loping" horse. The sturdy conformation, large and supple muscles, and long hair made this seem a lumber-ing gait; yet it was surprisingly rapid, and a startled bear could disappear quickly. Contrariwise, the same gait served for attack. The quickness of a wounded grizzly in closing with his opponent was testified by many California hunters armed with muzzle-loading rifles. In the few seconds available, the man could only try his pistol, draw his knife, or scramble up a tree. Mounted men, endeavoring to rope grizzlies, often had difficulty in escaping if the bear turned on them, because the beast could practically equal the speed of a horse for a short run.

The adults evidently did not climb to any extent—certainly not as black bears do. Young grizzlies, perhaps until about two years of age, are reported to have been fairly adept and to have climbed trees to obtain food for themselves, and possibly for the adults. The old ones could step up onto low and more or less horizontal large limbs of oak to seek acorns, but could not go high in the trees. Adams stated that "full-grown grizzly bears rarely climb, and rarely attempt to do so; but sometimes if they see the object of their pursuit climbing, they will attempt to follow" (Hittell, 1860 : 159).

## SWIMMING AND WALLOWING

The grizzly bear was not afraid of water and could swim well. In 1827 a boat crossing the channel near Los Angeles Island (now Angel Island) in San Francisco Bay came upon one swimming toward the island. The bear "approached the boat, intending to climb into it, when some soldiers who were in it . . . fired four balls at . . . close range, just as the bear was getting its claws upon the boat, and killed it stone dead" (Carter, 1929 : 239). In another instance, on the ocean beach near San Luis Obispo, two bears were encountered by a hunting party. The male was killed, whereupon the female took off "with long strides along the edge of the inlet toward the breakers."

She pushed through the heavy surf, disappearing and reappearing as it rolled over her; and in a few minutes we saw her swimming straight out to sea . . . [Then the grizzly turned and began] swimming slowly toward us, rising and sinking on the long waves . . . The bear paused and floated on the sea a while . . . At length she began to strike out boldly making straight . . . for [a member of the party]. She struck ground about one hundred yards from us. . . . And now she came on with a rush . . . (Anon., 1857 : 821.)

Bidwell (1897 : 76) in the 1840's saw a bear in the Sacramento Valley plunge in and swim across a stream without hesitation. While Grizzly Adams was ferrying his live animals across the Columbia River in 1853 (see chap. 9) his pet "Lady Washington and a black cub plunged overboard; but, being good swimmers, they followed the raft" (Hittell, 1860 : 177).

Bears enjoy wallowing in pools, and the grizzly was no exception. Baths during hot weather helped the animals to cool off and to avoid insect pests; and they enjoyed warm water. Paso Robles Hot Springs in San Luis Obispo County formerly were visited by grizzlies. Angel (1883 : 369-370) wrote: "A huge grizzly was in the habit of making nocturnal visits to the spring, plunge into the pool, and, with his forepaws grasping the limb [of a cottonwood that once extended low over the water], swing himself up and down in the water, evidently enjoying his bath, his swing, and the pleasant sensations of his dips in warm water, with unspeakable delight."

## VOICE

Bears are relatively quiet or silent creatures. Except when seriously disturbed or wounded, the sounds they utter are usually low—mainly conversational. In this respect they are in sharp contrast to some other large carnivores, among which the howling of wolves, the "songs" of coyotes, the cries of wildcats, and the roaring of Old World lions and tigers are proverbial.

According to Mills (1919 : 225), the grizzly of the Rockies is silent most of the time. "When he does say anything it is in a queer, but expressive language. He utters a choppy champ of a cough; he says 'Woof,' 'Woof,' with various accents; he growls eloquently; he grunts and he sniffs. The youngsters say something like 'eu-wow-wow,' and when forlorn give an appealing cry . . ." In the words of Seton (1929 : 42), "grizzlies cough, growl, grunt, roar, and sniff, in expression of various feelings."

The California contemporaries of the grizzly wrote little about its voice. Adams (Hittell, 1860 : 203) heard the "yelping of cubs" in the den of the Yosemite region where he captured the youngster that grew to be "Ben Franklin." He tells (pp. 164, 310) of older cubs running about "yelping and howling" when he killed their mothers. "Monarch," the San Francisco captive, was heard to give a quick short "*koff, koff, koff*" when his two little offspring in an adjacent cage put their paws between the bars of his enclosure. Their mother "often uttered a sort of choppy coughing sound to them." (Seton, 1929 : 35.)

As Adams (Hittell, 1860 : 202) was seeking the den of "Ben's" mother,

there seemed to be a snuffling underground, very faint at first, but growing louder and louder, until there was no mistaking it for the growl of a bear. . . . but the sound died away in a few minutes . . . [Later, Adams himself uttered a terrific yell, and a moment afterward] there was a booming in the den like the puffing and snorting of an engine in a tunnel, and the enraged animal rushed out, growling and snuffing, as if she could belch forth the fire of a volcano (*ibid.*, p. 205).

One early Californian attempted to phrase a sound of the grizzly in the wild. McCauley (1910 : 5) wrote that "again, and again, we heard a low, deep-muttered 'm-o-u-g-h, m-o-u-g-h,' as we moved about in the bush . . . on stooping down, I saw a great, shaggy monster rise up from his warm nest to a sitting posture, quite near me, giving at the same time that significant 'm-o-u-g-h —o-u-g-h,' so well understood by the Pioneer hunter . . ." Another grizzly sound evidently was the result of quickly exhaling. Browne (1862 : 604) said, "I had often heard of the peculiar 'snort,' or blowing sound, uttered by a grizzly when alarmed, and as the crackling of the bushes was followed by this infallible sign, there could be no doubt on the subject." "Lady Washington" and other bears in Adams' experience snorted (Hittell, 1860 : 114, 143, 328). His other companion, "Ben," uttered grumbling noises when he was hungry (*ibid.*, p. 333).

Grizzlies when newly caught or wounded were decidedly vocal. The captive "Ben," once injured by another grizzly, "went bounding off to camp, 'yelling at every leap' " (*ibid.*, p. 314). When Adams' trap closed on "Samson" (see our chap. 10), the hunter was wakened by the voice of the big bear—"It was the awfullest roaring and echoing in the mountains I had ever heard . . ." (*ibid.*, p. 302).

Wellman (in Grinnell *et al.*, 1937 : 86) said of a wounded grizzly, "The giant beast sank to the earth with a roar and a continuation of bawls which were thrown back by the granite walls of the canyon of the Merced like distant thunder." A grizzly killed above Pasadena by Giddings (Reid, 1895 : 129) gave, before dying, "one of the most terrific barks or grunts" he ever heard.

## TRAILS AND TRACKS

Many kinds of wild mammals make and use trails in going to and from the places where they feed, drink, rest, and sleep. In regions not seriously disturbed by man, such trails are followed by successive generations of animals until they become conspicuous well-beaten paths, evident even to the casual human observer. In the days when grizzlies were abundant in California the trails of the big bears were to be seen in many parts of the lowlands, particularly along streams and in streamside thickets and in the dense chaparral that clothed much of the foothill country. Today one may see trails of black bears and of deer in some parts of California. Those of the deer from feeding or resting grounds to watering places are often steep, whereas the routes used by the bears gain or lose altitude more slowly. The lesser gradients of man-made trails in national parks are more to the liking of black bears, and early morning travelers often see imprints of animals that plodded along the trail the previous night. It is likely that the grizzlies in hills and mountains were similarly averse to steep trails, although they could climb or descend abruptly when occasion demanded.

Near Calistoga on March 22, 1852, Bartlett (1854, 2 : 35) said, "There was no road, nor even a trail, save those made by wild animals . . . Our guide often directed our attention to the huge tracks of the grizzly . . ." Blake (1857 : 47–48), while riding in the Tejon region during 1853 near the Cañada de las Uvas, was on a trail that extended in a direct line over the hills and then "followed a long, narrow valley between the hills, among groves of oak. The path was very dusty, and apparently much travelled, but the mule was very unwilling to keep it. I soon found it to be a *bear trail*, and full of recent tracks of great size." The Tejon region then was notable for its large population of grizzlies.

Members of the State Geological Survey during 1860 to 1864 climbed many hills and mountains. Letters of the botanist, W. H. Brewer (Farquhar, 1930), often mention grizzly trails in the chaparral. On February 3, 1861, he wrote that after descending from the highest point of the Santa Ana Range the party "had to pass the same chaparral. Trying an easier way in one direction for a short distance, we found trails, but the traces of grizzlies grew so very numerous, that we took to the ridge again" (*ibid.*, p. 38). On September 3 of the same year, while on Black Mountain (Santa Clara County), the men "found tracks and traces of grizzlies, more abundant than we have seen them before—we were in paths where their fresh tracks *covered* the ground, but we did not meet any" (*ibid.*, p. 178).

In 1850 William Thurman hunted grizzlies in hills back of Sonora, Tuolumne County, where the chaparral was so dense that he and his companion could penetrate it only on bear trails (Kelley, 1903 : 86).

Newberry (1857 : 47–48), who was in the region northeast of Fort Reading, Shasta County, and around Lassen's Butte in 1855, said that the wide intervals between mature forest trees were covered by "a dense growth of manzanita, ceanothus, and low scrub oak. These thickets . . . are intersected in every direction by . . . well-beaten paths" of grizzlies.

Evidently these bear routes were extensive and probably led in various directions to give their users access to feeding and watering places; the shaded avenues under the shrub cover afforded shelter and possibly served as places for resting and sleeping (fig. 6). We may infer that the clear height of these routes was perhaps four feet—hump high for the bigger bears—and that obtruding branches of the brushy chaparral were soon broken away by frequent passages of the animals. The extent to which footprints showed on the surface would depend on the texture and slope of the ground. Many chaparral areas, at least today, have little soil under the shrubby plants.

Grizzly trails in the alluvium of the flat valleys were even more conspicuous and notable in character. P. B. Reading (1930*b* : 192) on coming into the head of the Sacramento Valley about November 1, 1843, said of the bears, "They are so numerous as to have worn large paths five or six inches below the surface of the earth."

In 1856, when Mrs. J. J. Stevenson was living within about

Fig. 1. Black bear, adult. Distinctive features are the slightly convex forehead, absence of shoulder hump, short foreclaws about same length as hind claws, and smooth-surfaced black (or brown) coat of uniform texture and color. Yosemite National Park, photographed Nov. 3, 1929, by Joseph Dixon.

Fig. 2. California grizzly. Nahl portrait now in Colton Hall, Monterey.

Fig. 3. Female grizzly and her two cubs. The young, perhaps nine months old, show parental characteristics of the hump, long rough fur, long foreclaws, and flat forehead. Yellowstone National Park, photographed September, 1929, by Joseph Dixon.

Fig. 4. Sociable grizzlies at a garbage dump. Typical of the habit of the big bears to gather where food was available. Yellowstone National Park, photographed Sept. 14, 1929, by Joseph Dixon.

Fig. 5. Habitats once occupied by grizzlies in California. *Above*—Hill slope in Cañada de los Osos, San Luis Obispo County. *Below*—Bald Hills above Redwood Creek, Humboldt County.

Fig. 6. A manzanita thicket about 12 ft. high, the kind that served as grizzly shelter; Fiddletown, Amador County. *Above*—Exterior appearance. *Below*—Interior of thicket with network of branches such as those under which the grizzlies made their trails. Photographed Feb. 21, 1954.

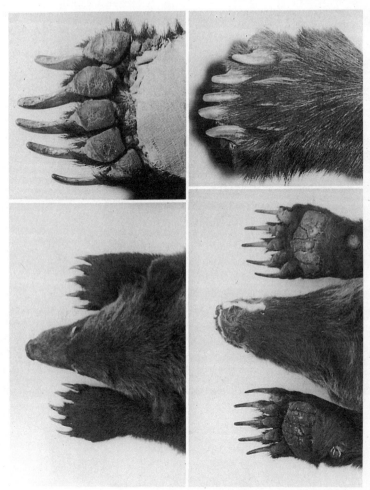

Fig. 7. Feet and claws of California grizzly, from museum specimens. *Left*—Cub (USNM 11809): upper and lower surfaces of head and forefeet. *Right*—Adult (MVZ 16616): lower surface of forefoot (5 in. wide) and upper surface of hind foot.

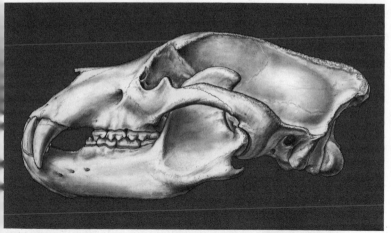

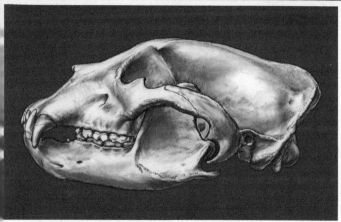

Fig 8. Skulls of Califor[nia] bears. *Above*—Grizzly (M[VZ] 46918), forehead conc[ave,] cheek teeth longer and m[ore] massive. *Below*—Black b[ear] (MVZ 29803), forehead m[ore] convex, cheek teeth prop[or]tionately smaller. Draw[n] by Karen Klitz, 1996.

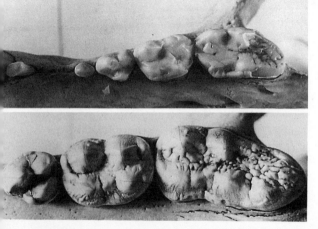

Fig. 9. Cheek teeth of bears. Black bear above, grizzly below; size of the last molar is the most distinctive character.

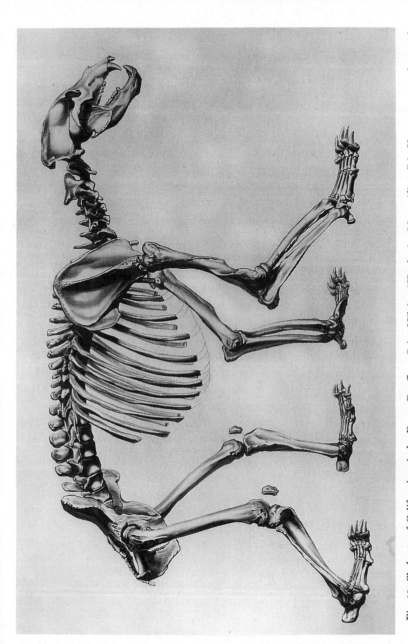

Fig. 10. Skeleton of California grizzly. From Geoffroy Saint Hilaire, "Zoologie—Mammifères," in *Voyage autour du monde sur la frégate La Vénus*, 1846. Specimen collected at Monterey, autumn, 1837; now no. A2754, Museum National d'Histoire Naturelle, Paris.

Fig. 11. The flexible and dexterous grizzly. Living examples in Fleishhacker Zoo, San Francisco. These individuals are as much "Californians" as many citizens of the state because they are "native-born" descendants (second or third generation) of a pair brought about 1934 from Yellowstone National Park.

Fig. 12. First published sketch of a California grizzly. Drawn by Ludovik Choris at San Francisco in the autumn of 1816 during the visit of Kotzebue's vessel, the *Rurik*. (Choris, 1822 : pl. 5.)

Fig. 13. Dagger made from femur of grizzly bear, decorated with *Olivella* shells. Canalino Indians, Ojai Valley, Locality 111 (no. 2776, Santa Barbara Museum of Natural History).

Fig. 14. A bear-and-bull fight on the range, 1849. The "observer" was J. Ross Browne, Secretary of the California Constitutional Convention later the same year. (Browne, 1862 : 748.)

Fig. 15. "Roping the Bear" at Santa Margarita Rancho (San Diego County) of Don Juan Foster. Painted by James Walker about 1876 from notes made in the 1840's. Oil painting on canvas about 24×42 in., in California Historical Society, San Francisco. The grizzly is being strangled by several nooses; usually the animal was roped on each foot.

Fig. 16. "The Pull on the Wrong Side." (Evans, 1873.)

Fig. 17. Bear-and-bull fight in an arena with horsemen. (Original in R. G. McClellan, *The Golden State*, 1872.)

Fig. 18. Bear-and-bull fight. Contemporary sketches of a spectacle at Mokelumne Hill, Calaveras County, in the early 1850's. Drawn by J. D. Borthwick (1857 : 296).

Fig. 19. "Pistol," 32-caliber, discharging five balls simultaneously. Used as a set gun for killing grizzlies. Owned by Albert Kennedy, Clear Lake Oaks, Lake County. Photographed March 1, 1953.

Fig. 20. Home-made steel grizzly trap. Owned by Alex Kennedy, Long Valley, Lake County. Length 84 in. Photographed March 1, 1953.

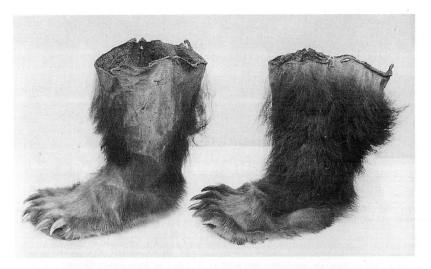

Fig. 21. Grizzly bear boots obtained in the 1850's by General Joseph Hooker at Agua Caliente, Sonoma County. Now in the possession of the Jansen family at Watsonville. Height 14 in. Photographed April 27, 1953.

Fig. 22. Relics of log grizzly trap, Seneca Creek, Santa Lucia Mountains, Monterey County. Length outside, 10 ft. 6 in.; width, 7 ft. 1 in. Photographed July 11, 1952.

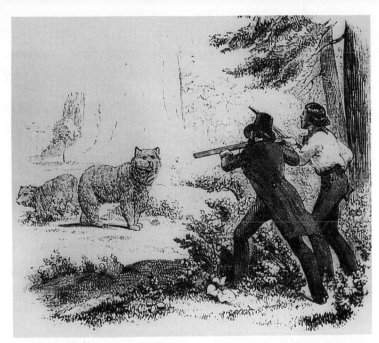

Fig. 23. A "scene of wonder and curiosity." An 1870 artist's conception of grizzly hunting in the "Frezno Grove" of mammoth sequoias. (From Hutchings, 1870.)

Fig. 24. "Peter Rescued from the Grizzly." One of the "elegant designs" credited to Charles Nahl, "the Cruikshank of California," in Alonzo Delano's *Pen Knife Sketches*. (Delano, 1853 : 71.)

Fig. 25. Grizzly Adams and "Ben Franklin," his trained grizzly hunting companion. Illustration by Charles Nahl (Hittell, 1860 : 186).

Fig. 26. Grizzly Adams and his bears in the menagerie at San Francisco about 1856. Illustration probably by Charles Nahl. (*Vischer's Pictorial*, 1870, pl. 121, copy in Bancroft Library.)

Fig. 27. Adams exhibiting under Barnum in New York, 1860. (Barnum, 1875 : 532.)

LIFE
OF
# J. C. ADAMS,
KNOWN AS

OLD ADAMS,
## OLD GRIZZLY ADAMS,
CONTAINING A TRUTHFUL ACCOUNT OF HIS
**BEAR HUNTS, FIGHTS WITH GRIZZLY BEARS,**
**HAIRBREADTH ESCAPES,**
In the Rocky and Nevada Mountains, and the
WILDS OF THE PACIFIC COAST.

NEW YORK. 1860.

PRICE, - - - - - TEN CENTS

Fig. 28. Adams in the dime novel. Cover of 1860 pamphlet in Huntington Library, San Marino, Calif.

Fig. 29. "Samson," the "1500-pound" California grizzly caught and exhibited by Grizzly Adams. Illustration by Charles Nahl. (Hittell, 1860 : 300.)

Fig. 30. "Bob," a yearling Mendocino County grizzly at the Albert E. Kent home in Chicago, summer of 1869. Earliest photographs of a California grizzly. (From William Kent, Jr., through California Academy of Sciences.)

Fig. 31. Transferring "Samson" from trap to cage in the central Sierra Nevada near ʌsemite Valley, winter of 1854–55. Adams atop the cage; Stanislaus and Tuolumne, ʌ Indian helpers; and a teamster with the oxen. Woodcut drawn in New York in 1860 ʌm description by Adams for the *New York Weekly* (May 31, 1860, p. 8).

Fig. 32. The taking of "Monarch." The grizzly was chained for the night between two trees, and by day was lashed firmly to the "go-down" sled hauled by four horses from the place of capture to the railroad. (Kelley, 1903 : 42.)

Fig. 33. "Monarch" in his final manifestation as overstuffed in 1911. Specimen in California Academy of Sciences, San Francisco, since 1953. Photographed Oct. 31, 1952.

Fig. 34. Stone grizzly carved in 1808 on washing trough at Santa Barbara Mission. Water from an octagonal fountain flowed through the bear's mouth into a large basin. The outflow passed through another bear spout of which only the distinctive front claws remain.

Fig. 35. Grizzly bear plaque. An example of foundry art showing resemblance to the Nahl illustrations of the 1850's but of unknown origin, purpose, and date. Bronze about 10½ in. long, owned by Francis P. Farquhar, Berkeley.

Fig. 36. A lively September 9 in the "gay 'nineties." Arch for parade of Native Sons of the Golden West in San Francisco, Admission Day, 1890, with the grizzly above all. Photograph from Roy D. Graves, San Francisco.

Fig. 37. Bear River (formerly Rio Oso) south of Marysville in north central California; once grizzly habitat, now just a name on a highway sign. Photograph by Bill Woodcock, 1996.

40 feet of the San Joaquin River, trails of both elk and grizzly were close to the house. A bear once stopped for about an hour to watch her while she was sewing. The grizzly trails there were a succession of pits 8 to 10 inches deep and 14 to 16 inches apart, made by the bears, perhaps generations of them, repeatedly stepping on the same spots. Because of the irregular surface, no other animals used these trails. Indians and other early white settlers reported similar trails under thickets of blackberries and vines close to the river. (Reported by Frank Latta, Bakersfield, April 20, 1953.)

The grizzlies, unlike other local native mammals, had nothing to fear from their animal associates and could walk slowly along these special trails. Seton (1909 : 1046–1047) reported similar trails elsewhere in grizzly country and added that the bears, in going up a bank or over logs, will put their feet "into the same tracks each time till they become a kind of stairway."

The actual print of the foot on the ground is similar to that of a black bear, but larger. In the words of Mills (1919 : 253) the big "hind foot leaves a track similar to the barefooted track of a man, while the track of the fore foot has the appearance of the grizzly's having walked upon the front of his foot,—the ball and toes,—with the heel [wrist] upraised."

The size of the hind foot is one of the few particulars given in the account of Vizcaíno's 1602 visit at Monterey—"a good third of a yard long and a hand [span?] wide" (Wagner, 1929 : 247). These agree with measurements of the 1900–1901 specimen from San Onofre Canyon: the sole without claws was 12 inches long and 8 inches wide (Merriam, 1914 : 190). During 1850, William Taylor (1858 : 153), en route between San Jose and Santa Cruz, measured, under the chaparral, a bear track that was 14 by 7 inches. Some other reports of the size of the foot imprint are the following: 1849, near Cottonwood, Shasta County, full 9 inches behind the toes (Kelly, 1851, 2: 98); 1849, near mouth of Feather River, 8 inches wide (Stillman, 1877 : 134); 1887, east of Wawona, Madera County, 13 by 10 inches (Grinnell *et al.*, 1937 : 84); 1891, Bear Meadow, Kings River Canyon, 8 inches wide across the toes (Muir, 1938 : 329); 1897, Mount Waterman, Los Angeles County, 13 inches (Arthur Carter, letter,

MVZ, Feb. 21, 1939); 1908–1911, Mount Bullion, Mariposa County, 17 by 9 inches or a little more [possibly in soft soil] (Grinnell and Storer, 1924 : 70); and 1924, Tulare County, over 12 by 6 inches (Fry, 1924 : 1).

Occasional early writers left statements not in accord with the preceding. Revere (1849 : 143) wrote, "The track of this huge plantigrado *measured* [italics ours] twenty-two inches in length, and eighteen inches across the ball." Either the bear had slipped wildly in soft dust—or the author did. Mills (1919 : 254) gave the largest track he measured in the Rockies as slightly more than 13 inches long by 7½ inches wide.

## SIGN

Both grizzlies and black bears in other places than California are wont to leave marks on trees by standing and clawing off the bark; various reasons have been assigned to the practice, none entirely satisfactory. Bears also rub against trees or logs as an alternative to scratching with their claws. In the Kern River region claw marks on a big manzanita were noticed on one occasion (Grinnell *et al.*, 1937 : 76), and in the Sierra Nevada during 1850 Bruff (1949 : 332) found where a grizzly, after feeding on an ox carcass had wallowed and then rubbed himself between the roots of a large pine. These are the only instances of this type of behavior by the grizzly we have found in the California literature.

The relatively large piles of grizzly droppings must have been conspicuous near feeding places and trails, but the only mention of them is that by Fray Juan Crespi, at Monterey Bay in 1769: "not far from this spot we saw manure like that of horses [tule elk?] and we saw bears' dung everywhere" (Bolton, 1927 : 27). Adams (Hittell, 1860 : 117) mentions that his captive "Lady Washington" was quiet at night except when, actuated by an innate love of cleanliness, she roused and removed to the end of her chain "to dispose of bodily waste." He says further, "My personal observations . . . have convinced me that cleanliness is as much a natural virtue in the ursine, as in the feline race." Frank Latta was told that dens of grizzlies in the floor of the San Joaquin Valley always had a dung pile one to two feet high at one place outside; the animals did not foul their shelters. Seton (1929 : 33) was told by J. M. Mackenzie of Bakersfield, who found many dens, that none contained dung, for the bears buried it near the den; at feeding areas, however, their droppings were scattered.

## DENS

Many kinds of terrestrial mammals make use of dens or burrows for shelter from the weather and for rearing their young. Ground squirrels, pocket gophers, and the like, regularly dig their own; other mammals, including the bears, are opportunists, using natural facilities if available but excavating when necessary. A naturally formed cavity or cave sometimes suffices; or the bear may remodel and enlarge it. If none is available the animal digs its own.

The best description of a California grizzly den is that by Adams (Hittell, 1860 : 199–201) of the place where he obtained his companion "Ben Franklin." In the spring of 1854, he went through Yosemite Valley "to the headwaters of the Mariposa River" (possibly toward Little Yosemite) and discovered a grizzly's den there. The scene "was a cañon-like ravine between two hills, densely covered with thickets of chaparral, with here and there a bunch of juniper bushes, a scrubby pine, or a cedar. A heap of fresh dirt in the thicket on one side, indicated the site of the den. It resembled the earth which a miner wheels out and dumps at the opening of a tunnel; and in size was as much as about fifty cart-loads." From concealment about a hundred yards distant, Adams saw

its character to be similar to that usually dug by the California grizzly;—in form something like an oven, having an entrance three or four feet in diameter and six or ten feet long, with a larger space, or den proper, rounded out at the extremity, intended for the lying-in place of the dam and the bedding of the cubs. A number of such dens I had seen in the Sierra, varying only according to their position and the quality of the ground in which they were excavated. The ravine here was rugged and narrow; and the den penetrated its steep, bushy side, about fifty feet above the bed of the stream . . . The entrance consisted of a rough hole, three feet wide and four feet high. It extended inwards nearly horizontally, and almost without a turn, for six feet, where there was a chamber, six or eight feet in diameter and five feet high, . . . and its entire floor was thickly carpeted with leaves and grass . . . (*Ibid.*, 1860 : 208.)

Another den found elsewhere by Adams when following a wounded grizzly was in "the thickest portion of the chaparral . . . marked by a heap of earth" (*ibid.*, p. 62). Adams said that a den was never occupied by more than one old bear (p. 208).

Dens were also used in the hills at lower elevations. There is mention of one in 1849 near Shingle Springs, El Dorado County (Kent, 1941 : 41); and another at Skunk Hollow, Santa Clara County, in 1879, is described as a "sort of a cave in the hillside, heavily bordered with chaparral" (N 91). In hills about Bakersfield, J. M. Mackenzie found many dens like those described by Adams (Seton, 1929 : 33). In the Santa Susana Mountains near Los Angeles the bears denned in caves screened by thickets in the bottoms of the canyons (Powers, 1872 : 278).

Alex Kennedy (interviewed in April, 1953) of Long Valley, Lake County, saw a number of dens when he was a boy, about 1880. The females dug dens in the earth on hillsides covered with chamise (*Adenostoma*), in which they spent the winters and gave birth to their cubs. These were horizontal caves, 18 to 20 feet deep, and were usually near the ridge of a hill and on the south-facing slope. Claw marks were still visible in the earth walls when Kennedy saw the caves. Years after a fire had removed chamise on a hill near Chalk Mountain, a tunnel about 3 feet in diameter and 15 feet in length was discovered; it was not man-made, and presumably had been dug by a grizzly.

Early residents of flatlands in the San Joaquin Valley told Frank Latta of Bakersfield (interviewed April 20, 1953) that on the borders of Tulare Lake and in similar places grizzlies dug their dens, in the manner of ground squirrels. The entrance was about 4 feet in diameter; at some distance below the surface the hole extended laterally for 10 or 12 feet. The odor at a grizzly den, probably from long occupancy, was such that a saddle horse could not be induced to go near the place.

We infer that dens were a necessity for female grizzlies in all parts of California because the small, blind, nearly naked, and helpless young had to be suckled a long time before they were able to travel on the surface. In the lowland valleys such dens probably served also as shelters during the midday peak of summer heat. The females, when not attending tender cubs, and the males in the lower, warmer parts of California, probably bedded down in any convenient place. The grizzly did not suffer from sleeping out, because it was well insulated by a rather heavy coat of hair which was often underlaid by a generous supply of subcutaneous fat. In Lake County, according to Alex Kennedy, the males did not use dens; in winter they made piles of brush, about three feet high, on which to sleep.

The females also bedded in comfort, according to Mrs. Anne B. Fisher (1945 : 194). An itinerant wool buyer of early days once stopped for the night under a great oak on the Rancho San Lorenzo in Salinas Valley.

While the woolgatherer cut mustard stalks and piled them high for his bed, a great she-grizzly had the same idea, in about the same place. They worked back to back in the bedmaking process, each intent on his own comfort and oblivious of the other until the man cut a last armful and the grizzly saw him through the thinned out brush. She stampeded the mules and overturned the wagon; then she treed the woolgathering merchant and kept him there until daylight!

## MIGRATION

Grizzlies were evidently present over much of their range in California through most or all of the year, but information about them in the higher Sierra Nevada is scant. At lower altitudes there were "episodes" between men and grizzlies in practically all months. Yet this does not necessarily mean that all the bears were strictly resident. A few reports indicate that some moved about with change in their food supplies. Of the Sacramento Valley one writer said, "the grizzly bear during the spring and fore part of the summer come down from the mountains and out of the thickets on the river and feed upon . . . [the clover] like herds of swine . . ." (N 2). John Work and his party found bears in abundance in the Valley in the winter and spring of 1832–33, but he made no mention of them in summer when he went from Stockton to Red Bluff (Maloney, 1943). Johnson (1849 : 177) wrote of the upper Delta region in April, 1849, that "the grizzly bears do not frequent the valley till later in the season, when acorns are ripe." Tyson (1850 : 71) made a similar comment.

In San Diego County during July, 1853, it was reported that bears had lately "descended from the mountains in great numbers to feed upon the sweet clover of the valleys" (N 24). By contrast, there was evidently an uphill migration in the San Bernardino Mountains, because it was said of Bear Valley (elevation about 7,000 feet) that "toward the first of September, when it [the clover] has attained its height, the valley is greatly resorted to by those animals . . ." (N 53).

There is one precise record of migration from a "tagged" bear. One evening in the fall of 1852 a bear was wounded with a large-bore rifle, by Martin A. Reager, at the place where the Glenn–Tehama County border meets the Sacramento River. Two years later it was killed up in the mountains at Alder Springs, about forty miles to the west. It was identified by a deformity of the front foot and by the huge ball from Reager's gun which was in the body. (Reported by Frank Reager, Sr., Jan. 1, 1954.)

## FOODS AND FEEDING

The California grizzly was an omnivorous opportunist—it ate almost anything and everything that was available. In this respect the big bear was somewhat like the house rat, the domestic pig, and even modern man. The grizzly, among the native animals of the state, was at the top of the food chain; it came first to dine on whatever was available, and the others "took second table." In most areas occupied by grizzlies in California there were no wolves; the mountain lion then as now probably foraged mainly on deer; the coyote had varied food habits but filled a lesser role; and only the badger could vie with the bear in digging rodents out of the soil. None of the herbivorous animals partook of so wide a variety of plant materials as the grizzly. Elk, antelope, and deer used grasses or browse; rabbits and hares ate leafy vegetation; and the rodents had various diets according to their kind but always more or less restricted in scope.

At all seasons there was a variety and abundance of food, both animal and plant. By all contemporary accounts, whales, water birds, fish, elk, deer, and antelope were common in many places. The land was well stocked with rodents. Resources of the plant kingdom were equally rich. Clovers and other pasturage grew to unbelievable height and density in both mountain and lowland valleys. Berry-producing vines and bushes screened numerous stream borders and hill regions that have long since been made bare by overgrazing, fire, and water manipulation. Oaks of various species then covered many valley flats and slopes now cleared for agriculture or towns and afforded acorns for several months each year.

Writers, from the Spanish period onward, often mentioned the food and feeding habits of grizzlies. There is far more in print on this topic than on others of equal importance. This is because early travelers and residents had frequent opportunity to see the bears while eating and also to interpret from the evidence where bears had foraged the objects of their search. Foods actually mentioned are listed below, but others undoubtedly were eaten.

| | | |
|---|---|---|
| Whales (carcass, offal) | Ground Squirrels | Yellow-jacket nests |
| | Field mice | Grubs (? in soil) |
| Deer (carrion, hunter kills) | Gophers | Wood-boring larvae |
| Elk | Lizards | Honey |
| Domestic livestock | Frogs | |
| | Fish (live, dead) | |
| Grasses | Tule roots | Serviceberries |
| Wild oats | Roots | Huckleberries |
| Wheat (in field) | Truffles | Wild plums |
| Corn (cultivated) | Dogwood "berries" | Wild cherries |
| Wild (sweet) clover | Elderberries | Grapes (wild and cultivated) |
| Red clover | Gooseberries | Nuts |
| Alfilaria (filaree) | Manzanita berries | Acorns |
| Brodiaea | Salmonberries | |

Grizzlies in the Rockies ate also tender shoots of many plants, spring flowers, mushrooms, rose hips, pine nuts, aspen tips, the bark of alder and aspen and cottonwood, the tips of branches from wild plum and berry vines, grasshoppers, bumble bees, ants and their larvae and eggs, snakes, trout, and salmon. One ate about a hundred pounds of potatoes in a garden. They only occasionally killed big game but got much as carrion. Some became cattle killers. No grizzly ate human flesh; bodies in grizzly areas have remained for days untouched. (Mills, 1919 : 63 ff.)

Grizzlies in California took "meat," large or small, fresh or putrid, as they could. Those living near the seacoast were attracted to the bonanza supply wherever a whale washed ashore—and the one-time abundance of whales in our coastal waters probably made this a not uncommon event. The first report of bears eating this food was by the Vizcaíno party at Monterey in 1602; a very large whale had gone ashore, "and the bears came by night to dine on it" (Wagner, 1929 : 247). Revere (1849 : 259) wrote that the carcass of a whale, thrown upon the beach, will attract a "regiment of bears"—and Kotzebue (1821 : 3, 41) used the term "countless troops." The fondness of bears for whale meat led to one of the most humorous bits of grizzly lore we have uncovered (see our chap. 11).

In the 1850's and later, when the Monterey beach was used by whaling crews as a place for rendering the fat, grizzlies fed on the offal that floated ashore. "Sometimes there would be twelve or fifteen there at one time." Both men and bears continued these practices—there were "lots" of bears—at least until 1875. Bears were roped (for bear-and-bull fights) on the beach at late as 1881. (Statements by early inhabitants to Mrs. Anne B. Fisher.)

Along the coast west of San Luis Obispo a big grizzly was seen eating dead fish washed into a marshy inlet (Anon., 1857 : 821). According to Davis (1929 : 224), bears angled with their paws for live or dead fish on the ocean shore of San Mateo County. Grizzlies in coastal Alaska regularly fish the inlets for salmon, and those in California probably had similar ways. The habit of cleaning up the carcasses of animals that washed ashore made the grizzlies the major beach sanitarians of early California.

We think it unlikely that grizzlies regularly killed the agile wild herbivores—antelope, deer, and elk—although occasional surprise attacks may have succeeded. Any individual that lagged behind the herd or became separated in a thicket, or one slowed in its reactions by injury, disease, or senility would have been easy prey. There are early stories (see chap. 5) that in the presence of a group of cattle a grizzly would tumble or roll in tall grass and lure them close, then spring on one.

Grizzlies made inroads on the mission and ranch cattle and also fed on the remains of those that were slaughtered. With the beginning of the American period, as ranches and farms started to keep livestock that was actually domesticated, the bears killed and ate cows, heifers, bulls, sheep, some horses, and particularly pigs—the grizzly, like the black bear, was fond of pork.

While Indians were taking deer for tallow near Mission San Francisco Solano in 1827, bears came to the hunting grounds to feed on the carcasses (Carter, 1929 : 241). Cattle killed in the matanzas proved equally attractive and afforded a vastly greater supply of food. In these feasts the grizzlies were joined by coyotes, vultures, and other scavengers.

The pioneer white hunters were plagued by grizzlies that appropriated their kills of deer, elk, or antelope while the men went off to obtain horses to carry home their game.

If a large carcass could not be consumed at one feeding, the grizzly, like the mountain lion, often buried it. In the late 1840's at Carmel, a hunting party killed several deer and left them afield at night to attract a bear, which they hoped later to shoot. Upon returning some hours after sunup they found a grizzly busily burying the deer. (Revere, 1849 : 40.) Grizzlies also nosed out and raided carcasses buried by the mountain lions. These two big mammals were enemies, as will be detailed later in this chapter. The grizzly, at times, might bury a person it had injured, or one who feigned death in hope of escaping injury. When fresh meat was not readily available, the bears worked over carcasses long dried. They sometimes came to crunch bones discarded outside the cabins of settlers in remote places.

Of lesser game sought by grizzlies, there is mention of the bears plowing up the soil in search of burrowing rodents—ground squirrels (Lyman, 1924 : 249), and pocket gophers. "They did not disdain to hunt and devour small field-mice among the tulé" (Revere, 1849 : 259). Likewise they tore apart logs to find wood-boring grubs (insect larvae). Xántus (1860b : 268) said that in getting at these larvae they "sometimes lift such huge trunks, that not even two oxen could move . . ." Nests of yellow jackets in the ground were taken as found. Wild colonies of bees in standing trees were beyond the reach of grizzlies, but apiaries on farms were raided. As early as 1860, Xántus reported from

Fort Tejon that grizzlies would "sack the bee-hives," reportedly eating the contents of six or eight at a time. Between 1879 and 1880, bears were shot at apiaries in Pasadena (Giddings in Reid, 1895 : 129). Dr. Loye H. Miller told us that in the summer of 1884, in San Antonio Canyon above Ontario, a bear entered a honey storage house by making a hole in the roof, knocked the spigots off honey tanks, and feasted, letting honey run all over the place. Another bear "caused considerable damage" by raiding beehives in Trabuco Canyon, Riverside County, in January, 1903 (N 101).

In summary, then, the animal food of California grizzlies ranged from whales to insects, a greater variety than partaken by any other animal. The livestock brought by white men proved a boon to the bear population, and campaigns to exterminate the grizzlies were undertaken mainly for the prevention of attacks on this resource.

The plant kingdom afforded the grizzlies an even greater variety of food than the animal sources. It was the opinion of some writers that the grizzly fed predominantly on "vegetable" materials, but we cannot now assess the relative contributions from these two supplies. The big bear, by its physical constitution, could take more widely of plant products than any other mammal. It could graze like a deer or cow, garner berry crops like a squirrel or robin, plow the soil for roots or bulbs (as well as insects), and reach to a height of 9 or 10 feet, well beyond the stretch of the tallest elk. For acorns or other tree crops young grizzlies could climb into the trees and the adults could walk up on large, low limbs.

Many early accounts tell of the grazing habits of grizzlies. The bears congregated locally to forage in wild pastures and made seasonal migrations to such places. Near Carmel Mission in the late 1840's, Revere (1849 : 39) came upon "a luxuriant field of wild clover growing up about as high as the bellies of our horses, which the bear had evidently been demolishing. Bruin, like some other animals, loves to be 'in clover,' and his choicest Apican morsels consist of the ripe sweet heads of that fragment grass, which he takes by way of dessert after venison." Near Bakersfield during the spring bears fed on red clover, alfilaria, and "purple flower grass" [brodiaea?] (Mackenzie, in Seton, 1929 : 26). Bear Valley in the San Bernardino Mountains gained its name from being the resort of many grizzlies that came to forage on the clover, which reached its height toward the first of September and was the "favorite food of the grizzlies" (N 53). The clover-eating habit was also noticed in valleys of San Diego County in the same season (N 24). In the Sacramento Valley the bears used clover during spring and early summer (N 2). Later, when there were far fewer grizzlies, a large one fattened on the luxuriant clover and wild oats of San Andreas Valley in San Mateo County (Evans, 1873 : 56–66).

Bears foraged in man's cultivated fields. There is an engraving of "Thieving California Grizzlies in a Wheat Field" (Pennoyer, 1938 : 86), the source and artist of which are unknown, showing the bears making their own harvest. A cornfield between Redding and Castle Rocks (in the early 1870's?) was nearly destroyed; the owner reported that grizzlies were very fond of ripening ears and would "tread down acres in a single night" (Boddam-Whetham, 1874 : 200). Some years earlier, near Port Wine (now La Porte, Plumas County), a man was killed by a grizzly when he tried to protect his corn patch from robbery (T.B.M., 1907 : 30–31).

Grizzlies dug for plant foods of several kinds. Many authors merely say "roots," although we think that bulbs, such as those of brodiaea, may have been among the foods so obtained. On June 9, 1862, W. H. Brewer (Farquhar, 1930 : 279), in the region south of Mount Diablo, wrote: "We found Mount Oso rightly named—Bear Mountain, of the Spaniards—for the whole summit had been dug over by bears for roots." At Fort Tejon, Xántus (1860b : 268) observed similar results: "This bear sometimes amuses himself with digging, like the pigs, and sometimes during a moon-light night, he will dig up many acres of lands, [so] that not one [blade of] grass is to be found on it." J. S. Hittell remarked (1863 : 110): "The roots which he [the grizzly] eats are of many different species, and it was from him that we learned the existence of a California truffle, very similar to the European tuber of the same name." Roots of the tule (*Scirpus*) also were reportedly sought (Cassin, 1858 : 14).

Berries in great variety were eaten by the grizzlies, as has been mentioned. Some of the crops were local and available for but a short season; others could be found in many places and over a considerable period. Berries of the manzanita (*Arctostaphylos*, many species) were in the latter category, and the bears ate quantities of them. Chaparral thickets that included manzanita bushes provided both food and shelter for the animals. When J. S. Newberry (1857 : 48) traversed parts of the Modoc lava beds in 1855, he found the mountain slopes "covered with bushes of service berry [*Amelanchier*], gooseberry of several kinds, plum and cherry trees, all loaded with fruit, upon which the numerous tracks proved many bears were feeding." On the Salmon River, Siskiyou County, during late summer, Wistar (1937 : 262) discovered a thicket of wild plums, covering less than an acre, every tree bending under an untouched crop of wild fruit. Upon returning "before sunrise next morning, we found only a scene of devastation and ruin, some grizzlies having visited it in the interim and broken and torn down every tree, smashing and destroying such fruit as they could not eat . . ." The sink of Cache Creek, Yolo County, was once overgrown with tules and blackberries; in early summer, when berries abounded, so did grizzlies in search of these fruits (McCauley, 1910 : 5).

More writers speak of acorns than of any other single food. One of them said: "When acorns are ripe the grizzly grows fat and heavy—his belly drags along the ground" (Anon., 1857 : 820). From Oregon to the Mexican line and from the seacoast to middle altitudes of the Sierra Nevada the California landscape has oaks in profusion, including the great and graceful valley oaks of the Sacramento–San Joaquin flats. In grizzly days, before the century of clearing, burning, and overgrazing, the country must have been more widely and densely covered by oaks than it is today. Acorns were a staple of grizzly diet wherever available, and the bears migrated locally to oak groves when the crop was ready. Indians also depended importantly on these nuts, but they sometimes avoided traveling among the oaks when grizzlies were at *their* harvest.

Joaquin Miller (in Muir, 1894 : 154) wrote that he had "seen this animal feeding under the oaks in Napa Valley in numbers together, . . . as composedly and as careless of danger as if they had been hogs feeding on nuts under the hickory trees of the Wabash." The grizzly did not depend only on the acorns that fell. Colton (1850 : 119–120) said that "instead of threshing them down like the Indian, he selects a well-stocked limb, throws himself upon its extremity, and there hangs swinging and jerking till the limb gives way, and down they come, branch, acorns, and bear together. On these acorns he becomes extremely fat, yielding ten or fifteen gallons of oil . . ." Stillman (1877 : 130) saw an oak stripped of acorns and the ground strewn with fresh leaves and little branches where a grizzly had been feasting. At Fort Tejon in 1853, Blake (1857 : 47) made his campfire of branches downed by grizzly harvest. This army post was "among large, umbrageous oak trees that bore large crops of acorns, and therefore had been a great rendezvous for grizzly bears . . ." One visiting bear stampeded all the horses and mules in camp (Edgar, 1893 : 26). Almost the only writer who mentioned the grizzly in Yosemite was John Muir (1917 : 194), who watched one eating acorns under a Kellogg black oak. In the Russian River Valley, according to Revere (1849 : 145), the bears came down to the plains during September to feed on ripe fallen acorns.

Frémont and his party of 1846 were witness to the harvesting of acorns by grizzlies in September in the Salinas Valley:

Where we were riding on the prairie bottom, between the willows and the river hills, some clumps of shrubbery hid for the moment an open ground on which were several of the long-acorn oaks [the valley oak, *Quercus lobata*]. Suddenly we saw among the upper boughs a number of young grizzly bears, busily occupied in breaking off the smaller branches which carried the acorns, and throwing them to the ground. Seeing us as soon as we discovered them, they started to climb down but, were apparently checked and driven back, running backward and forward in great alarm. Dismounting quickly and running into the open we found the ground about the trees occupied by full-grown bears, which had not seen us, and were driving the young ones back until the jingle of our spurs attracted their attention. In their momentary pause the young ones clambered down and scampered into the brush, as did some of the larger ones. (Frémont, 1887 : 571.)

Yet twelve, old and young, were killed. This community harvest of acorns presumably was duplicated in many parts of California when oaks and grizzlies abounded.

As a result of the bounteous and varied natural food supply, grizzlies often accumulated quantities of fat (see chap. 7). This reservoir tided the bears over any seasonal scarcity and was especially important to females suckling cubs.

## DRINKING

Grizzlies of California, and particularly those of the lowlands, had periodically to seek drinking water. The places where they drank were even mentioned by some early travelers. Bruff (1949 : 351), in the Sierra Nevada in 1850, found a rocky basin used as a drinking fountain by "a large grizzler." At Fort Tejon in 1853, "they frequently came to the water to drink, in the evening, just after sunset" (Blake, 1857 : 47). From Inskeep's Valley, Tehama County, then a remote place well populated by grizzlies, a newspaper (N 47) reported that "in certain hours of the day, they may be seen in gangs or herds, like wild cattle upon a Spanish ranch, repairing to the small stream which flows through the valley, for the purpose of slaking their thirst." In Shasta County, near Cottonwood, William Kelly (1851, 2 : 98) said: "stopping in a gully to look for water, I found a little pool, evidently scratched out by a bear, as there were footprints and clawmarks about it; and I was aware instinct prompts that brute where water is nearest the surface, when he scratches until he comes to it."

## REPRODUCTION

In the literature on California grizzlies there is no precise information about mating, birth, growth of the young, or other phases of reproduction. To outline this part of the life cycle, we summarize observations by Skinner (1936) in the Yellowstone region. The essential features doubtless were similar in California, but the time schedule and rates of growth in young here differed because of the marked contrast in climatic cycle and food supply. Mating, birth, and maturity may have been earlier in the lowlands of California.

In northwestern Wyoming, females bred first when about three and a half years old, and then every third or second season. The sexes met in mid-June; the pairs remained together about a month and then separated permanently. Each female lived alone through the remainder of summer; she seemed restless and traveled far, possibly seeking a well-drained den site. She excavated a retreat and retired in October, insulated from cold by a 3-inch coat of hair and fat 2 to 4 inches thick under the skin. The den was soon blanketed by snow except for a small breathing hole.

Young reportedly were born in mid-January[1] and usually numbered two or three but occasionally one or four. They had fine short hair but appeared naked, were about 9 inches long and weighed 8 to 12 (one 18) ounces. Forty days after birth, when the cubs were 12 inches long and weighed 2 pounds, the eyes opened and milk teeth were emerging from the gums. The female aroused about mid-April and soon was abroad, her fat exhausted, and ravenously hungry. The cubs then weighed 5 to 8 pounds and were the size of rabbits. They nursed until about August and then began taking solid food. At this time they weighed 20 to 40 pounds and possibly doubled in weight before going into hibernation in company with their mother. By the end of the second summer they weighed 100 to 200 pounds and then denned together but apart from their parent. In the third summer they began separate lives. (Skinner, 1936.)

Some details of mating and young in captive grizzlies are given by Seton (1909 : 1042–1045). The Central Park Zoo in New York City received a female in 1884 and obtained a male in 1891, from sources not stated; both were full grown on arrival. They produced a cub in 1899, and during July, 1900, mated many times. On January 17, 1901, a young female was born; it was 8½ inches from tip of nose to end of tail and weighed 1½ pounds (p. 66). The cub was blind, pale pink or flesh-colored, and covered with short, very fine gray hair. The ears were low, the openings not visible. The little one showed grizzly charac-

---

[1] Skinner (1936) gives the gestation period as 230 (to 266) days, but this disagrees with his statements in regard to the time between mating (mid-June to mid-July) and birth (mid-January)—about 199 days. The young of "Monarch," in San Francisco (Seton, 1909 : 1042) were born 158 days after the parents had mated.

teristics in the shape of head and jaw, hump on shoulders, and paws; but the tail was proportionately longer than in adults. It lived only a few days.

"Monarch," the San Francisco captive, mated with a Rocky Mountain female on July 19, 1904, and on December 23, (5 months and 4 days later) two cubs were born. They were kept

Newborn Montana grizzly cub in Central Park, New York City, Jan. 17, 1901.
Length, 8½ in.; weight 1½ lbs. (Seton, 1909 : 1043.)

hidden by the mother for several weeks. When seen by Seton on March 18, 1905—they were nearly 3 months old—the cubs were about a foot high at the shoulder and estimated to weigh 12 or 15 pounds (p. 67). The pelage was gray except for the ears, feet, and patches around the eyes, which were dark. The cubs were already eating meat, fruit, and bread and were very playful. The mother tried to keep them in the den, but they ran past her and she followed. She often uttered a choppy coughing sound at the cubs. As they were being fed, a heavy rain started and the cage bars slammed loudly. The female immediately raked and cuffed the two young *under her body*; straddling very wide, she sheltered them from the rain and guided them back to the den. The cubs often put their paws through the bars of "Monarch's" cage, and he would sniff at them loudly and utter short quick notes but made no attempt to molest his offspring. (Seton, 1909.)

The only other actual dates of birth recorded are of non-California grizzlies in the National Zoölogical Park, Washington, D.C.—January 15, 1913; January 9, 1914; and January 11, 1953 (E. P. Walker, letter, Nov. 25, 1953).

Early Californians held a curious belief that no pregnant grizzly was ever taken—at least four writers make a statement to this effect—although females accompanied by quite small cubs were met on various occasions. In the words of Colton (1850 : 120), "as soon as she discovers herself with young, she ceases to roam the forest, and modestly retires from the presence of others, to some secluded grotto. There she remains, while her male companion, with a consideration that does honor to his sex, brings

The offspring of "Monarch" in San Francisco, sketched by E. T. Seton, March 18, 1905, when they were about three months old. (Seton, 1909 : 1045.)

her food. She reappears at length with her twin cubs . . ." This ornate description glosses over the state of ignorance then pertaining to these phases of bear biology. Bears were in pairs for mating but not thereafter. It is now known that in some mammals after mating, development of the embryo starts, then ceases for a time, and later is resumed. This phenomenon is referred to as delayed implantation; the early stage remains essentially free in the uterus of the female, then becomes implanted and growth proceeds rapidly. This is known to be true of bears. The early stage (blastocyst) is microscopic in size and cannot be found except by careful laboratory study. Casual examination, as by a hunter, would not show any evidence of this first development. Female grizzlies in California may have become more reclusive or may have retired to their dens for the few weeks during which embryonic development proceeded before birth of the small-sized cubs.

Only one California record hints the time of birth. At Salmon Falls, American River, during December, 1849, a female was seen with one cub the size and color of a Maltese cat, its eyes not yet open (Sacramento *Union*, April 12, 1891). The number of cubs seen with females was most commonly two (at least 23 reports); but females with three were noted (3 reports), and several with only a single young one (5 reports); the last may have started with two cubs.

Grizzly Adams (Hittell, 1860 : 209) reported that the litter was frequently three but that the dam often devoured one cub, and always did so if it died or was deformed. The eyes of cubs opened in a week or ten days and the young cut their teeth when about two months of age (*ibid.*, pp. 71, 290). This hunter said that the grizzly had "a dangerous mouth at six months" (*ibid.*, p. 291), but that the grizzly of California did not have a full complement of teeth until about two years old. He made the further statement (which we have not seen elsewhere) that "every year a ring is added to its tusks,—the first ring being for the second year; and as the animal sometimes reaches the age of fifteen or sixteen years, a corresponding number of rings are found" (*ibid.*, p. 291).

Grinnell (1938 : 75), on the basis of a "fair agreement of testimony from various sources," wrote that "females bred first when they were two years old and bred thereafter every year." Since birth was a winter event, the term "two years" could mean either the second or the third summer of life. The third summer seems more probable to us; this would mean that breeding first took place when the female was about two and one-half years old and that she bore her first litter when she was three years old. The same author wrote that "cubs left their mother, to fend for themselves, when about eleven months old, females being then about two-thirds grown, males not more than one-fourth old-age size," but cited no evidence for these assumptions. Elsewhere in the literature, statements are made that females were seen with "yearling" cubs, which would indicate that breeding did not take place annually. Production of litters in alternate years seems more probable, when it is remembered that females had to provide quantities of milk for the young over a period of months.

The only description of a California grizzly cub we have found is not in scientific literature but in a delightful tale by Bret Harte (1875) of "Baby Sylvester." Taken when its mother was shot, the cub was reared by a Sierran miner.

His body was a silky, dark, but exquisitely-modulated gray, deepening to black in his paws and muzzle. His fur was excessively long, thick, and soft as eider-down . . . He was so very young, that the palms of his half-human feet were still tender as a baby's. Except for the bright blue steely hooks, half-sheathed in his little toes, there was not a single harsh outline or detail in his plump figure (*ibid.*, pp. 179–180).

Two larger specimens of California cubs are known, both being stuffed skins in the U. S. National Museum. No. 1440 (or 1444), collected in November, 1855, in Calaveras County, by J. S. Newberry of the Pacific Railroad surveys, is designated as "six months" of age; the specimen lacks feet. It now (1953) measures 620 mm. in snout-rump length, the ear from the crown is about 55 mm., and the longest hairs on the hump region are about 70 mm. (nearly 3 in.). The snout is pale brown, and the hind legs are nearly black. The top of the head, hump region, and sides are grizzled, and even the sides of the head and chin have some pale hairs.

The larger specimen (no. 11809) is labeled merely "California," without locality, date, or name of collector. A few measurements (in millimeters) are: snout to end of rump, 820; head (snout to back of ears), 230; ear from crown, 70; tail, about 60; girth (hair compressed), 530; longest hair on hump, 100–105; mid-back hair, 40–50; claws of the right middle front toe, 42; and corresponding rear toe, 40.

No detailed records have been found of den life of the California grizzly. A hint toward this phase is the account of a female black bear in Pennsylvania (Matson, 1954). She denned

by December 5, was out once (December 8–11), and gave birth
to cubs on January 3; the family left the den about March 21.
Her confinement was at least 106 days, and she passed no feces
for 78 days.

The female California grizzly with cubs was ever vigilant
and vigorous in defense of her offspring, a fact abundantly testified by many early hunters and others. Even at long range she
would usually take the offensive and only rarely would depart
quietly. Anyone coming upon her at close range was in trouble
immediately, even though he did not molest either the mother
or offspring. If a cub was injured or killed, the mother's rage
was enormous. Two ranchers in San Luis Obispo County, hidden
in a pit, shot a lone cub accompanying its mother. It made the
"most dolorous howls" before it died. "These pitiful cries . . .
enraged the old bear" and she "rushed in a circuit around her
slain cub, looking into the trees and leaping at them, . . . tearing
great pieces of bark and wood from them with her powerful
claws and terrible teeth, uttering frightful howls, as if nothing
but the destruction of something could appease her wrath. This
she continued during the night" (Angel, 1883 : 218).

## RELATIONSHIPS WITH OTHER ANIMALS

Since the grizzly was a common and important element of
the original fauna, we are curious as to how it got along with
the other wild species. Perhaps the most important question is
that of its relationships with the black bear, which, under primeval conditions, was reportedly uncommon in places where the
grizzly abounded. According to Dr. J. S. Newberry, who traveled widely in California during 1855 (reported by Townsend),
"the grizzly was met in many places, to the apparent exclusion
of the black bear . . ." J. L. Wortman said that "in many parts
of the West where Grizzlies abounded [in the 1880's] the black
bear occurs very rarely," and Townsend (1887 : 183) wrote in
the same period that "since the Grizzly began to disappear before the advancement of the settler, the other species has been
more numerous, leading to the inference that the Black Bear
will not be found in abundance where the larger species is well
established."

Old-timers, according to Grinnell (1938 : 71), frequently expressed the opinion that grizzlies would not tolerate black bears within their home territories. In support of this view he mentions that black bears expanded their range in one area favorable to their needs after the larger species was exterminated. His example is "the Tehachapi-to-Santa Barbara tangle of chaparral-clothed mountains, even up to the 1890's the metropolis of grizzlies, as these beasts vanished, blacks (or browns) came in from the southern Sierra, to the eastward, spread and multiplied, until, according to Forest Service reports they are now [1938] relatively numerous in Los Padres National Forest."

Alex Kennedy, a resident of Long Valley, Lake County, however, told us (in April, 1953) that both species lived in that region during the last half of the nineteenth century. No antagonism between the two was noted, and after the grizzly was exterminated the black bears did not become more numerous. A newspaper note from Tulare County when grizzlies were common stated: "The scarcity of acorns has driven the black bears, in large numbers, from their usual haunts in the mountains down among the settlements. At the upper settlement of Tule River, quite a number have been killed, besides two or three grizzlies." (Visalia *Sun*, in Sacramento *Daily Union*, Dec. 15, 1860.) This account suggests that the two species had a common forage ground. Among the oaks, black bears could climb to escape the terrestrial adult grizzlies.

Still another notion about grizzly–black bear relationships was the statement made by James M. Mackenzie of Bakersfield to E. T. Seton (1929 : 71) that "Monarch, the King of Kern [see our chap. 10] was always followed by from one to a dozen Blackbears, who played Jackal to the Lion, that is, they devoured the remains of a beef killed by the King, after he had feasted." No one else ever hinted at such a trait.

In summary, there is evidence that the two species in some places were relatively near one another. The black bear could forage in the absence of the grizzly and could escape by climbing a tree, if not by running, should the grizzly come close.

Three California accounts detail something of the antagonism between the grizzly and the mountain lion, or panther.

Livingston Stone (1883 : 1189) was told by the McCloud River Indians that the panther always killed the grizzly when the two fought. They said that the grizzly was afraid of the lion and that the latter would spring on the bear's shoulders and cut its throat. Stone saw places in the mountains where the ground had been torn up, evidence of a desperate conflict between a panther and a bear. The Indians said they had found bears killed by panthers but no panther a bear had killed.

An actual bear-and-panther fight in the central coast region was watched in the 1840's. Three hunters, originally seeking a female grizzly with cubs, had been grounded by the escape of their horses. Going cautiouly along a creek bordered by willows and grapevines, they approached a waterfall that plunged into a green, transparent pool over which a large tree had fallen.

With the sounds of the torrent came . . . the growls of two wild beasts, alternate and furious.

On the right hand, squatted on one end of the bridge, was a small, male grizzly, and opposite to him, at the other end, a full-grown panther, who was tearing up the bark of the trunk, and gathering and relaxing herself as if for a spring. The alternate roaring of these infuriated beasts filled the valley with horrible echoes.

We watched them a minute or more. The bear was wounded, a large flap of flesh torn over its left eye, and the blood dripping into the pool. My companion bade me shoot the tiger, while he [Colin Preston] took charge of the bear. We fired at the same instant; but, instead of falling, these two forest warriors rushed together at the centre of the bridge, the bear rising and opening to receive the tiger, who fixed her mighty jaws in the throat of her antagonist, and began kicking at his bowels with the force of an engine. At the instant both rolled over, plunged, and disappeared. We could see them struggling in the depths of the pool; bubbles of air rose to the surface, and the water became dark with gore. It may have been five minutes or more before they floated up dead, and their bodies rolled slowly down the stream. (Anon., 1857 : 823.)

Another natural fight between grizzly and mountain lion was described in the San Bernardino *Argus* of 1873 (Ingersoll, 1904 : 371):

Some hunters were witness to a desperate fight in the San Jacinto mountains, the other day, between a mountain lion and a bear. The fight is described as terrific. The superior strength of the bear easily enabled him to throw his antagonist down, but the latter used his paws and jaws so fearfully that the bear could not keep him under. Both animals were covered with blood. They fought till both were exhausted, when the lion dragged himself off to the jungle, leaving the bruno in possession of the field.

The grizzly sometimes appropriated game shot by hunters before the men could recover it—and the huge California condor was a competitor on occasion. Heerman (1859 : 29) wrote that he had "known four of them, jointly, to drag off, over a space of two hundred yards, the [dead] body of a young grizzly bear weighing upwards of one hundred pounds." That condors ate any flesh in company with grizzlies is improbable; they probably took what the bears left. The burying of carcasses, a practice of both grizzly and mountain lion, may have been an effort to circumvent the big scavenger birds. Heerman (*ibid.*) says, however, that a deer covered by shrubbery and heavy branches was found and eaten by condors. Conversely, the condor may have aided the grizzly. Soaring aloft at great height and for much of the daylight hours, one of the birds could locate a carcass. As it dropped down, other condors would converge toward the site. The earth-bound grizzly could find it only by his nose. Maybe the vulture congregations guided the bear to such a feast. When he arrived, however, the birds would flap their wings vigorously and take off to avoid being included in the bear's repast.

To the dominance of the grizzly there was one exception. "The only animal they dread is the little 'sorillo' . . . the . . . skunk . . . If the fragrant guest approaches while the grizzly gentleman is discussing his favorite meal of a long-buried carcass —which he has stowed away himself for the sake of the 'fumet'— the latter will retreat from his dainty repast with a reluctant growl, while master Sorillo quietly takes his place." (Revere, 1849 : 259.)

## LONGEVITY

Information on the span of life in grizzlies is available only for captives in zoölogical gardens. "Monarch," the California grizzly, reached the cage in San Francisco "late in October," 1889, when already of some size and age; he was killed, when decrepit, in May, 1911, after nearly 22 years in captivity. No detailed records on other California captives have been found.

Flower (1931 : 179), in a study on duration of life in mammals, had record of one grizzly in London—which had been given to the Zoological Garden by the Hudson's Bay Company and was probably from Canada—that lived from about 1813 until 1838, or 26 years as a captive. Two females from Montana in the London garden survived for 26 and 27 ½ years, respectively. Flower also mentioned a grizzly in the zoölogical garden at Cologne that lived for 30 years. Another is said (Anon., 1850a : 210) to have lived for 30 years in the Tower Menagerie, London, and then was moved to the Zoological Garden.

Information on grizzlies in the National Zoölogical Park has been provided by Assistant Director Ernest P. Walker (letter, Nov. 25, 1953). These records and those previously mentioned suggest that the big bears have a moderately long life span.

GRIZZLIES IN NATIONAL ZOÖLOGICAL PARK, WASHINGTON, D.C.

| Source | Sex | Date Received | Estimated Age on Arrival | Date Killed or Died | Duration of Life (Years) |
|---|---|---|---|---|---|
| Montana............ | ♂ | June 4, 1888 | ...... | Feb. 6, 1904 | 16+ |
| Yellowstone National Park..... | ♂ | June 30, 1894 | adult (730 lbs.) | May 9, 1913 | 19+ |
| | ♂ | June 28, 1902 | 2 (190 lbs.) | Mar. 31, 1906 | 6 |
| | ♂ | Sept. 13, 1902 | 4(?)ª | Feb. 2, 1919 | 21(?) |
| | ♂ | July 29, 1908 | 3½ (415 lbs.) | Mar. 27, 1922 | 17¼(?) |
| | ♀ | Aug. 3, 1910 | adult (325 lbs.) | Oct. 23, 1934 | 24+ |
| New Mexico......... | ♂ | Aug. 8, 1918 | 8 mos. | May 27, 1936 | 18 |

ªWeight, Jan. 31, 1903, 500 lbs.

Conditions in a zoölogical garden differ greatly from those in the wild. The animals have food, although not always of the most appropriate sort, and are sheltered from attack; but the usual cage is far more confining than the animal's normal environment, and the possibilities of disease or parasitism are greater. It seems unlikely that grizzlies in nature often attained ages as great as those of captives. Some skulls of wild grizzlies show indications of advanced age by the degree of tooth wear and the thickening and roughening of the sutures and ridges on the brain case, but there is no basis for estimating their actual age. Accidents and injury from attack by others of their kind probably resulted in the death of wild grizzlies long before they reached a life span approaching those of the sheltered captives.

## DEATH

The grizzly was so formidable that it probably had few real enemies. It seems likely that once a bear was fully mature, only the physical forces of nature, disease, or actual senility could cause its death. Of grizzlies in the Rockies, Enos Mills (1919 : 58–60) mentions individuals killed by a forest fire, a desert flood, a falling stone from a cliff, and a snowslide. He found one fat young grizzly which, though apparently healthy, had been frozen to death while hibernating during an extremely cold winter with little snow, and had a report of an old bear that perished similarly.

The end for one grizzly on the Salmon River in Siskiyou County, in 1849 or 1850, was thus described by Wistar (1937 : 262–263):

Down in the deep, gloomy bottom of one of its darkest and most secluded cañons, I once came upon a curiosity seldom found anywhere, in the shape of a complete and untouched skeleton of a grizzly, unfound even by the wolves and foxes. It was bleached clean and white, with just enough of the cartilaginous attachments remaining to hold all together. The position was one not unfrequently assumed by the animal in death, that is, prone on all fours, the head resting on the forepaws, something like a dog which waits impatiently for his master.

# 4 Grizzlies and Indians

When anyone but an anthropologist speaks of the Indians of California he almost invariably calls them "Diggers." This name, brought from Great Salt Lake and Humboldt Valley by white settlers, is one of the most inappropriate terms ever used to designate a geographical association of peoples. It is a derogatory misnomer, for it implies that the natives were primarily root-grubbers—which they were not. It expresses a contemptuous failure to appreciate the cultural complexity of men and women who lived and thought differently from Europeans; and, furthermore, it ignores the very important fact that the natives did not belong to a single homogeneous group.

## CALIFORNIA INDIANS

Actually, the Indians of California displayed a remarkable diversity of physical types, and they used at least 135 idioms that can be classified into no fewer than a half dozen major tongues. An example of this extraordinary linguistic profusion is indicated by a few of the names for bears listed in Appendix C.

No other region in North America the size of California contained so many different tribes. Most of the Indians of the state, however, did not recognize "tribes" in the sense in which we use the word. The highest political unit usually was the village, which the Spaniards called a *ranchería*. Between rancherias, even in the same area, there was often considerable diversity of beliefs, customs, and habits; and generalities about the aboriginal life tend to obscure this provincialism. Nonetheless, for the sake of intellectual order, ethnologists have had to erect a system of tribal groupings; although this may seem false to the natives, it is based on real morphological, cultural, and linguistic relationships. The locations of tribes mentioned in this chapter are given in Appendix C.

Because the material culture of our sophisticated civilization more or less insulates us from nature, it is difficult to become fully aware of the intimacy with the out-of-doors that was a necessary part of the life of Indians. All aspects of nature—alive, inanimate, or imagined—had a profound effect on their existence. Even the smallest natural objects and the most inconspicuous changes in the seasons or the weather were full of meanings that are lost to us. Wild animals in particular were invested with extraordinary significance, and for size and ferocity the most impressive species encountered in California was the grizzly. It was the only truly dangerous animal. Whether an Indian met the bear in its actual form, as known to the white man, or saw it with imagined and usually evil attributes beyond our comprehension, it was an integral part of his environment and one of the factors that shaped his life. In this chapter we sketch the extent of this influence on the aboriginal Californians and explore the different ways in which the various tribes reacted to the presence of the beast. Their responses provide us also with a few clues regarding the natural history of the grizzly.

The information in anthropological literature on which we have to rely is tantalizingly meager. In fact, there is practically none for the coastal region south of San Francisco Bay where grizzlies were most abundant and no black bears were present. The Spaniards who colonized that region left relatively few ethnological records, and the tribes they introduced to Christianity expired or forgot their ancestral habits before modern-day researchers were afield.

## ATTITUDES TOWARD GRIZZLIES

The grizzly was certainly the one formidable animal in the environment of the Indian almost throughout the area of California. Before the coming of the Europeans, the big bears rather than man dominated the scene, because the natives lacked adequate weapons and were afraid of grizzlies, whereas the bears had little to fear from any living creature. They were at the top of the food chain and could treat native man with contempt. That fact alone explains many attitudes of the Indians toward the great carnivore. He was their hereditary enemy (Goddard, 1903 : 5) and the most evil and odious being of which they could conceive (Powers, 1877 : 240). His ferocious disposition, according to the Yokuts, was clearly evident even in death when the muscular fibers bristled erect as his flesh was cut with a knife (Kroeber, 1925 : 526).

Because of this belief in the inherent wickedness and ferocity of the grizzly, the most effective curse a Wintun could hurl at another man was "May the grizzly bear eat you!" or "May the grizzly bear bite your father's head off!" (Powers, 1877 : 240). By contrast, the black bear was considered sacred and lucky. It would run from the Indians, and they could hunt it for its flesh and pelt without fear.

In the cosmology of the Luiseño, grizzlies, along with stinging weeds and rattlesnakes, were the avengers whom the great God Chungichnish invoked upon those who disobeyed his commandments (C. G. Du Bois, 1908 : 76). The Wiamot, or Chingichnich of the Juaneño, is said to have announced that after his death he would ascend to the stars and from there he would watch people and send bears and other terrible things to punish the faithless (Kroeber, 1925 : 638). The Pomo had a hedonistic conception of heaven and believed that the good Indian, after death, would enjoy its delights:

in some far, sunny island of the Pacific—an island of fadeless verdure; of cool and shining trees, looped with clinging vines; of bubbling fountains; of flowery and fragrant savannas rimmed with lilac shadows, where the purple and wine-stained waves shiver in a spume of gold across the reefs, shot through and through by the level sunbeams of the morning—they [good Indians] will dwell forever . . . [and] the deer and the antelope will joyously come and offer themselves for food; and the red-fleshed salmon will affectionately rub their sides against them, and softly wriggle into their reluctant hands (Powers, 1877 : 154–155).

But the ghosts of the wicked Indians had to stay behind in the bodies of miserable and tormented grizzlies, forever roaming the wilderness to be hated and loathed by all who saw them. The Wintun likewise believed in this unhappy palingenesis and would

not partake of the flesh of the big bears for fear of absorbing
lost souls (*ibid.*, p. 240). It was bad luck, too, to dream about
grizzlies, for one who did would then sing bear songs, throw fire
around in a hysterical manner, grunt, and walk on all fours
(Loeb, 1932 : 40).

## HAZARDS FROM THE BEARS

The terror that the grizzlies inspired among the Indians is
not surprising in view of the superstitions associated with the
beasts, as well as their natural ferocity. A Shasta expressed the
firm conviction that "The biggest man is scared of a grizzly. He
will cry and tremble. Anyone who had had trouble with a grizzly will just bawl and cry. If you just hear one, it scares you to
death. You may not know you are shaking until you light your
pipe and your hand will just be shaking. Nothing else has that
power." (Holt, 1946 : 311). When Dr. Pickering traveled in California with the United States Exploring Expedition during 1841
he noted that, because of the grizzly, Indians "kept on the hills
and other high ground, very carefully avoiding the favorite resorts of this animal" (Cassin, 1858 : 14).

We have little evidence of the frequency with which grizzlies inflicted bodily injury or death upon the natives, and some
of it, like the foregoing, is only circumstantial. For example, the
Wintun forbade its young men to eat the sinews of deer because
of the belief that the tendons would shine at night and reveal the
youths to grizzlies (Cora Du Bois, 1935 : 9). Yet a man who had
the fortitude to withstand the agony of fire would hold his face
in flame and absorb a shower of sparks, which, thereafter eman-
ating from his countenance, might induce a grizzly to take flight
(*ibid.*, pp. 12–13). The Pomo did not allow a man to go hunting
alone until he was twenty-five or thirty years of age, for fear of
bears (Loeb, 1926 : 181).

When the explorer William Kelly (1851, 2: 93) was in the
Sacramento Valley north of Red Bluff in the autumn (of 1849?),
he recorded that the Indians wanted to remain with his party all
night, "giving us to understand (as they went away reluctantly)
by signs and noises, that they were afraid of being attacked by
the bears [which were abundant here]." A similar attitude was

related by James O. Pattie, who in 1828 when crossing the Sierra Juarez of Lower California on foot suffered such anguish from an inflamed leg that he had to stop while the remainder of his party went on to the Mission of Santa Catarina for a horse. In his journal he says that "the Indians gave me the strictest caution against allowing myself to go to sleep in their absence. The reason they assigned for their caution was a substantial one. The grizzly bear, they said, was common on these mountains, and would attack and devour me, unless I kept my guard." (Flint, 1930 : 266.)

While in the land of the Costanoans near what is now Crystal Springs Lake, south of San Francisco, Fray Pedro Font, who accompanied the Anza expedition of 1776, wrote in his diary that there were many grizzlies abroad and that "they often attack and do damage to the Indians when they go to hunt, of which I saw many horrible examples" (Bolton, 1930 : 349, 362). At Mission San Luis Obispo in 1772, bears were said to have been very destructive, and "not a few of the Indians showed that they had been lacerated and maimed by their terrible claws" (Hittell, 1885, 1 : 347). José Longinos Martínez, in California during 1792, wrote that the bears killed many Indians, "within a short time I have seen two dead gentiles, victim of this ferocious animal" (L. B. Simpson, 1938 : 34–35).

In replying to an *interrogatorio* on the Indians, submitted by the Spanish government in 1812 to the civil and ecclesiastical authorities of California, the fathers of Mission Santa Ynez wrote of the Chumash Indians that "in gathering their fruits and similar wild seeds, they were continually in danger of being attacked by the bears . . ." (Engelhardt, 1932 : 15). Similarly, at Mission San Buenaventura, the padres noticed that during ceremonies for rain the Chumash took precautions to insure "that no bear might catch them" (Engelhardt, 1930 : 34). The San Francisco *Californian* of March 15, 1848 (N 1) reported that three young Indians at the Rancho of Señor Armijo in Suisun Valley had been killed by a bear. J. Quinn Thornton (1855 : 90), who visited California the same year, commented that grizzlies "sometimes attack and devour the savages." J. S. Hittell (1863 : 110) wrote that during seasons of scarcity they "break into the huts of Indians and eat them." Townsend (1887 : 184) saw a Shastan whose patella had been bitten off by a grizzly, leaving a "cancerous sore" that under "barbarous Indian treatment" had not healed in twenty years. Chever (1870 : 130) stated that he had encountered "Indians bearing the scars of conflicts with grizzly bears." Strong (1926 : 60) tells of the killing of two Cahuilla women by a grizzly about 1875 in the San Jacinto Mountains. From the foregoing evidence we may conclude that the grizzly injured or killed Indians sufficiently often to keep them in a state of respect and terror.

It was on this basis that the early-day poet Joaquin Miller explained the little mounds of stones formerly found on old Indian trails that led from one tributary to another along the headwaters of the Sacramento River:

It was the custom for each Indian as he passed the place where one of his people had been killed or maimed by one of these monsters, to pitch a stone or pebble onto the spot. And thus from year to year the mound of stones was formed. No doubt some sentiment of pity or respect lay at the bottom of the custom; but back of that lay the solid and practical fact of a warning to all unwary passers-by that the grizzly bear had been there and probably at that moment was not many miles away (Muir, 1894 : 150).

Early in the Spanish period of California, a sure way to earn the gratitude of the natives was to destroy grizzlies. The Indians were delighted that at last human beings had come into the land who could successfully dispute the domination of the bears, and they often showed their pleasure with gifts. Such was the experience of the founders of Mission San Luis Obispo a few months after Don Pedro Fages and his troops had been in the vicinity slaughtering grizzlies. According to Fray Francisco Palóu:

Although in that region there was no village of any kind, the Indians soon began to arrive . . . it had been only three months since the soldiers had been there at the time of the slaughter of the bears (for which they were very thankful, as the land had been rid of these fierce animals, who had killed many of the Indians, of whom not a few of those who were still alive showed the terrible scars of their dreadful claws), they were glad to show themselves delighted that our people had come among them to live. They began to visit the Mission with great frequency, bringing little presents of venison and wild grains to the Father. (Palóu, 1913 : 137.)

Some Indians lived where both grizzly and black bears were present; they evidently feared only the grizzly. For example, Susan Little, the one Indian living on the Hupa Reservation in 1952 who had had personal experience with the grizzly, recounted the following incident (to Robert Talmadge). In 1860 when she and some playmates were gathering hazelnuts in the company of an elderly woman, they heard bear sounds in the bushes. The children, not being afraid of black bears and knowing nothing of the then rare grizzly, wanted to sneak up on the animal and frighten it away with sudden yells. But their guardian, the instant she got a glimpse of the bear, ordered them to freeze in their tracks—"Stand like sticks," she said quietly.

Susan remembered the beast as "a great big brown with a great red mouth and big red eyes." (The Hupa called grizzlies "great big browns" to distinguish them from "little blacks and browns.") The animal was tearing up a thicket of hazel bushes. The children were so terrified that they wanted to run, but the old woman kept them still, even for a long time after the grizzly had gone.

## GRIZZLY HUNTING

One result of the respect the Indians had for grizzlies was the reluctance of many tribes to trap or hunt the beasts. Seth Kinman, a famous early-day bear hunter of Humboldt County, told Powers (1877 : 101–102) about the reaction of a company of Wiyots who by mistake caught a grizzly in a rope noose set for elk:

One day an Indian came running to his [Kinman's] cabin with all his might, desperately blown after a hard six-mile stretch . . . Panting and puffing, and in a drip of perspiration as if he had just emerged from the sweat-house, he made out to reveal his errand by pantomine some time before he recovered his wind. Kinman quickly caught down his rifle and they ran back together. Arrived on the spot he found an enormous grizzly bear snared in the noose, frantic with rage, roaring, lunging about, dragging down the bushes and saplings with the pole, and throwing himself headlong when suddenly brought up by some tree. The Indian would not venture within rods of him. Kinman slowly approached and waited for the mighty beast to become a little pacified. He waited not long though, lest the rope might chafe off, and presently drew up and sent a bullet singing into his brain. The great brute fell, quivered, then lay quiet. But it was only when Kinman approached and stamped on his head with his heel that the cowardly Indians were assured; and then from all the forest round about there went up a multitudinous shout. From a score of trees they scrambled down in all haste. Not more than a dozen had been in sight when Kinman arrived on the ground, but now scores collected in a few minutes, gazing upon the enormous brute with owl-eyed wonder, not unmixed with terror.

Not all Indians, however, were as reluctant as the Wiyots to tangle with their great enemy. When Jedediah Smith was exploring the Sacramento River close to the mouth of Chico Creek during 1828, he found circumstantial evidence that the Maidu, at least, had the temerity to shoot arrows at bears:

In the evening several of us went out hunting for there was considerable sign of Bear Deer Elk and Antelope in the neighborhood. Mr. McCoy and J Palmer killed a large Grizly Bear in tolerable order and on opening him found nearly in the center of the lights a stone Arrow head together with about 3 inches of the Shaft attached to it. The men brought that part of the lights containing the arrow into camp. The wound appeared perfectly healed and closed around the arrow. 3 indians who came with us to camp were busily employed on the share of the Meat alloted to them and on the entrails of the Bear. They filled themselves so completely that they were puffed up like Bladders. (Sullivan, 1934 : 76.)

There were several reasons why some Indians undertook the dangerous task of combating grizzlies. For one thing, aborigine and beast were competitors for the same kinds of food, a state of affairs that was bound to bring conflict at certain times of year. The early journals have numerous accounts of grizzlies coming down out of the mountains to the lowland valleys in the spring and early summer to feed upon clover (*Trifolium*) "like herds of swine" (N 2). It is less well known that Indians did also. John Work, who explored the Sacramento Valley in 1832–1833, was astonished to see the Maidu "spread over the plain and gathering & eating different kinds of herbs like the beasts" (Maloney, 1943 : 336). Other early observers also remarked on the novel sight of numbers of Indians on all fours cropping the heads of succulent clover like so many beasts of the field. In the autumn, the acorn was the principal item in the

diet of both man and bear. T. R. Peale in 1841 observed that
both "Indians and bear . . . thrash down the acorns, which is
almost as effectively done by one as by the other" (Cassin, 1858 :
14). When grizzlies came to rob the caches that the Wintun
had hidden in the forks of trees, the Indians would try to drive
them back to the hills by setting fire to the grass or by going
after them with spears.

The Kato had a more pressing reason for hunting the grizzly.
They needed the hide for the "bear man," who made war on
the human enemies of the tribe (Loeb, 1932 : 45). The Cahuilla
preferred to leave the beasts strictly alone. If they met one they
tried to persuade it to be peaceful. But if a man-killer disregarded
these diplomatic overtures the various clans joined forces to hunt
him down (Strong, 1926 : 60). The Wailaki killed to avenge the
victim of an attack (Loeb, 1932 : 88).

Apparently the westside Monos, who in a sense were not
truly Californians because they were an isolated offshoot of the
Shoshonean stock, were the only people who made a sport of
killing grizzlies. It was their opinion, on the basis of considerable
experience, that grizzlies were easier to hunt than black bears.
A black bear always looked about suspiciously while he ate and
therefore was difficult to approach, they said, but a feeding grizzly cared only about filling his stomach. (Gayton, 1948 : 262.)

The Monos had a morbid interest in bears, and much of
their conversation revolved around the beasts (Gifford, 1932 :
50). Of them Powers (1877 : 397–398) said:

They are not such a joyous race as the Californians . . . Their business is with war, and fighting, and hunting; hence they have more
taciturnity, more stern immobility of feature, than the Californians.
. . . They pursue and slay the grizzly bear in single-handed combat,
or in companies, with bows and arrows, but the Yokuts hold that
animal in mortal terror.

## HUNTING METHODS

Actually, the Yokuts sometimes tried to kill grizzlies by
shooting arrows at them from the safety of a nest constructed
high in an oak under which bears fed on moonlight nights
(Kroeber, 1925 : 529). The Salinans, however, instead of taking
to the trees, hid in carefully excavated holes from which, with

some chance of survival, they could harass a grizzly who came to the bait (Mason, 1912 : 124).

A safer variation of the pit method was used by the Costanoans, at least during Mission times:

They dug a large hole, about five or six feet deep, directly under the branch of a tree, covered it with brush and a light coating of earth, and made all smooth on top. From the branch would be suspended a quarter of beef. Bruin would scent the meat, and, approaching without suspicion, would fall headlong into the pit. Shooting with bows and arrows, the Indian, having come out of his place of concealment, would presently kill the bear. (Davis, 1889 : 312.)

Torchiana (1933 : 416) adds clubs to the weapons reputed to have been used during such skirmishes.

The Kato and Pomo, who in some respects were almost as bold and battle-loving as the Monos, conferred the title of "chief grizzly bear hunter" on their war leader. It was his duty to go alone into a grizzly's cave and drive the beast out with a club while his warriors stood by to greet its emergence with a hail of arrows. (Loeb, 1932 : 45.)

Besides these rather fragmentary accounts of the hunting of grizzlies by Indians, we have detailed descriptions of two distinct techniques.

Before a hunter of the Atsugewi or Shasta started on the trail of a grizzly, he danced the war dance as though he were setting out in pursuit of a human being. When he found the animal's cave, he blocked the entrance and signaled for help. Hearing his call, the other Indians held a conclave at which a shaman sang and, receiving portents of the future, foretold the outcome of the approaching battle. Then they marched to the cave. There, two or more of the bravest held poles across the entrance in the form of an X. By that time the grizzly was angry and usually needed little persuasion to rush out; but if for any reason he was hesitant, they argued with him in kin terms. When he appeared and struggled to climb over the barrier, they pushed up the poles, forcing him against the roof of the cave. After holding him for a few seconds in that vulnerable position, they shot him with poisoned arrows. (Garth, 1953 : 134; Holt, 1946 : 311.)

Frank Latta told us of an incident in which Indians used

this method of hunting bears. Long ago a female grizzly gave birth to young in a shelter under Hospital Rock near Sequoia National Park. In an effort to capture one of the cubs, several Indians went to the cave with long poles, which they pointed at the entrance, forming a semicircle like the spokes of a half wheel. As the bear came storming out, the men tried to hold her at bay to permit a youth to slip behind her and grab a cub. They were no match for the beast, though, and, while one man was being scalped and the others knocked about, an old Indian grandmother called out "Haliwanshee!," which means, "Damn fools, get out of here quick!" Many years later that vigorous expression in the native tongue was suggested to the Department of Highways as a euphonious name for the new mountain road into the Park, but when the officials obtained the translation they chose the prosaic "Generals Highway" instead.

The second technique of hunting grizzlies for which we have a definite account was a specialty of the Maidu. The Indians who shot the arrow that Jedediah Smith found in the "lights" of a bear were presumably of this tribe.

Grisly bears were hunted only by those who were very fleet of foot, and renowned hunters. The grisly was never attacked except by a number of men together, and in the foot-hill region in the following manner: Four or five men would go in a party, and all but one would hide behind trees or rocks in the vicinity of the bear. One man then went as near the bear as possible, and shot once, or twice if he could. He then ran, followed by the bear, toward the place of concealment of one of the other hunters. Slipping behind the tree or rock, the first hunter would stop; and the fresh runner would instantly jump out and run toward the place where another man was concealed. The bear would follow this second runner, and as he passed the tree or rock, the first would again shoot at him. The second runner would similarly change places with the third man, who, running toward the fourth, would lead the bear away again. Thus each hunter had time to rest, and to shoot several arrows, while the other men were taking the attention of the bear. By thus changing off, they tried to tire out the bear, and fill his body full of arrows until he finally succumbed. It was always, however, dangerous sport, and not infrequently several of the hunters were killed. (R. B. Dixon; 1905 : 194–195.)

## CELEBRATING VICTORIES

A victory over the big hereditary enemy was an important occasion, as might be expected. Thomas Jefferson Farnham (1849 : 381), who traveled in California during the middle of the last century, wrote: "It is seldom that the Indians, with their imperfect weapons, venture to attack this formidable animal; and whenever one is killed by them, the occasion becomes a matter of great rejoicing, and the fortunate victor is ever after held in great esteem by his comrades."

Long ago, when the Cahuilla near Palm Springs conquered a man-killer from the San Jacinto Mountains, the desert air rang and throbbed all night with the singing and dancing of the people around the prostrate form of the monster (Strong, 1929 : 76). The Wintun, possibly as a cathartic against the fear engendered by a live grizzly, indulged in hysterical hilarity while the hide of a dead one was being fleshed. They threw pieces of the meat at one another, and anyone who was hit had to plunge into the river. Later, by the glow of campfires, the animal's head was put in front of a young and vigorous male dancer, who pantomimed the killing of the bear with a split-stick rattle. He dodged back and forth, striking at the silent head, while the rest of the company sat around in a circle nodding their approval. (Cora Du Bois, 1935 : 12.) The Wailaki hung the hide outside the sweathouse where all the people could beat it with their fists (Loeb, 1932 : 88). When the Luiseño were so fortunate as to kill a grizzly, they erected a cairn at the spot (Sparkman, 1908 : 199).

## USE OF THE GRIZZLY

At least five tribes ate flesh of the grizzly. One of these was the Nomlaki, who, however, attached more importance to preserving the pelt (Goldschmidt, 1951 : 401); another was the Patwin (Kroeber, 1932 : 277). The Salinan did not relish old bears but considered the cubs a delicacy (Mason, 1912 : 121), and the Klamath baked the paws in ashes and then skinned them (Spier, 1930 : 160). The Atsugewi ate a grizzly only if it was known not to have killed a man (Garth, 1953 : 134).

Among the tribes having a horror of grizzly flesh, the Wintun and Yurok believed that, since bears devour Indians, to feast on them would be to practice cannibalism (Cora Du Bois, 1935 : 12; Kroeber, 1925 : 526). Apparently *lex talionis*, the law of retaliation, did not apply. The Yurok said simply that they did not eat grizzlies because grizzlies ate them (Gibbs, 1853 : 128). Other tribes known to have included the big bears on the list of taboo foods are the Valley Maidu (Curtis, 1914 : 107), the southern Maidu (Kroeber, 1925 : 409), the Pomo (Gibbs, 1853 : 113), and the Luiseño (Sparkman, 1908 : 199). A few Gabrieliños ate grizzly, but in general it was rejected on superstitious grounds (Hugo Reid, 1926 : 11). A girl of the Kawaiisu who was frightened by a grizzly while under the narcotic influence of the juice of Jimson weed, thereafter was forbidden bear meat (Kroeber, 1925 : 604).

Like the tribes of the Rocky Mountains that came in contact with grizzlies, the Luiseño in California saved the claws to make necklaces which were worn at certain dances (Sparkman, 1908 : 210). Similarly, the bear shaman of the Shasta performed his esoteric rites with a collar of claws locked round his neck (R. B. Dixon, 1907 : 486). According to the priest in charge of Mission San Carlos, in his reply to the Spanish *interrogatorio*, the Costanoans suspended a grizzly tusk from the neck of any man who killed a bear. This was done "in token of an heroic feat and bravery" and thereafter he was respected. (Engelhardt, 1934 : 132.)

Among tribes such as the Kato and Owens Valley Paiute, with whom the institution of bear shamanism involved physical impersonation of the beast, the hides were saved to clothe their "bear men" (Loeb, 1932 : 41; Steward, 1933 : 322). The Chumash of the Santa Barbara coast made the pelt into "a little cape like a doublet reaching to the waist," which was worn only by the master of a fishing launch (Bolton, 1930, 4 : 252, 259). When Vizcaíno discovered Monterey Bay in 1602, the Costanoans living there presented the Spaniards with "skins of bears" (Wagner, 1929 : 247). Two centuries later, in Suisun Valley, Padre José Altimira noted that the Indians, probably Wintun, greatly esteemed the bear skins he gave them (Anon., 1860a : 115).

The strangest of all uses, however, was associated with human burial. Both the Wintun and the Nomlaki prized the pelts of grizzlies and black bears as shrouds for the bodies of the dead (Cora Du Bois, 1935 : 11; Goldschmidt, 1951 : 304). The fur side was put inward and thus touched and adhered to the flesh of the corpse. The bearskin was the Nomlaki's most valued possession. The Cahuilla included the perforated pelvis of a grizzly among the ceremonial objects with which the dead were mourned. Called *yuuknut*, which is translated as "frightening," it was said to be very powerful; blown on by a chief, it produced a whistling sound that could kill irreverent persons. Then, along with eagle feathers, it was rolled in a reed matting to make the death bundle, known to the Indians as "the heart of the house." (Strong, 1929 : 128.)

In the mounds of prehistoric peoples that once lived on the Santa Barbara coast, archaeologists found long, sharp daggers fashioned from the femurs of grizzlies (Orr, 1942 : 81; Gifford, 1947 : 124). Their grooved handles were beautifully inlaid with beads of *Olivella* and irridescent rectangles of abalone shell (fig. 13). They had been made by an ancient and warlike tribe, who, like the modern Monos, apparently considered the grizzly to be a worthy object of the chase. Similar daggers, however, were found in the mounds of a later but prehistoric race of Indians who were not hunters. The conclusion has therefore been advanced (by P. C. Orr), that these more timid people learned to make the daggers by imitating those found in the burial sites of their aggressive predecessors.

The only hint we have of Indians keeping grizzlies as pets comes from the Portolá expedition of 1769, as told in the diary of Father Crespi. The expedition was then a few miles north of San Luis Obispo.

We pitched our camp on a high spot in the cañada, and here about sixty Indians from a ranchería which they said was not far away came to visit us. They regaled us with a few measures of gruel, and they were in turn given beads. They brought along a bear cub, which they were taming. They offered it to us but it was not accepted. From this incident the soldiers took occasion to name the place La Cañada del Osito [now Santa Rosa Creek]. (Engelhardt, 1933 : 8)

The Spanish explorers had already experienced the nature of the adult grizzlies and wanted none of the young!

## TALKING TO GRIZZLIES

Throughout California the Indians entertained a strongly developed feeling of kinship with wild animals and particularly with the grizzly. Whether they had the cold nerve to hunt the beast or whether because of timidity they tried to stay out of its way, they felt the bear was more closely one of them than any other animal. This was true despite the fact that by nature it was the most evil and odious being of which they could conceive. A reason for this sense of relationship doubtless lay in the manlike attitudes of the bears; they have many gestures, movements, and tricks that make them appear almost human. Furthermore, an enraged grizzly stands on its hind legs and fights with its fists like a man. Such a terrifying phenomenon was cause enough to make an Indian believe that he was looking at close kin, though a monster; and many tribes insisted that at least some of their people could commune with bears. According to Joaquin Miller (Muir, 1894 : 152), the Shasta had the notion that a grizzly would talk to a human being if the person would only sit still long enough to listen to what the bear had to say instead of running away in great fright. One wrinkled old woman in particular was held in great respect because daily she hobbled over a long trail to a heap of rocks at the edge of a thicket, where, so she said, she talked with a grizzly.

The Cahuilla relied for safety on an ability to converse with the big bears. A person who met one in the mountains called it *piwil* (great-grandfather) and said in a soothing tone, "I am only looking for my food, you are human and understand me, take my word and go away." The bear would rise, brandishing its great paws in the air; then, if it intended peace, it would drop to all fours and scratch dirt to one side. (Strong, 1929 : 115–116.)

One story of the Cahuilla is about an Indian who attended a bear-and-bull fight at the pueblo of Los Angeles. The grizzly, a cowardly individual, was getting the worst of the battle, being repeatedly knocked down. Then the Cahuilla man whispered to him, "You must fight and defend yourself, they are going to kill you." Whereupon the bear charged the bull and broke its neck. (*Ibid.*, p. 116.)

Even a female and cubs, the most dangerous and unpredictable of ursine groups, could be influenced by the proper words spoken in a diplomatic manner. Once when a party of Cahuilla encountered a she-bear with young near the modern town of Beaumont, the oldest and most respected man stepped forward and told her that they meant her no harm and that since she was a relative of theirs she should not bother them. The grizzly looked at them, understood, and went peacefully on her way. (*Ibid.*, p. 76.)

When a man met a bear in the mountains, the object of conversation by the Indian obviously was to assuage the beast. A Spanish-Indian half-breed, who accompanied the famous Bandini brothers on a grizzly hunt with rifles near Cobblestone Mountain, Ventura County, in 1873, had a different objective— but only because he was under the protection of the guns. Thus, in a sense, he perverted and degraded the ancient power that had come to him from his mother. When the bear was sighted, he demanded to be allowed to hold brief intercourse with it before the shooting began. Confronting the beast, who stopped eating berries to look at him, he called out, *"Que hay vale que estás haciendo aqui? Eso mirame bien; soy tu tata"* ("Well old pard what are you doing here? Look at me well; I am your daddy"). The bear seemed to resent this last insinuation, for he rubbed his claws on the ground angrily. The half-breed then launched forth on a long tirade reflecting on the bravery of all bears from time immemorial, and then, indulging in personalities, he made the most unkind and unwarranted allusions to the grizzly's own pedigree. At last, picking up a stone, he threw it at the bear with the remark, *"Tu no eres hombre y me retiro"* ("You are no man and I retire"). At this last insult the bear charged, and thirteen bullets were required to kill him. (Bandini, 1893 : 316–317.)

## SPIRITUAL KINSHIP

Closely related to the idea of kinship with animals and the notion that in former ages—and to some extent even now—all creatures were simply human beings clothed for the moment in fur, feathers, or scales, there is the concept of a world of spirits that haunted the living as well as the nonliving. Among California Indians, ceremonies connected with this religious belief were few but highly developed for a people of rudimentary political and material culture. One of the major dances of the Patwin, the *silai*, or grizzly dance, was essentially a spirit dance. (Kroeber, 1925 : 378, 384.) In the Maidu ceremony, the impersonator imitated the action as well as the appearance of the grizzly and cried "wuk-wuk" (*ibid.*, pp. 433, 435). On the rolling, oak-covered hills of the southern Coast Range, the Salinan performed their dance to the spirit of the bear in August—if there was likelihood that the trees would bring forth a full crop of acorns. The chant was "hau-wa-ya," repeated again and again. (Mason, 1912 : 178.) Among the Pomo, two stuffed raccoons with their tails removed were tied under the forelegs of the bear impersonator to represent cubs (Loeb, 1926 : 376). In the Utzumati or grizzly ceremony of the Miwok, the performer bound curved flakes of obsidian to his fingers to simulate the claws of bears. These objects, found in ancient burials of the San Joaquin delta, were long a mystery to antiquarians and were given the ambiguous designation of "Stockton curves." (Kroeber, 1925 : 450.) The Shasta sang a grizzly-bear song to their dogs in the hope of increasing the dogs' power of scent and ability to frighten game (*ibid.*, p. 294).

The grizzly, like other animate and inanimate objects of nature, was endowed with a spirit to be honored with dances and rituals, according to the beliefs of many Indians. It is not surprising, therefore, to learn the natives also entertained the idea that after death they were likely to be transformed into bears. The palingenesis that was the fate of evil men of the Pomo has already been mentioned. The Chumash, living near the "Valley of the Bears," a land of long grassy swales studded with oaks, where grizzlies were unusually plentiful, evidently believed that all who died there became grizzlies. The only hint of this is in the reply of Fr. Luis Martínez of Mission San Luis Obispo to the *interrogatorio* of 1812:

Among the pagans there is the notion about their eternity in which they change their nature; for those that died here on this soil, they are wont to say, were transformed into bears, and they went to live in some mountains two leagues from this Mission. I myself have known a Christian of this Mission who at twelve o'clock in the day time was chased by a bear. The people presumed that it was of that class of bears because of the little fear the horse manifested, and because the bear kept running after him. In the other rancherias they likewise have the same ideas. (Engelhardt, 1933 : 54.)

In the mythology of the California Indians there is a profusion of tales regarding animal spirits (Gifford and Block, 1930), but, strangely enough, the grizzly is poorly represented. The Wintun spoke of thunder and lightning as being a pair of destructive twins born of Grizzly Bear Woman, and they explained the rippling of the moon's reflection on river and pond as caused by a grizzly who eternally must run round and round the lunar orb (Cora Du Bois, 1935 : 75). Other tribes had tales that concern adventures of grizzlies; but in the main the bears seemingly did not appeal to the imagination of the Indians as much as smaller beasts, such as the coyote.

Anthropologists have speculated why the grizzly was not so favored a character for tales as the clever and mischievous coyote. The answer seems to be, in part, that the bear, the most impressive of all animals, came to have a far more important role than that of a mere mythological character from the past. It became instead a sinister and living force in the everyday native religion of the Indians and thereby exercised a profound and direct influence in the superstitions of almost every man, woman, and child in most tribes. Long before any anthropologist had seen the natives of California or had begun systematic interpretation of their beliefs, General Bidwell (1897 : 83–84) recorded that "the grizzly bear was looked upon by the valley Indians with superstitious awe, also by the coast Indians. They were said to be people, but very bad people, and I have known Indians to claim that some of the old men could go in the night and talk with the bears."

## SHAMANISM

Kroeber (1907 : 327), in discussing the religion of the Indians of California, points out that it "was very similar to that of savage and uncivilized races the world over. Like all such peoples, the California Indians were in an animistic state of mind, in which they attributed life, intelligence, and especially supernatural power, to virtually all living and lifeless things." They also held to the ideas and practices of shamanism—a belief that the supernatural is responsive to certain men, the shamans, who alone have the power of communion with the spirits. In California, the primary function of the shaman was to cure disease (Kroeber, 1925 : 851). All pathological conditions of the body were thought to be caused by foreign, hostile, animate objects, which the shaman could suck out through appropriate rituals and actions. Yet the shaman was not entirely devoted to doing good; he had also the capability of great evil. Depending on whether his inclinations were malevolent or beneficent, he could cause disease or remove it; his powers were equal in bringing about death or in warding it off. The art of doctoring, therefore, was indissolubly associated with the practices of witchcraft.

Throughout most of California there were three specialized classes of shamans: bear shamans, rattlesnake shamans, and rain shamans; of these the first was the most varied in its manifestations and the most extraordinary.

The bear doctor [shaman] was recognized over the entire state from the Shasta to the Diegueño. The Colorado river tribes, those of the extreme northwest, and possibly those of the farthest northeastern corner of the state, are the only ones among whom this impressive institution was apparently lacking. The bear shaman had the power to turn himself into a grizzly bear. In this form he destroyed enemies. The most general belief, particularly in the San Joaquin Valley and southern California, was that he became actually transmuted. In the region of the Wintun, Pomo, and Yuki, however, it seems to have been believed that the bear doctor, although he possessed undoubted supernatural power, operated by means of a bear skin and other paraphernalia in which he encased himself. Generally bear shamans were thought invulnerable, or at least to possess the power of returning to life. They inspired an extraordinary fear and yet seem to have been encouraged. It is not unlikely that they were often looked upon as benefactors to the group to which they belonged and as exercising their destructive faculties chiefly against its foes. In some tribes they gave exhibitions of their power; in others, as among the Pomo, the use of their faculties was carefully guarded from all observation.

Naturally enough, their power was considered to be derived from bears, particularly the grizzly. It is the ferocity and tenacity of life of this species that clearly impressed the imagination of the Indians, and a more accurately descriptive name of the caste would be "grizzly-bear shamans." (Kroeber, 1922 : 303.)

## GRIZZLY SHAMANS OR BEAR-MEN

When dealing with native accounts of the behavior of grizzly shamans, there is much difficulty in separating fact from belief; the two were inextricably intertwined in the Indian mind. Regardless of the amount of truth in tales of the often terrible deeds performed by persons who had the power to impersonate the grizzly or to commune with its spirit, the institution of the bear-man was very real to the aborigine. It was one manifestation of the fear and awe with which he regarded the great, subhuman-appearing beast that lorded itself over the California scene.

After studying the Nisenan, Beals (1933 : 390) wrote that "the belief in these bear-men must have been very strong, for it survives vividly in the minds of old people from whose memory many other things have faded. And one and all are agreed that many Indians were killed by these creatures. Nearly all had seen them in their youth in human form." An Indian reputed to have the power of turning himself into a grizzly was still living near Banning in southern California as late as 1920 (Gifford and Block, 1930 : 68). Among the Patwin, at least, bear doctors were of either sex (Loeb, 1933 : 220).

The savageness of the bears gave rise to a specialized class of shamans who, in turn, implanted greater fear in the minds of the Indians by attributing additional evil qualities to the grizzlies. The variety of form and function manifested by the grizzly shaman in different tribes was so great that this character has been described by a number of terms in the anthropological literature. Besides "grizzly-bear shaman," other typical names were "bear shaman," "bear doctor," "human bear," "bear man," "man bear," "werebear," and "were-bear shaman." All referred to the grizzly and none to the relatively innocuous black bear.

In some tribes, persons who had the grizzly power did not perform directly through the medium of a spirit. Hence, in a strict sense, they were not true "shamans." Those who practiced no curative medicine could not properly be called "doctors"; and those who did try to effect cures could hardly be designated as "werebears." Perhaps, then, the least restrictive and best term for the group as a whole is "bear-men"—if it is used with the understanding that the institution probably originated in shamanism (Kroeber, 1925 : 259). From the beneficent bear shaman or doctor who practiced curative medicine there evolved, through a series of intermediate forms, the malevolent and marauding *gauk buraghal*, who was concerned chiefly with plunder and killing for personal gain. When white men began coming to California, all stages of bear-men were represented among the different tribes. The institution was well developed in all areas where grizzlies were particularly numerous but was not present in the southern desert regions, where bears did not occur, or in the extreme northeast and northwest (among tribes such as the Yurok and Lutuami), where grizzlies were relatively scarce.

Among the Hupa of the northwest there was a faint suggestion of the bear-man in the *kitdonghoi*—a person of secret and evil proclivities who might be seen at night rushing about and throwing out sparks, and who could turn himself into a wolf or bear (*ibid.*, p. 136).

Since the doctors of the Hupa effected cures through live objects which resided in their own bodies and not through communication with guardian spirits, the concept of a true bear-doctor was an impossibility for the people of that tribe (*ibid.*, p. 137). Their relatives, the Wailaki, however, had an inspirational bear shamanism, for certain doctors became possessed involuntarily by the spirits of grizzlies. Then these spirits spoke through the medicine men to the patients. (Loeb, 1932 : 84.)

Shamans among the Shasta, whose guardian spirits were grizzlies, functioned primarily to cure bear bites. This fact suggests that the caste of grizzly shamans, as opposed to ordinary doctors, may have originated from a frequent and specialized need to treat persons who had been injured by the beasts. When practicing his power on the victim of an attack by a grizzly, the bear-doctor of the Shasta wore a collar of grizzly claws, and, before he effected a cure by pulling the animal's tongue from the patient's wound, he danced and growled and jumped about like a bear. (R. B. Dixon, 1907 : 486; Kroeber, 1925 : 303.) The Shasta doctor went no further in the exhibition of his supernatural powers than to simulate a grizzly; but in some other tribes there was actual transmutation of the body. Bear hair began to grow through the skin—a process said to have been very uncomfortable—and gradually the shaman turned into a grizzly, which, if met in the woods, was indistinguishable from the real animal.

Sometimes a Yuki man sought the power by bathing in the root hole of a wind-thrown tree. On emerging, he growled and scratched the bark in the manner of bears at their dances. Others began to get their power by dreaming of bears or by being carried away by real grizzlies. Some, as youths, had been taken into the woods by young bear-women and had lived with them and thus received their power. They sometimes would be gone from their people for many months. On returning to the tribe, these novices came under the tutelage of older shamans, who taught them to sing and to develop their strength in lonely places. Finally, when they were able to commune with the spirit of the grizzly by changing into the bestial form at will, there came upon them the power to cure bear bites, to destroy enemies of the tribe, and to kill any of their own people against whom they held a grudge. (Kroeber, 1925 : 200–201.)

The grizzly shamans of the Yuki had definite functions and powers, but those of some of the other tribes did not. The bear-men of the Yokuts did not cure disease (*ibid.*, pp. 516, 517) but seemed to be mainly exhibitionists. They came to the public dances naked and painted black except for headbands of eagle down and necklaces of claws or loincloths of bear skin. Beginning their exhibitions, they held their feet together, leaned their bodies forward, and let their hands droop. Then, jumping in short, stiff leaps to the rhythm of their growling song, they curved their fingers to look like the claws of grizzlies and finally leaped forward as if to seize a foe.

These bear-men of the Yokuts reportedly at times transformed themselves bodily into grizzlies and acquired the faculty to survive repeated killings. One who was killed and buried emerged from the ground in the form of a bear and walked away. Another, a Chumash, while in the animal form, it is said, had the misadventure of being roped by a Spanish vaquero and made to fight a bull at Mission San Luis Obispo (*ibid.*, p. 517).

A Cupeño bear-man who lived during mission times was reputed to have transformed himself, while attending ceremonies, in order to frighten people, and to have utilized his power also for killing calves on the range (Strong, 1929 : 252). The Wintun tell the story of a white man who married a Nomkensus woman. One spring, unbeknown to him, she turned herself into a grizzly and went to graze on clover. Another man, seeing a bear in his fields, wounded it with rifle fire, but it got away. Later the woman returned to her husband and showed him a bullet wound in her shoulder. (Cora Du Bois, 1935 : 5.)

Another tribe which believed in bodily transformation was the Nisenan. According to their most common account, a man who had the power and wished to change into a bear would rub himself with a particular herb to be found only in pools. Almost immediately bear hair began to sprout through his skin and he took on the appearance and characteristics of a grizzly. To become a man again all he had to do was to jump into the water. (Beals, 1933 : 391.) This was similar to the belief of the Tachi that a man desiring the power bathed at night. "At last a bear appeared in his dreams and instructed him. After many years, not before middle life, he reached the power of becoming a bear at will. He swam in a pool, emerged as the animal, and went on his errand. To resume human shape a plunge into the same pool was necessary." (Kroeber, 1925 : 514–515.)

The bear-men and bear-women of the Nisenan could not be killed. They not only were friends of the grizzlies, but they were inherently vicious and did no good for their tribe. During the summer, they went about in groups, as real grizzlies do, and they attacked whomever they met. (Kroeber, 1929 : 275.) The westside Mono believed the grizzlies that carried away women were really this kind of werebear, and that the children by such unions were half bear and half human (Gifford, 1932 : 51).

## THE BEARSKIN SUIT

The Indian belief that certain men, for good or evil, could transform themselves into grizzlies obviously belongs to the realm of fancy. More difficult to disprove is the often defiant belief of some tribes, particular those of central California, that bear-men encased themselves in the actual skins of grizzlies for the purpose of taking human life more easily. In this garb, they roamed the country, slaying and robbing whomever they en-countered, whether friend or foe. (Curtis, 1924 : 7–9.) They were greatly feared, and other Indians always tried to elude rather than to fight them; for by their malevolence, rapidity, fierceness, and resistance to wounds they were capable of inflict-ing greater injury than real bears (Kroeber, 1907 : 331). This belief obviously was a variation of the widespread werewolf idea, a belief that even today has a powerful hold on the imagination of many civilized people in the Old World.

According to the northwestern Maidu, the validity of the existence of such creatures could be ascertained by anyone will-ing to go up to a grizzly and pull the skin down over its head. Frequently a man would be found inside. (Loeb, 1933 : 200.) A Pomo who attempted thus to rob two presumed werebears of their power made the mistake of attacking real grizzlies. He was a great warrior, however, and managed to come out of the en-suing fight alive. (Barrett, 1917 : 458.)

The hide-wearing werebears differed from the beneficent grizzly shamans of some tribes by practicing no curative medi-cine—or at least none when they were in animal form. Nor did they have the spirits of bears as guardians. Their power resided wholly in the skin suit and its associated paraphenalia, and ap-parently it was transplanted to the wearer through an elaborate ceremony performed in the manufacture and subsequent don-ning of the apparatus. (*Ibid.*, p. 458.)

The Pomo said the art was learned from an older person. A natural cave or secluded spot in a deep canyon or on a rugged mountainside would be selected. Near by, the conspirators would prepare an elliptical clearing where dancing connected with the initial wearing of the suit could be performed. Then they fash-ioned the hide of a grizzly into the magical apparatus, operating only at night when the moon had not yet risen or the sky was overcast. If the moon suddenly emerged from behind a cloud, they ceased their work and hurried away. (*Ibid.*, p. 447.) Before the finished suit could be worn by a Kato man he had to purify himself by swimming and by anointing his body with the strong-smelling oils of the leaves of California laurel trees (Loeb, 1932 : 41).

Since the magical power, which was called into being by the ceremonies, resided in the bearskin, the suit had exceptional value to the bear-man and was always kept in a secret place. Sometimes when it had been left hanging on a tree in a lonely part of the woods and was full of its power, it might be heard to cry like a baby, or again to hoot like an owl. (*Ibid.*)

The following composite description of this extraordinary apparel is taken from several writings (Barrett, 1917; Curtis, 1924; Kroeber, 1925; Loeb, 1932): A close-fitting basket woven of oak twigs covered the head of the wearer. Each eye opening was provided with a disk of abalone shell having a small hole and sometimes was armed with pieces of obsidian for the purpose of piercing an enemy who in his fright might happen to dash into a bear-man. Both nose openings were plugged with pitch to resemble the secretions from the nostrils of a bear, or one was kept clear so that a grunting sound could be made through it. The tongue was a piece of abalone shell attached by a loose string; it simulated that shining, lolling organ of the grizzly.

For armor, the body was wrapped in belts of beads. The body, the appendages, and the basket on the head were covered with sections of hardened grizzly hide. This casing, which was tied together to make a tight fit, was supported on the inner side, along the backbone and ribs, by sticks of yew or wild plum. Small twigs were woven together in such a way that when attached to the feet they left the track of a bear. Two small baskets half filled with water were suspended under the armpits. As the bear-man moved, their swishing sounded like entrails slopping about within the abdomen of a walking grizzly. Oak galls were used by the Maidu for this same purpose. The Nisenan said that the rattling of the galls sounded more like a bear growling. (Beals, 1933 : 392.) As a further attempt at deception, sometimes a basket was carried containing crushed manzanita berries, which, when dropped in little piles, were meant to resemble the excrement of grizzlies.

The hide-wearer's weapons consisted of an elkhorn dagger six to ten inches in length, a flint knife, and a crooked yew staff for hooking the ankles of enemies fleeing through the brush. Some bear-men had sharp pieces of flint sticking out of their elbows, and when they forced a man down they kept him pinned with their forefeet and ground him to pieces, grunting the while. (Loeb, 1932 : 41.)

"Incased in a complete bearskin, his body wrapped with belts of beads serving as armor, and with a horn dagger concealed on him, this nefarious being roamed the hills in search of his human prey, not sparing even people of his own town, if he owed them the least grudge or if their spoliation seemed sufficiently profitable" (Kroeber, 1925 : 259).

## THE BEAR-MAN IN ACTION

In a few tribes, the suit-wearing bear-man was not entirely the nefarious being common in most places. Because of his reputed influence on the real grizzlies, he was important to the Nomlaki, Yuki, and Yokuts. He aided them in their bear hunts. Among the Kato and the Yuki he participated also in warfare. In fact, the bear-man of the Kato was held in high respect because he killed only enemies of the tribe (Loeb, 1932 : 41). As a youth, by virtue of being a swift runner of great endurance, he was selected and trained in the art of wearing the disguise. Before going to war he entered a brush house and danced in his bearskin. This dance "could be witnessed by everyone in the village. One or two doctors first pranced around in their bearskins. They waved their legs in the air, and grunted *hu hu hu*, like bears. Then one of the pair rushed out and obtained a huge log which he brought back under one of his arms. He allowed this to fall upon the floor of the dance house, where he worried it and pounced upon it as if it were an enemy." (*Ibid.*)

When the bear-man was on a war expedition, he was accompanied not only by six or seven warriors but also by women. The story is told of one who, having had a hard day trying to waylay an enemy, neglected to remove his fur skin before having intercourse with his wife; soon afterward she gave birth to a bear cub (*ibid.*). Once during a war between the Kato and the Yuki, a majority of the latter were routed, but two of their bear-men, who were shamans of unusual ferocity, went and remained with the Kato (Kroeber, 1925 : 158).

Among the Patwin, Maidu, and Pomo, the bear-man was truly nefarious. Even the Pomo name for the creature, *gauk buraghal*, which Loeb (1926 : 335) translated as "man bear," has a sinister sound. The various *gauk buraghal* of the Pomo country kept in touch with one another. Whenever two or more met at some hidden spring or other rendezvous in the mountains, they would gloat over their past triumphs. Sitting round the campfire, they would tell about people of their own tribes whom they had killed because of personal grudges, and about others whom they had done away with to obtain plunder. At the same time they would reveal the mountains which they planned to make the scenes of their future operations. Two or more often hunted together, deploying over a wide area. When one sighted a prospective victim, he would climb the nearest ridge to indicate the direction of the prey. The others, closing in, would hide beside trails along which the unsuspecting person was likely to pass. (Barrett, 1917 : 461, 457.)

They did not actually attack their victim until he was close enough to be taken quickly. Then they assaulted him suddenly and fiercely in the manner of a grizzly. The objective, apparently, was to throw him into a fit of terror and thereby gain an immediate psychological advantage—a feat not difficult to accomplish, considering the Indian's fear of grizzlies and even greater horror of the *gauk buraghal*. They destroyed his eyes as soon as possible by slapping or clawing them; but if he struggled too well to be blinded, then they stabbed him to death with the dagger held close to the bearskin's mouth to give the impression of biting. After he was dead, the killers ripped open his belly with a manzanita stick or other rough object, gutted and disemboweled him, and scattered his flesh over the ground as an enraged bear might do.

Having performed a murder, only four of which were permitted by the Pomo in one year, the bear-man gathered up the plunder and hurried to his hiding place, where he divested himself of suit and paraphernalia. Then he returned to his village, and there, unidentified as the criminal, lived peacefully with the fellow members of his tribe—until he again felt an urge to exercise his evil function.

## THE BEARSKIN SUIT—REAL OR IMAGINED?

Whether the bearskin disguise existed outside the imagination of the Indian and was actually worn for the abetment of murderous deeds is a debatable question. Models of such suits have been prepared for museums, but no ethnologist has discovered an original. Few students of the subject are as skeptical as Loeb (1926 : 335), who stated that the *gauk buraghal* of the Pomo "was probably an entirely mythical character in much the same manner as our own werewolf."

Kroeber (1925 : 260) concluded that some Indians probably possessed the disguises, but that "their feats of slaughter existed chiefly if not wholly in the imagination of themselves and their public." He argued that it would be impossible for a man to travel far on all fours and then fight successfully while encumbered with heavy wrappings and armed only with a dagger.

Barrett (1917 : 443–445) pointed out that belief in those creatures was deeply rooted in the native mind; even as late as the 1890's, when the white man had suppressed most aboriginal practices, the Pomo were convinced that the *gauk buraghal* still lived. In the same line of thought, Curtis (1924 : 68) contended that since the Indian imagination is not particularly inventive, firmly held traditions usually can be credited with at least a modicum of fact. He concluded that "it is highly probable, though not susceptible of proof, that certain men . . . actually on occasion wore bear-skin suits for the sake of the bear-like qualities to be gained thereby." Beals (1933 : 390) pointed out that because the bear-man had a tremendous psychological advantage over a lone Indian, to whom even a real grizzly was an object of dread, he was in a better position to commit a murder, if he so desired, than was an ordinary person. Beals concluded that "there seems good evidence that bear shamans . . . actually dressed as bears and in this disguise actually killed people."

The truth of the matter can never be known. Now that both grizzlies and the reputed bear-men have gone from the

California scene, all we can do is speculate regarding the possible existence of the latter.

Many "doctors" were inclined to be evil and selfish, and, considering the natural fear ordinary Indians had for real bears, shamans could easily have hit upon the idea of simulating a grizzly to terrify and to kill people more easily, while at the same time disguising their own identities. The animistic mind of the Indian was full of horrors that do not afflict us. When a native was suddenly confronted by a grizzly, he probably was too paralyzed by fear to resist. In fact, as Europeans and Americans learned later, the best way in such circumstances for an unarmed man to save himself was to pretend to be dead—a method of no avail against a bear-man. Our opinion, therefore, is that shamans who encased themselves in skins and performed murderous deeds may have existed. If, then, they are to be added to the list of creatures that roamed the woods and fields until a century ago, we wonder at the wild scene which must have transpired whenever a grizzly met a *gauk buraghal*. The lord of the California landscape was not likely to be tolerant of impersonators.

# 5 *Grizzlies and Spaniards*

Toward the end of the sixteenth century, the sea front of California presented an uninviting aspect to the scurvy-ridden Manilla galleons which, loaded with the riches of the Orient, regularly rode the westerlies across the north Pacific and thence south to the Mexican port of Acapulco. Except for the brief records of Cabrillo and one or two other early explorers, California was an unknown land lying far beyond the frontiers of New Spain and inhabited only by savages and wild beasts, many kinds of which had never been seen by Europeans.

Then the English pirates Drake and Cavendish burst into the Pacific and terrified the treasure-swollen galleons. Brought to the attention of the Council of the Indies, sitting at Madrid, this resulted in suggestions being sent to Mexico City, headquarters of the viceroyalty of New Spain, that a port of call be established on the California coast to give warnings of pirate danger. Little was done, however, until 1602 when the viceroy, the Conde de Monterey, delegated Sebastian Vizcaíno, mariner and merchant, to explore the western shores. He was instructed to make a detailed map of his findings and above all to look for a likely port where a settlement would provide the Manilla galleons with haven from Englishmen and respite from scurvy.

On December 16 of that year, Vizcaíno believed that he had come into such a port in the sweeping *ensenada* which a half century earlier Cabrillo had called the Bay of Pines. Ignoring the findings of his predecessor, Vizcaíno maintained that the port was the prize discovery of his voyage, and in honor of the viceroy he named it Monterey.

The official diarist of the expedition, Father Antonio de la

Ascensión, was equally excited. He described the practical advantages as well as the beauties of the site (Wagner, 1929 : 246–247): ". . . immense number[s] of great pine trees, smooth and straight, suitable for the masts and yards of ships . . . good meadows for cattle . . . fertile fields for growing crops . . . springs of good water . . . forests of great scarlet oaks . . . large live oaks . . ." But his conclusion that "it is a very good port and well protected from all winds" was a gross exaggeration, for Monterey is little better than an open roadstead. This hyperbole combined with a flattering account of the harbor, communicated by Vizcaíno to the Council of the Indies, gave rise to a legend about the port of Monterey, which was to haunt the Spanish Crown for fully a century and a half.

While at Monterey, Father Ascensión observed also that the Indians were friendly and that the land abounded with game, such as ducks, geese, doves, quail, condors, elk, deer, and rabbits. He saw California grizzlies and noted that they came to the shore at night to feed on the carcass of a whale which had been washed up on the beach by the breakers, that they were large animals, and that their impressive tracks measured "a good third of a yard long and a hand wide."

Father Ascensión's account is brief. In our history, though, it is important. Written on board a Spanish vessel anchored off the unexplored wilderness surrounding Monterey Bay, it is the first information obtained about the species of bear which, three and a half centuries later, was designated the official animal of the second most populous commonwealth of the United States of America (see chap. 12).

After the ships had set their gleaming sails to the northwesterly breezes and swept south to civilization, a long hiatus in exploration followed. Balked by a new viceroy, Vizcaíno and Ascensión died without realizing their ambitions for a California settlement. For more than one hundred and sixty years the Spanish Crown contemplated the splendors and advantages of that sweeping harbor far to the north but took no further action.

Then, coincident with a Russian threat from Siberia, an officer of force, vision, and enthusiasm arrived in New Spain. He was José de Gálvez, and he was armed with the authority of visitador-general. Using the Russian advance as an excuse, he determined to rediscover the port of Monterey and then colonize California with presidios, missions, and pueblos. From the already impoverished missions of the desert wastes in the peninsula of Lower California, he drafted such necessities as priests, altar furniture, foodstuffs, branding "forceps," and domestic animals, including mares with their stallions. Part of his sacred expedition, as it was known, was dispatched by sea; the remainder went by land. A rendezvous was held at San Diego, where Vizcaíno had reported a harbor, but the meeting there in the summer of 1769 was not auspicious. Scurvy had taken a frightful toll of the sailors, and the hardships of desert travel from Loreto to San Diego had reduced the size of the land party.

Nonetheless, the intrepid commander of the expedition, Gaspar de Portolá, marched north in search of the port of Monterey. Although the comparatively few persons able to accompany him were stout-hearted, they were physically emaciated. In his diary he referred to them as skeletons.

Winding its way north, the party found the grassy fields, cool and refreshing sea breezes, and watercourses along the coast of southern California a pleasant change from the stark aridity of the peninsula they had left behind. The priests professed to be shocked by the nudity of the male Indians, who flocked from their *rancherías* to gaze at the white men, but they actually were delighted to find so many souls in dire need of salvation.

The soldiers, being of a more practical mind, were on the watch for food. On September 2, 1769, when they were north of Point Conception and crossing a swampy canyon with a salt lagoon at its mouth, they saw tracks of bears—grizzlies. Accordingly they went hunting, each man armed with a lance, broadsword, and musket; and that afternoon they succeeded in killing one of the beasts.

Miguel Costanso, a distinguished cartographer and engineer, recorded in his diary (Teggart, 1911 : 56): "It was an enormous animal: it measured fourteen palms from the sole of its feet to the top of its head; its feet were more than a foot long; and it must have weighed over 375 pounds. We ate of the flesh and found it savory and good." Father Juan Crespi named the lagoon near which this first bullet-killed California grizzly fell "Lake of the Holy Martyrs . . ." Because the next bear killed was very lean, the soldiers named the pond (in what is now San Luis Obispo County) where it was taken *El Oso Flaco*. (Bolton, 1926 : 169; 1930, 4 : 268 note.)

In spite of Costanso's comments about the large size of the first grizzly seen at close hand, his modest estimate of the weight indicates the bear was of moderate size. The second animal undoubtedly was unhealthy, for it was thin at the time of the year when most wild beasts are fat. In short, then, these two bears were not entirely typical of their race, and the Spaniards, having dispatched them with ease, were totally unprepared for the resistance they encountered when bear hunting a few days later.

Near the site of the present town of San Luis Obispo, they came into a broad, somewhat marshy valley receiving waters from shaded canyons of live oak to the south and bounded on the north by pinnacles and buttresses topped with rock and brush. In this valley they were astounded to see numbers of bears scattered over the lush, moist fields and busily engaged in digging up roots. (Smith and Teggart, 1909 : 31, 33.) Here was meat in abundance for Portolá's hungry men. A grand bear hunt was planned.

That evening after the tired soldiers and missionaries had partaken of juicy steaks roasted over red-hot coals of oak, all the diarists of the expedition noted the happenings of the day. The account written by Costanso is the most detailed:

In this canyon we saw troops of bears; they had the land plowed up and full of the holes which they make in searching for the roots they live on, which the land produces. The natives also use these roots for food, and there are some of a good relish and taste. Some of the soldiers, attracted by the chase because they had been successful on two other occasions, mounted their horses, and this time succeeded in shooting one. They, however, experienced the fierceness and anger of these animals—when they feel themselves to be wounded, headlong they charge the hunter, who can only escape by the swiftness of his horse, for the first burst of speed is more rapid than one might expect from the bulk and awkwardness of such brutes. Their endurance and strength are not easily overcome, and only the sure aim of the hunter, or the good fortune of hitting them in the head or heart, can lay them low at the first shot. The one they succeeded in killing received nine bullet wounds before it fell, and this did not happen until they hit him in the head. Other soldiers mounted on mules had the boldness to fight one of these animals. They fired at him seven or eight times and, doubtless, he died from the wounds, but he maimed two of the mules, and, by good fortune, the men who were mounted upon them extricated themselves. (Teggart, 1911 : 59–61.)

The commander of the troops, Lieutenant Pedro Fages, was impressed by the skill and boldness of the bears. He wrote: "They are ferocious brutes, hard to hunt; they attack the hunter with incredible quickness and courage . . . he can only escape on a swift horse. They do not give up unless they are shot either in the head or in the heart." (Priestley, 1937 : 39–40.) Three days later the explorers refused to accept a grizzly cub presented by the Indians! (Bolton, 1927 : 188).

The valley where the Spaniards learned about the ferocity and tenacity of life of California grizzlies, and where the bears were thrown into a terrible fury by the bullets of the glittering strangers, was named by the soldiers *La Cañada de los Osos* (Smith and Teggart, 1909 : 75). The gentle priest, Father Crespi, called it *La Natividad de Nuestra Señora* (Bolton, 1927 : 185), but the name given by the troops is the one that stuck. Today the "Valley of the Bears" in San Luis Obispo County provides abundant pasturage for cattle and horses and fertile soil for crops (fig. 5).

In fact, the agricultural riches of modern-day California belie the extreme difficulty experienced by the early explorers in finding sufficient food. When Portolá reached the vicinity of Monterey Bay, which he did not recognize because there was no great harbor as described by Vizcaíno, "twenty and more leather-jacket soldiers" out of twenty-seven were ill with scurvy, "twelve wholly incapicated and the rest not far from it."

After a fruitless reconnaissance farther up the coast, the disappointed leader decided to return to San Diego. Monterey had not been found. Winter and the rainy season had set in. Each man was rationed to a single tortilla for breakfast, two for lunch, and two for supper, along with whatever game—mostly sea gulls—that could be shot. Under these circumstances, they were glad to leave the wintery wilderness of the north, and they eagerly looked forward to another feast in the "Valley of the Bears."

Although Father Crespi later wrote to Father Francisco Palóu at Mexico City that he had seen "bear's dung everywhere" in the vicinity of Monterey (Bolton, 1927 : 27), and Fages included "bears" in his list of land animals (Priestley, 1937 : 60), apparently the expedition had actually met few if any grizzlies north of San Luis Obispo. But the soldiers expected to repeat their previous adventures with the beasts in the south.

Costanso described their disappointment on December 27:

We were very desirous of arriving at this canyon, having the intention of killing some bears, thinking that we would find as many as on the previous occasion. Preparations were made for the hunt very early in the morning; the soldiers took horses and led them by the bridle in order that they might saddle and mount them when necessary. On beginning the march, however, a very heavy rain set in and lasted, without stopping, all day and the following night. (Teggart, 1911 : 143.)

The explorers halted in the Cañada de los Osos, but there was no bear hunt (Smith and Teggart, 1909 : 75). Even Mass had to be foregone the following day, much to the disappointment of Father Crespi, "because we are in a muddy place, all wet, and unable to move from one spot" (Bolton, 1926 : 247). Four days later, New Year's Day, 1770, the good intentions of the ecclesiastic were rewarded and he was able to record in his diary that at last

the Divine Provider did not fail us, and ordered that a bear should appear in the road with three cubs following her. Some of the soldiers thereupon changed to horses that were accustomed to the ferocity of these animals, and they succeeded in killing the mother and one cub, with which there was a great feast. Its meat is not in itself unsavory, and to-day it tasted better than if it had been a fatted calf. (Bolton, 1927 : 261.)

Costanso agreed: "The meat of these animals has a very good flavor and taste, . . . at that time it seemed better than the best veal" (Teggart, 1911 : 147).

Father Crespi in a letter to Gálvez dated February 9, 1770, said that "there are bears in abundance in the whole country from near San Diego up to the last region explored at San Francisco" (Bolton, 1927 : 45). Yet south of Point Conception the "Divine Provider" issued no further orders to grizzlies; as the explorers approached the end of their journey, they had to sacrifice a pack animal a day for food. They reached San Diego "smelling frightfully of mules."

Nonetheless, as soon as a ship brought supplies from Mexico, the indomitable Portolá turned north once again in a final search for the port of Monterey and the elusive, mysterious bay, which Vizcaíno and his pilot had likened to a lake in the shape of an O.

One evening when most of the members of the second expedition were camped on the Carmel River, Fages, Crespi, and a soldier took a long walk along the beach where more than a century and a half earlier Father Ascensión had seen grizzlies eating carrion whale. Suddenly all three "perceived that the bay was locked by points Año Nuevo and Pinos, in such a manner that it resembled a round lake like an O." With one voice they cried out, " 'This is the port of Monterey that we are seeking, for this is the letter described by Sebastian Vizcaíno.' " (Cleland, 1944 : 73.)

At once Portolá moved his party to the shore of the bay. On June 3, 1770, he established a presidio, even though there was no good harbor for supply ships, and Father Serra founded Mission San Carlos Borromeo. Then Portolá, having discharged his assignment, appointed Fages to be governor, and quit California forever.

Soldiers and padres in the new settlement were dependent on supply ships from Mexico, for nearly all crops that were planted failed and much of the livestock that was to be used for breeding had to be slaughtered for food. In 1772, ships on which Monterey and the new Mission of San Antonio de Padua had relied for succor did not appear. The situation became critical. There was only a small quantity of milk and vegetables on hand.

In these desperate circumstances, the thoughts of Fages turned once again to the "Valley of the Bears" and to the battle he and his men had waged there two and a half years earlier. Accordingly, late in May, the governor and thirteen soldiers marched from Monterey to the *Cañada de los Osos*. Father Palóu in his *Noticias* stated that the troops "remained there for three months, eating bear-meat and sending loads of it jerked to the others" (Bolton, 1926 : 357).

During that period, which Palóu (1913 : 132) termed the time of the "killing of these wild beasts," "twenty-five loads, or about 9,000 lbs. of bear meat" were sent on muleback to the two northern missions, San Carlos and San Antonio (Engelhardt, 1933 : 13). Until late in August, soldiers, padres, and neophytes lived on little else than the jerked flesh of grizzlies. But the settlements were saved.

The Fages bear hunt is one of the most dramatic and colorful incidents of the early history of California. Unfortunately, there is no detailed account of it by any of the participants. All we know is that Fages wrote to the viceroy at Mexico City, when his men had been in the field against the bears almost a month, saying that they had "succeeded in killing thirty to help us to much food." The fact that in 1772 Fages had to go "fifty leagues southward" (Engelhardt, 1934 : 35) to "make a killing of bears" (Bolton, 1926 : 357) may be an indication of the relative scarcity of grizzlies in some of the coastal areas of California before the ranges were stocked with cattle, sheep, and horses. Years after Fages' bear hunt, grizzlies were abundant at Monterey and San Antonio.

The numbers and distribution of wild animals cannot be accurately determined from the journals of the early explorers, for there is no way of knowing how many times a species was seen without being recorded. Fages, for example, mentioned grizzlies only once in his record of an expedition—actually a pursuit of deserters—which he undertook into new territory in 1772, yet undoubtedly he saw many bears. After crossing the southern end of the San Joaquin Valley, he wrote the first description of that great inland plain, which then abounded in water, calling it a "labyrinth of lakes and tulares" and remarking on the "plentiful game, such as deer, antelope, mule deer [tule elk?], bear [grizzly], geese, cranes, ducks" found there. (Bolton, 1931 : 213.)

In March of that same year, Father Crespi was a member of an expedition that went from Monterey to explore the harbor of San Francisco. The party saw "many antelopes and tracks of bears" in the San Benito Valley, and bears and deer at an estuary that is now part of the city of Oakland (Bolton, 1926 : 322, 338–339). At Strawberry Creek, near the western side of what is now the Berkeley Campus of the University of California, "the soldiers succeeded in killing a bear, so that they had fresh meat to go on with" (*ibid.*, p. 340). Yet Father Palóu recorded no bears in the diary he kept of an expedition from Monterey to San Francisco in November, 1774 (Bolton, 1930, 2 : 395–456).

When another explorer, Juan Bautista de Anza, broke a trail from Sonora to California that year, his expedition recorded grizzlies only at San Antonio Creek near the site of the town of Ontario. The party "halted for the night at a fertile arroyo which came from this sierra, and was thickly grown with cottonwoods, willows and sycamores. It was given the name of Los Osos, because of several bears which were seen here and then ran away." Father Francisco Garcés, a member of the party, noted in his diary, "Here there are many bears and sycamores." (Bolton, 1930, 2 : 95, 346.) The date was March 21.

Two years later, Anza led a second expedition. He entered coastal southern California by the San Carlos Pass, twenty miles west of what is now the junction of Riverside, Imperial, and San Diego counties, and then journeyed north to Monterey and San Francisco and returned. Father Pedro Font, who accompanied him, kept a voluminous diary (Bolton, 1930, 4 : pages cited below). Since Font was a master observer and interested in compiling detailed descriptions of the various natural features and phenomena seen along the route, we can assume that he recorded at least a majority of the meetings with grizzlies. His diary provides one of the most reliable indications of the numbers of these animals then in the southern Coast Ranges.

On January 3, 1776, the party reached the *Puerto de los Ossos* (near San Dimas, Los Angeles County) without having noted any bears, although the first expedition apparently had seen some at that site: "The other time they called this pass the Puerta de los Osos because here they saw a number of bears" (p. 173). During the ensuing two months they traveled to Mission San Gabriel, to the site of the city of Santa Ana, and to San Diego, but there is no word of bears.

The first mention by Font of an encounter with grizzlies

was on February 29. The expedition then was north of Point
Conception and near the mouth of the Santa Ynez River. He
notes that north of the point the character of the country was
sharply different, "thickly covered with flowers, and green with
a great variety of grasses . . ." (pp. 264–265). The party passed
"some sand dunes or sandy hills" and "a fairly wide flat between
two hills and half closed in by a pool of water which is there,
having no exit toward the sea, but not very miry. In this flat we
saw a band of six bears, of which many large ones are found in
those lands" (p. 226).

Anza and Font reached the "Valley of the Bears" in San
Luis Obispo County in March, but there is no report of their
seeing grizzlies. In fact, from the vicinity of Point Conception
to the beginning of the San Francisco Peninsula—a region where
in later days vast numbers of grizzlies were killed—Font recorded
none of the beasts. Then on March 27 in the hills west of Mill-
brae he "saw many bears, but although the men chased them
they were not able to kill any" (p. 330).

Two days later, in the vicinity of Crystal Springs Lake, a
few miles south of San Francisco, they were more successful:

We traveled through the valley . . . and crossed the arroyo of
San Matheo where it enters the pass through the hills. About a league
before this there came out on our road a very large bear, which the
men succeeded in killing. There are many of these beasts in that
country, and they often attack and do damage to the Indians when
they go to hunt, of which I saw many horrible examples. When he
saw us so near the bear was going along very carelessly on the slope
of a hill where flight was not very easy. When I saw him so close
and that he was looking at us in suspense I feared some disaster. But
Corporal Robles fired a shot at him with aim so true that he hit him
in the neck. The bear now hurled himself down the slope, crossed
the arroyo, and hid in the brush, but he was so badly wounded that
after going a short distance he fell dead. Thereupon the soldiers
skinned him and took what flesh they wished. . . . The commander
took the hide to give as a present to the viceroy. The bear was so
old that his eye teeth were badly decayed and he lacked one tooth,
but he was very fat, although his flesh smelled much like a skunk or
like musk. I measured this animal and he was nine spans [nine quarters of a vara, or over six feet] long and four high. He was horrible,
fierce, large and fat, and very tough. Several bullets which they fired

at him when he fled they found between his hide and his flesh, and the ball which entered his throat they found in his neck between the hide and the muscle with a little piece of bone stuck to it. (Bolton, 1930, 4 : 350.)

Anza called it "a monstrous bear . . . whose very fat flesh was taken advantage of by those who like it" (Bolton, 1930, 3 : 132). The fact that the hide of such an ancient, smelly beast was kept as a present for the viceroy indicates that the expedition had not killed other grizzlies.

From the San Francisco Peninsula the explorers went to the east side of the Bay, and on April 1 near the site of the town of San Leandro "came to an arroyo with little water but with a very deep bed grown with cottonwoods, live oaks, laurels and other trees, crossing it at the foot of the hills by making a detour. Before crossing it we saw on a slope four bears, which, according to all accounts, are very plentiful through here also, for we saw several Indians badly scarred by bites and scratches of these animals." (Bolton, 1930, 4 : 361–362.) Turning east, the expedition passed the Livermore Valley and on April 5 entered the rugged hills to the south: "All this country which we crossed this day and the next is very broken, and is the haunt of many bears, judging from the tracks which we saw" (*ibid.*, p. 415).

The last mention by Font of grizzlies was at Carmel when Father Cavaller of the Mission of San Carlos "gave Señor Ansa . . . some bear skins, some eight I think" (*ibid.*, p. 454).

From the foregoing evidence supplied by the various expeditions, there is every reason to believe that grizzlies occurred from near San Diego to San Francisco in all the areas explored by the Spaniards. Probably, though, they were not as abundant as they became afterward. Later records indicate that a diarist in the first half of the nineteenth century taking the route of Anza and Font would have found more frequent occasions to record grizzlies.

In the days of Anza and Portolá, the number of grizzlies was in balance with the environmental conditions then prevailing. The concentration noted by Portolá and Fages in the "Valley of the Bears" was presumably a feeding group assembled to dig up the choice tubers growing in that damp soil. Throughout

California in those days, roots, nuts, berries, and flowers undoubtedly were staple foods of grizzlies, together with some
carrion.

Portolá's expeditions and the settlements he helped found
did not immediately change the natural environment of grizzlies.
The expeditions were purely military and ecclesiastic; there were
no civilian colonists. By 1773 the two presidios and five missions
built of logs and tules were in a miserable condition. The missions were so poor that they could offer few inducements to the
natives. The padres at San Diego tried for a year before they
succeeded in persuading a single Indian to accept the faith and
take instruction in tending animals, tilling fields, and making
adobe brick. At the viceregal palace of faraway Mexico there
was talk of abandoning the whole California project, which had
to be supplied at great expense by a dangerous sea route. Then
through the meeting of two dynamic men in Mexico, forces
were set in motion that were destined to change the face of California irrevocably.

After Captain Juan Bautista de Anza, commissioned by the
viceroy, Antonio Bucareli, had searched out a land route from
Sonora to Monterey, breaking a trail across the deserts and
mountains, Bucareli authorized him to lead colonists and livestock along the same route. The second expedition consisted of
240 persons, 695 horses and mules, and 355 cattle. It reached
Mission San Gabriel on January 4, 1776. At last enough people
and livestock had come into the land to make for success in
colonization.

In 1773, according to Father Palóu's report to the viceroy
dated December 10, the missions had owned only 205 cattle;
313 goats, sheep, and pigs; and 107 horses (Mora, 1949 : 33).
Sixty years later they recorded 396,000 horned cattle; 321,500
goats, sheep, and pigs; and 61,600 horses (Bancroft, 1888 : 339;
but see Engelhardt, 1913 : 653, 655). The great ranchos—about
thirty in all—also had quantities of domestic animals, and un-
counted hordes had gone wild. In short, the colonization of California had been a huge success. The livestock had multiplied at
a phenomenal rate. The missions—twenty-one in all—were pros-
perous, powerful, and wealthy.

By the Secularization Act of 1833, an outgrowth of the Mexican revolution, the missions lost control of their lands and neophytes. But the pastoral development which was changing the face of California was in no way retarded. The crumbling of the missions merely ushered in the golden era of the ranchos.

In 1784, Governor Fages had obtained sizable land grants for some officers of the presidios, because, as he wrote, "The cattle are increasing in such manner, that it is necessary in the case of several owners to give them additional lands" (Cleland, 1941 : 30). During the Spanish period, not more than thirty such grants were made; but after Mexico threw off her allegiance to the crown, the lavish generosity of the new provincial government brought some eight million acres into the possession of about eight hundred grantees. Each rancho, an empire unto itself, grazed thousands of cattle, sheep, and horses; supported hundreds of Indian servants; and was an economically independent, self-sustaining community.

Under the benign influence of a mild climate and ignorant of economic competition or want, the Spanish Californians were able to devote themselves to the "grand and primary business of the enjoyment of life" (Cleland, 1941 : 44).[1]

In the preceding chapter we touched on the relations between the Indians and the grizzlies. In the present one we shall attempt to trace the manner in which the Spanish colonists reacted to the presence of the bears, and how those animals in turn were affected by the development of missions and ranchos. The grizzly, which by virtue of its temperament, size, and numbers was an important element in the life of the savages, was not

---

[1] The ranchos began to decline after the discovery of gold when the demand for meat created new and inflated standards of value. This "fatal discovery," turned the paradise of the rancheros into "a great, roaring Pandemonium, a hell on earth. Every canyon and every foothill swarmed with greedy gold-hunters, who squandered their money like water. The vaqueros suddenly saw every bullock in their countless herds become a skinful of silver, and all the yellow marrow of his bones was like fat gold. . . . They would have been more than human if they had not fallen before such temptation. . . . [They] became the most prodigal spendthrifts. Recklessly they gambled away their princely riches." (Powers, 1872 : 284–285.)

Then, with the great drought of 1863–1864 the pastoral era, "the day of the unfenced ranchos, of enormous herds of half-wild cattle, of manorial estates, and pleasure-loving *paisanos*, came to its inevitable close" (Cleland, 1941 : 183).

an animal to be ignored even by the relatively sophisticated new-comers from Mexico and Spain. Before the Spaniards had been long in California, their many experiences with the beasts gave them a heritage of bear lore, which was not forgotten even after the close of the pastoral era. Writing near the end of the last century about early-day life in the Santa Clara Valley, Guadalupe Vallejo stated that "innumerable stories about grizzlies are traditional in the old Spanish families, not only in the Santa Clara Valley, but also through the Coast Range from San Diego to Sonoma and Santa Rosa" (Vallejo, 1890 : 191).

The grizzly, being of a conservative nature, did not adapt itself at once to the somewhat more potent weapons of the new-comers. Instead of becoming wary and secretive, it attempted to treat the Spaniards with the same contempt it had shown for Indians. If, for example, a pueblo lay along its path, a grizzly would not hesitate to amble through the town, scattering the people in all directions. Quite often a serenade party in the pueblo of San Jose, "singing and playing by moonlight, was suddenly broken up by two or three grizzlies trotting down the hill into the street, and the gay caballeros with their guitars would spring over the adobe walls and run for their horses, which always stood saddled, with a reata coiled, ready for use." (*Ibid.*, pp. 190–191.)

At such times the grizzlies, intent on their own business, probably were not dangerous; but they could be provoked so easily into a rage that their mere presence always constituted a hazard. When James O. Pattie visited Monterey during the 1820's, he observed that "numberless grey bears" found their homes in the oak and pine woods and "are frequently known to attack men" (Flint, 1930 : 352). At Sonoma, Señora Vallejo, the wife of Mariano Vallejo, kept a small rifle as protection against bears. Although it must have given her more psychological comfort than actual protection, she said she had occasion to fire it at a grizzly that invaded the courtyard of her house. (Sanchez, 1929 : 49.)

According to Guadalupe Vallejo (1890 : 191), many herdsmen and hunters were killed by the beasts. In the main, though, the bears were not as great a hazard to the Spaniards as they were to the Indians. One reason was that the dons and vaqueros generally were mounted whenever they left their homes. But even being astride a horse was no absolute assurance of safety from bears. William Heath Davis (1929 : 303–304) described the experience of Don José Martínez who, while on horseback near his mansion at Pinole, was attacked unexpectedly by a bear. The animal ripped the ranchero's trousers and tore off a shoe and stocking before the horse could get away.

Duhaut-Cilly, who visited San Francisco in 1827–1828, saw "a soldier bearing recent and indisputable proofs" that the grizzlies "are not always of a very peaceful disposition." The man had been on horseback when he "suddenly found himself face to face with the bear, two steps away." Not prepared for a fight, "he had tried to escape the danger by turning back; but the animal had immediately thrown itself upon the crupper of his horse and stopped him short." When the beast "buried its claws in his right side and in his face," the soldier had cried out in agony and terror, and a courageous priest, by making "a good deal of noise," had frightened away the bear. (Carter, 1929 : 146.) The Spaniards and later the Americans learned that not infrequently a sudden, loud sound was the best means of intimidating a hostile bear.

Surprisingly, in view of the importance of bear meat during the days of exploration and settlement, the descendants of the original colonists loathed and would not eat it. They looked upon the flesh with the same disgust with which they regarded pork —they did not use lard even in cooking. (Bancroft, 1888 : 368, 370.) The chief value of grizzly carcasses was in sale of the pelts to foreign merchants. "Skins of bears" were among the articles included in a trade contract negotiated between a British firm and the prefect of the missions in 1822 (*ibid.*, p. 466). By 1834, five hundred bales of furs from different species of wild animals were exported annually (Cronise, 1868 : 61). The priests obtained grizzly pelts through the efforts of their own hunters (Vallejo, 1890 : 187).

The trade was concerned primarily with hides and tallow from cattle and wild ungulates; bear skins were of small consequence. Bears may have been killed for their oil; Farnham

(1849 : 329), in telling about Spaniards roping elk, deer, and antelope for hides and tallow in the San Joaquin Valley during the 1840's wrote: "The grisly bear inhabits the mountain sides and upper vales. These are so numerous, fat and large, that a common-sized merchant ship might be laden with oil from the hunt of a single season." Both Farnham and Powers (1872 : 276) saw bear skins used as coverings for the saddletree.

By 1825 cattle were so numerous on the California ranges that seemingly they could be slaughtered without end. The Spaniards used only the very choicest cuts for food. There had been a time, however, when depredations by bears were considered a hindrance to the growth of the herds. In 1796, for example, the priests of Mission Santa Barbara in a special *Nota* (Engelhardt, 1923 : 64) wrote that during the previous year "cattle, horses, mules, and sheep" had increased but slightly because of the "serious damage" done by predators, which included bears.

At Monterey during the decade 1801–1810, grizzlies killed and ate cattle every day in full view of the herdsmen, who were helpless to prevent these bold and audacious outrages (Bancroft, 1886, 2 : 142–143). One beast in particular distinguished itself by devouring five mules and seven cows. She-asses could not be turned loose but had to be confined in the presidio for safety. Ordinary fences gave no protection against the big predators, although barriers made from the horned skulls of thousands of slaughtered cattle are said to have been effective at times (McKittrick, 1944 : 173).

In 1805, an estimated four hundred head of livestock belonging to the Rancho del Rey were killed by bears. When Governor Sola arrived at Monterey from Mexico in 1816, he was told by the commandant that grizzlies "were continually coming down from the high mountains and destroying cattle, and that the inhabitants had no means of exterminating them" (Bancroft, 1888 : 424). Rojas (1953 : 47) said that when vaqueros camped on the range were awakened at night by "the bawling of frightened cattle," they knew that "the great *oso pardo* . . . was coming down the *cordones* [ridges] driving the cattle in panic before him." Even in the decade 1811–1820, bears, along with Indians and soldiers, made serious inroads on the numbers of domestic animals (Bancroft, 1886, 2 : 418).

We are of the opinion that grizzlies usually were unable to capture the fleet-footed antelope, deer, and elk which originally inhabited the range. How, then, were they able to be so successful in their attacks on cattle, which in the short span of forty years had become an enormous new supply of food? Domestic stock—even the half-wild breeds imported from Mexico—were so much less alert and swift than the native ungulates that they could probably be taken with relative ease. An extraordinary technique used by grizzlies is mentioned so often in the writings of early-day observers that it may be an explanation of the bears' success with cattle.

The California grizzly, although usually described as a fearsome beast, also had a playful nature; taken as a cub, it often grew up to be a droll and sportive pet. When the naval officer Joseph Warren Revere explored the Russian River valley in the late 1840's, he had a rare opportunity to observe a grizzly totally ignorant of the presence of man. The great beast was lying on its back and "in the most unceremonious and at-home sort of manner playing with his paws, which were sportively elevated in the air." (Revere, 1849 : 142.) Cattle are curious by nature. If they saw grizzlies playing like the one described by Revere, they very likely would draw closer and closer. Conceivably the bear might learn how to lure them within striking distance. At any rate, Colin Preston, a hunter who settled near San Luis Obispo in 1845 and reportedly killed two hundred grizzlies in a single year, made the following observation:

In the midst of a plain covered with wild clover, which is deep and close at that season (you can pluck the clover heads with your hand without bending from the saddle), I perceived a movement, and saw that it was a grizzly of enormous size rolling in the clover, with his paws playing stupidly in the air. The cattle on a hill-side not far distant were watching this movement, and a bull advanced toward it, drawn, it seemed to me, by curiosity . . . The herd followed him, grazing as they went. He forced his way through the tall clover until he came within fifty yards, and bellowed, tearing up the earth. The bear moved less, only now and then rolling a little to stir the field. The curiosity of the bull now changed into anger; he came slowly up, snorting and bellowing, and at length stepped suddenly forward, and plunged at the bear, who caught him in his powerful arms and held him down.

There was fifteen minutes of struggling and roaring, and the two immense beasts rolled over and over, crushing flat a wide area of the field. The herd gathered around, rushed upon them, and bellowed with rage and terror; but the bear never slackened his hold until the bull, exhausted, ceased to strive. Then up rose Bruin, light as a cat, and, striking out as a cat strikes, broke at one blow the shoulder of the bull. He fell as if dead, and the herd ran to the hills, groaning. (Anon., 1857 : 818.)

James Capen Adams, who knew more about the big California bear than any other man, saw a heifer succumb to a similar ruse. About sundown he heard a tremendous commotion among some cattle south of Corral Hollow, in 1855. A huge grizzly was

rolling and tumbling in the grass, while the cattle were gathering around him, and bawling as if crazy. . . . I had frequently heard of the sagacity of the grizzly in decoying cattle within his reach. . . . The bear was in the long grass, rolling on his back, throwing his legs into the air, jumping up, turning half somersets, chasing his tail and cutting up all kinds of antics, evidently with no other purpose than to attract the attention of the cattle. These foolish animals crowded around him; some bulls running up as if to make a lunge, and then turning aside, and all bawling violently. At last a young heifer . . . lowered her head and ran up, to thrust her horns into him. In an instant the bear rose upon his hind legs, and, making a leap, caught the heifer around the neck, and fixed his jaws in her nose. She made a jump to get away; but the bear, with a peculiar jerk of his head, threw her upon her side, and without loosening his hold, turned his entire body upon her. He then let go his hold upon the nose and seizing her by the neck, tore it open; the blood gushed in torrents from the severed arteries; and in a few moments she was dead. (Hittell, 1860 : 323–325.)

The other cattle alternately approached and were driven off while the bear was eating the fallen heifer.

Even horses were not immune to the attraction of a playful bear with ulterior motives. According to Don Eduardo Duarte, "In those early days the grizzly bears were very plentiful, and when they wanted to kill a horse they would lie on their backs in a grassy place and throw their huge legs in the air, to attract the inquisitive herds. Then, when they came close, the bear would jump to his feet and, before they could escape, he would strike one down with his paw." (Coolidge, 1939 : 17–18.)

A variation of this technique was for the bear to simulate a shaggy ball rolling downhill. Thomas Clark, who came to the Ojai Valley in 1868, saw three grizzlies—a huge male, a female, and a half-grown young one—undertake the killing of a steer by this trick.

It all took place in full daylight, and on an open plain, the bears paying absolutely no attention to Clark, who was about one hundred and fifty yards away.

The cattle were grazing in an open place, and a few moments before the attack was made Clark saw the bears slowly shuffling up a narrow arroyo. Then they separated. In a moment a bear rolled out of a fringe of willows that ran down like a point in the little plain. This rolling bear was all doubled up and bounding along like a football rolling towards the cattle, which instead of fleeing, pricked up their ears and watched the strange spectacle.

Suddenly at angles from either side of the plain the two other bears rushed forth, and almost before one could tell what had happened the larger of the two had reached the great steer that was intent on watching the rolling bear. The steer was totally unprepared for the attack. Bruin with a paw as heavy as lead felled the steer to earth. In five minutes from the time that Clark had first seen the grizzlies they had devoured a large part of the steer. (Kingsley, 1920 : 24.)

As a final word about this strange and ingenious method of luring livestock to their deaths, we quote again from Revere (1849 : 257–258) on the subject of the California grizzly:

He . . . commits great havoc among the cattle, his plan of operations being to roll himself up in a ball, and then, like an eminent statesman, to "set the ball in motion, solitary and alone." He selects for this pastime an open meadow, and while engaged in this ground tumbling he cuts all kinds of monkey-shines. He well understands the failings of the cattle, who are as curious as mother Eve, and will rush from all parts of the rancho to see the fun. . . . Well, the cattle will surround the bear in a wondering and gaping circle, until Cuffee—who is all the while laughing in his paw at their simplicity—seizes upon the first fat cow that comes within the grasp of his terrible claws, and . . . walks off with his prize, who thus pays the expense of the performance.

Not all cattle went to their deaths in such a docile and stupid manner. The bulls in particular were willing and eager to challenge the grizzlies' taste for beef. "Bulls used to come to . . .

[Fort Ross] with lacerated flesh and bloody horns after encounters with bears" (Bancroft, 1886, 2 : 639). J. Ross Browne, who served as recording secretary for the California Constitutional Convention of 1849, wrote the most vivid account we have seen of such an encounter on the range. In the summer of 1849, when Browne was journeying north of Soledad, he was thrown from his mule and then chased by cattle. Gaining refuge in a tree, he became the captive spectator to a wild battle between bear and bull (fig. 14).

A fine young bull had descended the bed of the creek in search of a water-hole. While pushing his way through the bushes he was suddenly attacked by a grizzly bear. The struggle was terrific. I could see the tops of the bushes sway violently to and fro, and hear the heavy crash of drift-wood as the two powerful animals writhed in their fierce embrace. A cloud of dust rose from the spot. It was not distant over a hundred yards from the tree in which I had taken refuge. Scarcely two minutes elapsed before the bull broke through the bushes. His head was covered with blood, and great flakes of flesh hung from his fore shoulders; but, instead of manifesting signs of defeat, he seemed literally to glow with defiant rage. Instinct had taught him to seek an open space. A more splendid specimen of an animal I never saw; lithe and wiry, yet wonderfully massive about the shoulders, combining the rarest qualities of strength and symmetry. For a moment he stood glaring at the bushes, his head erect, his eyes flashing, his nostrils distended, and his whole form fixed and rigid. But scarcely had I time to glance at him when a huge bear, the largest and most formidable I ever saw in a wild state, broke through the opening.

A trial of brute force that baffles description now ensued. Badly as I had been treated by the cattle, my sympathies were greatly in favor of the bull, which seemed to me to be much the nobler animal of the two. He did not wait to meet the charge, but, lowering his head, boldly rushed upon his savage adversary. The grizzly was active and wary. He no sooner got within reach of the bull's horns than he seized them in his powerful grasp, keeping the head to the ground by main strength and the tremendous weight of his body, while he bit at the nose with his teeth, and raked stripes of flesh from the shoulders with his hind paws. The two animals must have been of very nearly equal weight. On the one side there was the advantage of superior agility and two sets of weapons—the teeth and claws; but on the other, greater powers of endurance and more inflexible courage. The position thus assumed was maintained for some time— the bull struggling desperately to free his head, while the blood streamed from his nostrils—the bear straining every muscle to drag him to the ground. No advantage seemed to be gained on either side. . . .

As if by mutual consent, each gradually ceased struggling, to regain breath, and as much as five minutes must have elapsed while they were locked in this motionless but terrible embrace. Suddenly the bull, by one desperate effort, wrenched his head from the grasp of his adversary, and retreated a few steps. The bear stood up to receive him. I now watched with breathless interest, for it was evident that each animal had staked his life upon the issue of the conflict. . . . Rendered furious by his wounds, the bull now gathered up all his energies, and charged with such impetuous force and ferocity that the bear, despite the most terrific blows with his paws, rolled over in the dust, vainly struggling to defend himself. The lunges and thrusts of the former were perfectly furious. At length, by a sudden and well-directed motion of his head, he got one of his horns under the bear's belly, and gave it a rip that brought out a clotted mass of entrails. It was apparent that the battle must soon end. Both were grievously wounded, and neither could last much longer. The ground was torn up and covered with blood for some distance around, and the panting of the struggling animals became each moment heavier and quicker. Maimed and gory, they fought with the desperate certainty of death—the bear rolling over and over, vainly striking out to avoid the fatal horns of his adversary—the bull ripping, thrusting, and tearing with irresistible ferocity.

At length . . . the bull drew back, lowered his head, and made one tremendous charge; but, blinded by the blood that trickled down his forehead, he missed his mark, and rolled headlong on the ground. In an instant the bear whirled and was upon him. Thoroughly invigorated by the prospect of a speedy victory, he tore the flesh in huge masses from the ribs of his prostrate foe. The two rolled over and over in the terrible death-struggle; nothing was now to be seen save a heaving, gory mass, dimly perceptible through the dust. A few minutes would certainly have terminated the bloody strife . . . when, to my astonishment, I saw the bear relax in his efforts, roll over from the body of his prostrate foe, and drag himself feebly a few yards from the spot. His entrails had burst entirely through the wound in his belly, and now lay in long strings over the ground. The next moment the bull was on his legs, erect and fierce as ever. Shaking the blood from his eyes, he looked around, and seeing the reeking mass before him, lowered his head for the final and most desperate charge. In the death-struggle that ensued both animals seemed animated by supernatural strength. The grizzly struck out wildly, but with such destructive energy that the bull, upon drawing back his head, presented a horrible and ghastly spectacle; his tongue, a mangled mass of shreds, hanging from his mouth, his eyes torn completely from their sockets, and his whole face stripped to the bone. On the other hand, the bear was ripped completely open, and writhing in his last agonies. Here it was that indomitable courage prevailed; for blinded and maimed as he was, the bull, after a momentary pause to regain his wind, dashed wildly at his adversary again, determined to be victorious even in death. A terrific roar escaped from the dying grizzly. With a last frantic effort he sought to make his escape, scrambling over and over in the dust. But his strength was gone. A few more thrusts from the savage victor, and he lay stretched upon the sand, his muscles quivering convulsively, his huge body a resistless mass. A clutching motion of the claws—a groan—a gurgle of the throat, and he was dead.

The bull now raised his bloody crest, uttered a deep bellowing sound, shook his horns triumphantly, and slowly walked off, not, however, without turning every few steps to renew the struggle if necessary. But his last battle was fought. As the blood streamed from his wounds a death-chill came over him. He stood for some time, unyielding to the last, bracing himself up, his legs apart, his head gradually drooping; then dropped on his fore knees and lay down; soon his head rested upon the ground; his body became motionless; a groan, a few convulsive respirations, and he too, the noble victor, was dead.

During this strange and sanguinary struggle, the cattle . . . had gathered in around the combatants. The most daring, as if drawn toward the spot by the smell of blood or some irresistible fascination, formed a circle within twenty or thirty yards, and gazed at the murderous work that was going on with startled and terror-stricken eyes; but none dared to join in the defense of their champion. No sooner was the battle ended, and the victor and the vanquished stretched dead upon the ground, than a panic seized upon the excited multitude, and by one accord they set up a wild bellowing, switched their tails in the air, and started off at full speed for the plains. (Browne, 1862 : 747–749.)

Such natural bear-and-bull fights, although spectacular and gory, had little effect on the population of bears. And the military campaigns launched from time to time by the Spaniards against the animals that were eating the mission cattle were hardly more effective in this respect than the attempts by the bulls to defend their own herds. In 1792, grizzlies were so plentiful and injurious in the environs of Mission Santa Cruz that soldiers of the guard were ordered to hunt them for target practice (Torchiana, 1933 : 193). Seven years later, "troops of Purísima made a regular campaign against the bears of that region" in an effort to lessen the destruction of livestock (Hittell, 1898, 2 : 560).

In 1801, Raymundo Carrillo, commandant of the garrison at Monterey, reported that although the vaqueros had roped thirty-eight during the past year, depredations by grizzlies continued. He proposed to ambuscade the troops at a place baited with carcasses of old mares. (*Ibid.*) For this campaign against the bears, Governor Arrillaga authorized the use of a thousand precious cartridges (Bancroft, 1886, 2 : 143). More than fifty grizzlies were killed at Monterey during that year (*ibid.*, p. 142), but the bear population continued to grow. Thirty mares were slaughtered and their carcasses poisoned with *yerba de puebla,* but the bears were not seriously affected (*ibid.*, pp. 143, 418.)

In time other methods of control were tried. At Mission Santa Clara, for example, in the 1820's several hunters often were posted on a special platform built some fifteen feet from the ground, in a tree. According to the noted traveler Duhaut-Cilly:

men are kept there armed with rifles, each one loaded with two bullets. Twenty paces from the tree is a horse, dead several days, the decay of which begins to make itself manifest. The bears, which, they say, have a very acute sense of smell, are drawn thither from a long way; and as they come, they are shot with great ease by the hunters. Padre Viader, president of Mission Santa Clara, a modest and truthful man, assured me he himself had killed a hundred in this way. (Carter, 1929 : 153.)

A bear-killing device perfected by the neophytes at Santa Clara was a large, strong box made of rails and having a doorway. A *reata* was coiled on the floor round a hunk of meat. When a grizzly stepped within, the rope was yanked to snare him and the door was slammed shut. Then the Indians attacked with spears. (W. H. Davis, MS, Huntington Library.)

At Monterey, about 1824, Pattie observed that "great numbers" of grizzlies were taken by stratagem, the Spaniards

killing an old horse in the neighborhood of their places of resort. They erect a scaffold near the dead animal, upon which they place themselves during the night, armed with a gun or lance. When the bear approaches to eat, they either shoot it, or pierce it with the lance from their elevated position. Notwithstanding all their precautions, however, they are sometimes caught by the wounded animal; and after a man has once wrestled with a bear, he will not be likely to desire to make a second trial of the same gymnastic exercise. (Flint, 1930 : 352.)

At one time the governor appointed expert bear hunters in different parts of the country. One of the most famous of these men, Don Rafael Soto, "used to conceal himself in a pit, covered with heavy logs and leaves, with a quarter of freshly killed beef above. When the grizzly bear walked on the logs he was shot from beneath." (Vallejo, 1890 : 191.)

This extraordinary and seemingly dangerous technique was employed because bears could seldom be killed from a distance with the type of rifle available to the Spaniards. Duhaut-Cilly also described the pit method of getting close to and underneath a grizzly to make certain that the shot would be lethal. According to him, the hunters "dig a deep pit, covered over with a strong hurdle of boughs, on which they put some flesh of the kind to allure bears; and keeping themselves below, they kill them with thrusts or rifle shots." (Carter, 1929 : 153.)

The Spaniards, by temperament, were not hunters like the Americans who came later; nonetheless, their records indicate they destroyed a surprisingly large number of bears. The grizzlies, having intimidated the Indians for centuries, at last were confronted by superior creatures. Yet the government hunters were no more effective in reducing the bear population than the troops had been. In spite of the many bears killed by guns, lances, ropes, and traps, their numbers in the southern Coast Ranges apparently increased markedly during the first half of the nineteenth century. They did not decline until after the great drought of 1863–1864 had ended the Spanish pastoral era and the American hunters had been at work for some years.

The herds of cattle, horses, and sheep, which were sculpturing every hill in the coastal region of southern California with tier upon tier of terrace-like trails, provided a powerful stimulus to the reproductive capacity of the grizzlies. Having a new and abundant food supply, the bears increased in number despite attempts to destroy them.

The natural deaths among the enormous herds gave the grizzlies a large additional food supply; and time and again the bears also availed themselves of opportunities to feast on the countless animals killed by the Spaniards and left unused either near the settlements or on open lands.

One reason for these frequent slaughters of livestock was the pressing need to save the range during severe droughts, such as occurred in 1809–1810, 1820–1821, 1828–1830, and 1840–1841 (Bancroft, 1888 : 337–338). Wild horses, for example, were so numerous that on the approaches to the pueblos they would eat up the grass and spoil the pasture for domestic animals (*ibid.*, p. 346). Bidwell (1897 : 75) recorded horses in herds "twenty miles long" on the west side of the San Joaquin Valley; and every rancho supported great numbers from which candidates for the *caballada*, or band of highly trained animals, were selected (Cleland, 1941 : 79).

To reduce this overpopulation of horses, the government periodically organized rodeos, or roundups, during which the herds were driven into specially built corrals. The branded stock was then sorted out, and the wild individuals or *mesteños* were turned loose to be lanced by mounted vaqueros. Thus thousands were killed (Bancroft, 1888 : 346), and the pasturage was saved for cattle (Cleland, 1941 : 79). Another method was to drive the unwanted horses over precipices until there were piles of mangled bodies below the cliffs (Bancroft, 1888 : 347). Guadalupe Vallejo (1890 : 189) said that "in 1806 there were so many horses in the valleys about San José that seven or eight thousand were killed. Nearly as many were driven into the sea at Santa Barbara in 1807, and the same thing was done at Monterey in 1810." Even tame ones were slaughtered when the numbers became too great. On one occasion, the government threatened Juan José Nieto with forfeiture of his ranch if he allowed his horses to continue increasing (Bancroft, 1886, 2 : 418). And in "the great drought of 22 months between the rains of 1828 and 1830, during which the wells and springs of Monterey gave out, . . . it was estimated that fully 40,000 head of horses and neat cattle perished throughout the province" (Bancroft, 1888 : 338). Cattle also were slaughtered at special *desviejars* merely to get rid of the older stock (*ibid.*, p. 341). With so many dead bodies of domestic animals to be had without effort, grizzlies were able to enjoy almost perpetual feasts and multiplied as they had never done before.

Likewise grizzlies were attracted to the private butchering grounds of the missions and ranchos. Guadalupe Vallejo (1890 : 190) recalls that

every Mission and ranch in old times had its *calaveras*, its "place of skulls," its slaughter-corral, where cattle and sheep were killed by the Indian butchers. Every Saturday morning the fattest animals were chosen and driven there, and by night the hides were all stretched on the hillside to dry. At one time a hundred cattle and two hundred sheep were killed weekly at the Mission San José, and the meat was distributed to all "without money and without price." The grizzly bears, which were very abundant in the country . . . used to come by night to the ravines near the slaughter-corral where the refuse was thrown by the butchers.

Deer carrion also became available to the bears in greater abundance than previously, for the Spanish Californians pursued deer on horseback and caught many with the *reata*. Then, as related by Duhaut-Cilly, "the flesh of these dead animals, from which the fat had been removed, remaining abandoned on the hunting ground, bears, attracted by this prey, come from all sides to feed upon it" (Carter, 1929 : 241).

When the hide and tallow trade between the missionaries and rancheros on one side and the Yankee and British merchants on the other flourished along the California coast, enormous numbers of cattle were killed for their skins and fat. Many "hundreds were slain in a day," and there were "immense heaps of carcasses" (Yount, 1924 : 43). In 1834, the missionaries, who anticipated secularization, reportedly disposed of some 100,000 cattle for the hides and tallow (Cronise, 1868 : 61), leaving most of the meat to be eaten by wild animals.

The slaughtering of livestock for hides and tallow was known as the *matanza*, or sometimes as the *nuqueo* if it was done by breaking the necks of the cattle with knives. The *nuqueadores* would ride

at full speed over the fields, armed with knives. Passing near an animal, one gave it a blow with the knife in the nerve of the nape of the neck, and it fell dead. These nuqueadores . . . were followed . . . by dozens of peladores, who took off the hides. Next came the tasajeros, who cut up the meat into tasajo and pulpa; and the funeral procession was closed by a swarm of Indian women who gathered the tallow and lard in leather hampers . . . A field after the nuqueo looked like Waterloo after the charge of the old guard. (Bancroft, 1888 : 340.)

Then "bears came to the place at night to feast on the meat that was left after the hides and tallow were taken" (Davis, 1929 : 80).

Davis, who had personal experience with the hide and tallow trade, related that his wife's father, a Spanish Californian, once with the help of ten soldiers and a relay of horses "lassoed and killed forty bears in one night" at the *matanza* ground near the site of the town of Mountain View.

It was in the killing season, and the bears smelling the meat, had come down from the mountain to partake of it. My father-in-law said this was the most exciting event of his life, and that they were so interested in dispatching the bears they forgot all danger. The animals were lassoed by the throat and also by the hind leg, a horseman at each end, and the two pulling in opposite directions till the poor beast succumbed. The fun was kept up until about daylight, and when they got through they were completely exhausted . . . (Davis, 1889 : 97–98.)

This love of excitement was deeply rooted in the character of the Spanish Californians, who, although they abhorred ordinary manual labor, would exert prodigious energy in dangerous sport. Perhaps the most prominent and thrilling activity in which they indulged extensively was the pursuit of grizzlies with the lariat, known as the *reata* (fig. 15). This they did extensively and intensively, at any season and at any time of day or night, and on the least provocation or excuse. It was a risky sport and, in the words of one Anglo-Saxon contemporary, "worthy the entertainment of an emperor, when it is conducted by the natives of this country, and after their own fashion" (Garner, 1847a : 370). Don and vaqueros, old and young, were equally enthusiastic adherents. Even some of the ecclesiastics tucked up their robes and participated: "Father Real . . . was often known to go with young men on moonlight rides, lassoing grizzly bears" (Vallejo, 1890 : 187).

Because the average vaquero rarely carried a rifle or a pistol, firearms being almost entirely the property of the military and special personages, the use of the *reata* as a weapon was developed to an extraordinary perfection. Made of pliable rawhide plaited in four strands to the thickness of the little finger, about sixty feet in length, and having the far end tied in a running knot, this instrument, like the horse, was an indispensable part of the equipment of every male Californian. A common saying was, "*La reata es el rifle del ranchero.*" The strongest were made from hides of cattle that had starved to death in winter, but they turned black when braided and had to be tallowed constantly (Rojas, 1953 : 30). Duhaut-Cilly observed that

when one of these men desires to use his rope against a man or an animal, he holds it coiled in his hand; he goes at a gallop to within fifteen paces of his enemy, while making the fatal cord turn above his head like a sling; and at the favorable moment, he unrolls it in throwing it with so much skill, that he never fails to bind by the neck, body or legs, the individual he threatens, dragging him instantly with great cruelty over the ground, at his horse's utmost speed (Carter, 1929 : 152).

The horses which the rancheros rode had been carefully trained to assist their masters in the subjugation of wild cattle, human enemies, and savage beasts. They were beautiful and intelligent creatures, for like their Mexican progenitors, they "were partly of Moorish or Arab blood, small, finely formed, agile, and capable of almost incredible endurance" (Cleland, 1941 : 79). Although innumerable feats of skill had to be performed in the ordinary course of managing the cattle herds, the severest test of the quality and training of a horse was the animal's behavior when attacking a grizzly.

One American, William R. Garner, who witnessed mounted Californians bear hunting with the *reata*, wrote in 1846 from Monterey that "every motion of . . . the horses, which seem as though they were doubly proud when they feel the strain of the lasso from the saddle, and appear to take as much delight in the sport as the riders themselves, is grand beyond my power of description." He went on to say that

this method of hunting the bear is one of the noblest diversions with which I am acquainted. . . . It requires an extraordinary degree of courage for a man to ride up beside a savage monster like the grisly bear of this country, which is nearly as active as a monkey, and whose strength is enormous. Should a lasso happen to break, . . . the bear invariably attacks the horse and it requires very often the most skilful horsemanship to prevent the horse or its rider from being injured. It requires also great skill to know when to tighten the lasso, and to what degree, to prevent it from being suddenly snapped by too sudden a strain. The rider must have his eye constantly on that of the bear, and watch his every motion. Sometimes, either through fear, carelessness or inadvertence, a man may let go his lasso. In this case, another, if the bear takes off, (which he is likely to do,) will go as hard as his horse can run, and, without stopping his speed, will stoop from his saddle and pick the end of the lasso from the ground, and, taking two or three turns round the loggerhead of his saddle and checking his horse's rein, again detain the bear. (Garner, 1847a, 371.)

This same writer also gave a detailed account of the preparation for the contest and the subsequent happenings:

Whenever a Californian wishes to catch a bear, and which at any time he is ready to undertake for the sake of the diversion, he . . . looks well over the ground for about two miles all around the spot where he intends to lay his bait. This is done for the purpose of reconnoitring every step of ground that he thinks he may have to ride over, for the purpose of ascertaining if there are any squirrel holes or ravines, and likewise to form a judgment which way the bear will be most likely to run from the bait, on her being surprised. At least one of his companions accompanies him on this excursion.

They then go and catch a mare, . . . or if this is difficult, a stray horse will answer the purpose. As soon as they have lassoed their victim, they take it to the place previously selected for laying the bait. On this spot they strangle the animal, and then let out its entrails, that the bear may scent it at great distance. They then cut off one quarter of this animal, and drag it all over the ground for a half or three quarters of a mile round the spot, then take it back and leave it with the carcass, always covering it over with some grass or bushes that the birds [condors, vultures] may not devour it before the bear makes his appearance.

The bait being left in perfect order, and the ground well reconnoitred, they go away, and do not trouble the bait the first night, because, if the bear comes the first night, he will be sure to return if he is not troubled, and most likely with two or several more. Consequently the second night is the best hunt. The owners of the bait

then invite, in secrecy, four or five choice friends. They do not invite too many, because, through too much excitement amongst many
persons, or eagerness to get the first chance to throw the lasso, the
bear gets wind that all is not right, and being a very cunning animal, if he once begins to suspect that the enemy is near, he keeps
so good a watch that it is impossible to catch him. . . . It is not the
fleetest horse that is considered the best for this employment. It requires a tame, lively horse, with a good government in the mouth,
and a strong back.

Everything being prepared, men, horses, saddles, and lassoes,
they all start at sun-down or dusk, and keep carefully to windward
of the bait, which must be placed on a piece of ground clear from
rocks, trees or bushes, and near or within about eight hundred yards
of one of these, for the purpose of hiding themselves, that the bear
may not see them when he is approaching the bait. A horse that has
been catching bears three or four times, will keep a strict watch for
the approach of the bear at the bait, and will invariably let his rider
know—not by any noisy motion, but by deep suppressed sighs, and
pricking up his ears. Whenever one or more of the horses do this,
the men who have been lying by on foot, mount as quietly as possible, and when all are ready with their lassoes in their hands ready
to swing, they put spurs to their horses, which at that moment is
very little needed, that noble animal appearing to all intents and purposes to be as anxious as his rider to capture the savage animal. The
horse, being swifter than the bear, if the plan has been well laid, is
sure to overtake him before he can get to any bush. The foremost
rider throws his lasso, and seldom fails of catching the bear, either
by the neck or round the body or one of its legs. Should he miss,
there are several more close at his heels to throw their lassoes. As
soon as the bear finds himself fast he rears and growls, taking hold
of the lasso with his two fore paws. At this crisis the lasso must always be kept tight; if not, the bear will extricate himself immediately.
(*Ibid.*, p. 370.)

Perhaps the most remarkable feature of the struggle was
the sagacity and skill of the horses and their own delight in mas-
tery of the bear:

The horse, from the very moment the bear is lassoed, keeps his eye
on every movement, and appears to do . . . all in his power to protect and defend his rider as well as himself; as it often happens, that
. . . the bear will entangle the horse's legs with the lasso, . . . if it is
a horse that has been used to lassoing bears, he will with the greatest
agility clear himself, without the least motion from the bit. I have
several times seen a horse, when the bear has been approaching him
from before, instead of turning round to run away or to run on one

side, wait until the bear got close to him, watching him all the time with a steady eye, and all of a sudden take a leap right over the bear, and then turn suddenly round and face him again. This feat of course is only done by such horses as are well acquainted with bear-hunting. . . . From the moment a horse sees the bear, it matters not at what distance, he begins to tremble, and his heart beats so loud that his rider can distinctly hear it. But . . . as soon as the horse feels by the strain of the lasso that the bear is lassoed, his fear leaves him, and he is from that moment in the highest glee. If the bear is a very large one, two or three more persons will throw their lassoes on him, because an old bear will be very apt to take the lasso in his mouth and bite it off, or bring such a strain on it as would break it.

The bear being now well secured, with three or four lassoes on him, the horses, arching their necks and snorting with pride at their prize, walk away with the savage animal, which is rearing, plunging, and growling. (*Ibid.*, pp. 370–371.)

Descendants of the Spanish rancheros hunted bears with the *reata* even during the early years of the American period. An article in the San Francisco *Daily Alta California* of March 7, 1875, tells in a gilded style of "a thrilling encounter with a grizzly bear, in which Governor [Romualdo] Pacheco figures prominently":

Governor Pacheco has among his accomplishments,—and they are many—one, possessed, we believe, by no other Governor in the United States. He can lasso, and get away with a wild, grizzly bear; and we saw him do it, in May, 1852, on the Rancho de los Osos (Bear Ranch), in San Luis Obispo, then the residence of Governor Pacheco's mother. Away up in the mountains, among the wild cats, the grizzly bears take their morning naps, after their nightly prowling about in search of any stray calf, pig or other small game. Early one morning the enormous print of a grizzly's foot was seen in the earth close by the dwelling of the Governor's mother, and in a few minutes Romualdo and two or three others were in the saddle and off for the mountains. When the tall wild oats, half-way up the mountain, were reached, the party had not ridden more than two minutes among the tall, dry wisps, when the horses started, snorting loudly, and instantly a huge grizzly stood erect, with a terrific presence, high above the dry, wild oats . . . with shaggy, fur overcoat, his eyes gleaming fiercely, his cruel teeth and red mouth unpleasantly conspicuous. Every man and every horse seemed for the instant petrified as if, while every nerve and every muscle and wary sense was at the utmost tension, they had suddenly looked upon Medusa. In a second's time, Pacheco had spurred forward, swinging his lasso. The bear began sparring warily . . . But Pacheco's lasso shot out like an arrow, and clasped about the huge fore-foot, when the horse (who saw every movement, and was just as wide-awake as Pacheco), sprang the other way, and the lasso being fast to the pommel, the bear was instantly thrown to the ground, when two other men, quick as lightning, had thrown their lassos, and caught the hindfeet; then another rider caught the loose fore-foot, and the four horses took their positions like cavalry animals trained by some noiseless signal, and slowly marched down the mountain's side, two horses in the van and two in the rear, dragging Ursa Major quietly down the grassy descent, the rear horses keeping just taut-line to prevent the bear from getting any use of his terrible hind claws. Nahl has painted some of these California lassoing scenes, that have been as near justice to such exciting tableaux as could be done by the painter's art, but nothing could portray the intensity of excitement and action brought forth at such a moment. Pacheco was . . . twenty-one years old, and the handsomest man we ever looked upon. . . .

When he first realized the sudden presence of the terrible enemy, and stood erect in his stirrups, his face gleaming with the glory of youth, fearlessness and excitement, his great, black eyes sparkling, his white teeth tightly pressed upon his nether lip, perfectly still for a moment, he was the most glorious object in nature.

In no longer time than the sight of this could be just taken in, he sprang forward, his long, dark hair tossed wildly for a moment, and then, he had captured the bear . . . The captors slowly took their prisoner down to the house, where a long, heavy piece of timber lay upon the grass. Fastening the bear's hindfeet to the timber with the strong lassos, and the fore feet to a strong, deep-driven stake, they stepped away to a respectful distance, their eyes upon the ferocious creature, and their hands upon their saddle-pommels. . . .

The bear lay with his head between his huge paws, covering his eyes, save occasionally, when he would furtively lift his eyes, like a sulky child, to look at his captors; then covering his eyes again, remain a moment, and steal another look. Soon he gave heavy sighs, and some one said, "He is dying!" We expressed surprise to learn that the bear was wounded.

"He is not wounded," they replied, "but his heart breaks—he dies of rage." And, in a few moments he had breathed his last, and was dragged away some distance from the house, and left.

Pacheco pointed to the sky. We looked, and saw a hundred carrion crows, whose watchful eyes had seen the feast long before it was half way down the mountainside; and before we were a hundred yards from the dead bear, its body was completely hidden by the sable, flapping wings of the hungry undertakers.

The roping of bears as practiced by the Spanish Californians was a refined skill and there were many tricks to the art that took years of experience to master. One Yankee visitor to California, having acquired some dexterity at throwing the *reata*, brimming with confidence, elected to pit his newly acquired skill against a grizzly. Noosing one which he happened upon in a wooded glen, he reined back his mount in full expectation of toppling the bear flat onto its face,

when the *rieta*—which is attached always to the pommel of the saddle —came up taut. Judge of his astonishment . . . when that bear quietly assumed a sitting position, took hold of the *rieta*, and commenced to draw it in, hand over hand! The hapless descendant of the Pilgrim Fathers stuck to the horse and saddle until he saw the slack all drawn in, and the bear and the horse coming rapidly together [fig. 16], . . . then hastily descended and hunted a tree, abandoning the horse to the underwriters. He had learned only half of the trick. (Evans, 1873 : 66.)

A similar incident was described by J. M. Letts at Santa Barbara in 1849:

. . . one of the *Bruin* family had taken up his residence on a *rancho*, not far distant. . . . Now, said Bruin had been a quiet neighbor, and had taken nothing excepting the appurtenances of said *rancho*, and had a most religious adversion to any additional *ties* between himself and neighbors. When said neighbors approached and attempted to . . . [lasso him], Bruin, as dignified people will do, stood up and looked them in the eye. Six lassos were simultaneously thrown. He caught three of them, and, hand over hand, hauled the horses in, and with one stroke took off from one of them his entire haunch. The riders cut their lassos, and, . . . took the longest kinds of steps toward the mission-house. (Letts, 1852 : 137–138.)

Bell (1927 : 257) described the unhappy fate of a man who noosed a bear round the middle with a rope attached to his saddle. The grizzly sat down and, paw over paw, quietly drew in horse and rider like a fisherman nonchalantly bringing up a deep-sea catch. Finally, the horse in its frantic struggles to escape overturned the saddle and dumped the man into the receptive arms of the beast. Seton (1929 : 68) tells of cowboys in California who had to release a grizzly because even after it was roped round the neck the animal could drag two horses with ease.

A few experienced bears developed such cunning in avoiding the lassos that they could not be caught. One in particular committed outrages at Mission San Juan Bautista for many years until finally shot by an American, who was paid a fee of twenty cows (Revere, 1849 : 258).

Although the Spaniards usually went in groups of not fewer than four persons when they challenged the grizzly with the *reata*, some men became so adept at the sport that they took on the beasts singlehanded. The usual technique was for the rider to throw the rope and then, aided by the brilliant maneuvers of his horse, entangle the grizzly about the trunk of a tree; the animal was then killed with the "broad-bladed *machete*, or Mexican hunting knife" (Vallejo, 1890 : 191). Bancroft (1888 : 434–435), though, tells of another method. Manuel Lario, after getting a rope round a bear, galloped away with the beast in full pursuit. Approaching a tree, the intrepid and ingenious Spaniard hurled his end of the rope forward and over a horizontal limb, caught it again as it came down on the other side and, continuing ahead at full speed, suddenly was brought up with a jerk as the astounded bear dangled from the branch.

Ramón Ortega, who was born in the early 1840's and spent most of his life in the mountainous sections of Ventura County, was reputed to be the greatest bear hunter of southern California (Gidney, Brooks, and Sheridan, 1917 : 362–363). He recounted having lassoed seventy grizzlies in five years on Rancho Sespe and some two hundred during his lifetime. Once on the Potrero Seco he roped fifteen in a single day. On the Sespe he killed forty in little more than a month.

The most dauntless of the Californians who pursued the grizzly was Don José Ramón Carrillo, a member of the aristocracy who matched his skill with the beasts only for sport. Riding alone one day he gave chase to a bear that in trying to escape crashed into a pit about six feet deep. Before Carrillo could rein up his horse, he too went over the embankment—man, horse, and bear were thrown together in wild confusion. The grizzly attempted to climb out but was unable to do so. Thereupon Carrillo, perceiving that in such close quarters he would be highly vulnerable if his antagonist should decide to fight, placed his arms under the bear's hind parts. Exerting all his strength, he gave the animal a heave and boosted it out of the pit. (Davis, 1889 : 497–498.)

Carrillo was most famous, however, for the duels with bears in which he used a light-handled sword. One such encounter in Sonoma County was described by Davis (1929 : 302–303). Another, in southern California, is more vividly described by Bell (1930 : 106–107):

These grizzlies stand on their hind legs and spar, fence, parry and strike like a skilled fencing master or prizefighter. . . . Ramón Carrillo . . . overtook a huge grizzly in the Encino Valley, challenged him and fought him singlehanded with a light sword.

This Ramón was a desperado, a man who fought for the love of it and who was never defeated until finally riddled with American buckshot. General Covarrúbias with a party which included Ramón Carrillo was journeying from Santa Bárbara to Los Angeles when they sighted the bear, out on the San Fernando plain. They surrounded the Old Man of the Mountains, who promptly stood up like a man and offered defiance.

"Stand back, please, señores," requested Carrillo, dismounting and drawing his sword. "Allow me to fight a personal duel with this grand old gladiator." . . .

Ramón advanced like a dancing master, flourishing a rapier-like blade which he always carried. Bruin stood on the defensive, staring with angry astonishment. Deftly and with a smile . . . the young Californian, with all the skill and grace of a trained bullfighter, danced around the grizzled giant and got in his stinging, maddening thrusts here and there. The grizzly rushed time and again with terrific roars, but the man waited only long enough to sting the huge menacing paws with rapier point and then sidestepped to safety.

The excitement of the picturesque mounted audience grew almost beyond control, and the "vivas!" first for the caballero and then for the bear, drove the animal duelist almost frantic. With utmost coolness and always laughing, Ramón Carrillo fenced with that grizzly for one hour. When all concerned seemed to be tiring of the sport he stepped in and with a quick thrust to the heart laid the splendid brute low.

# 6 Bear-and-Bull Fights

The Spanish Californian had a temperament too emotional and fiery for him to be content with a quiet and humdrum existence. Craving excitement and color and being free to devote himself to the "grand and primary business of the enjoyment of life," he seized upon every opportunity for dangerous sport, social gaiety, and devotional pageantry. He was adept at arranging spectacles—from the solemn Mass in his missions to all-night dances in his adobes and thundering rodeos on the range—in which color and drama, and not infrequently blood and violence, combined to arouse the emotions. But of all these forms of excitement, nothing so impressed and horrified the Anglo-Saxon visitor—and was least understood by him—as the fights between grizzlies and wild bulls that were staged during fiestas and on feast days and Sundays at the missions, pueblos, and presidios.

These fights were an outgrowth of the sporting heritage and temperament of the people. In Spain, ever since the Latin races had first gone to the Pyrenees, battles between bears and bulls had been a form of amusement (Kingsley, 1920 : 22). The Californian colonists, therefore, would have been untrue to their ancestry if they had failed to avail themselves of an exceptional opportunity to continue this tradition. Grizzlies, present in abundance, could be captured with the *reata*, and cattle were numerous. Natural conflicts with bears on the range, as described by Browne (see our chap. 5), were proof of the inherent antipathy of the two species. Under these circumstances, it was inevitable that in California the bear-and-bull fight reached a higher stage of development than anywhere else in the world. The people were intensely fond of the sport and would go great distances to wit-

140

ness a contest. Pío Pico, for example, stated that "for the purpose of being present at bull-baits he frequently rode in one day from San Diego to Los Angeles" (Bancroft, 1888 : 346).

The earliest formal fight of which we have record was held at Monterey in 1816 in honor of the newly appointed Governor of California, Pablo Vicente de Sola. At that time, however, the techniques for such a contest were already well established, and it is certain that fighting between bears and bulls had thrilled the people of the capital for a number of years previously.

## A FIGHT ON THE RANGE

Some of the first fights probably were held on the range for the benefit only of the vaqueros. The Reverend Walter Colton, Alcalde of Monterey in the late 1840's, describes in his diary a fight which, except for the paraphernalia of his aristocratic friends—servants, bottles of ale, and guns—doubtless was typical of these informal conflicts.

THURSDAY, OCT. 28. The king of all field-sports in California is the bear-hunt: I determined to witness one, and for this purpose joined a company of native gentlemen bound out on this wild amusement. All were well mounted, armed with rifles and pistols, and provided with lassoes. A ride of fifteen miles among the mountain crags ... brought us to a deserted shanty, in the midst of a gloomy forest of cypress and oak. In a break of this swinging gloom lay a natural pasture, isled in the centre by a copse of willows and birch, and on which the sunlight fell. This, it was decided, should be the arena of the sport: a wild bullock was now shot, and the quarters, after being trailed around the copse, to scent the bear, were deposited in its shade. The party now retired to the shanty, where our henchman tumbled from his panniers several rolls of bread, a boiled ham, and a few bottles of London porter. These discussed, and our horses tethered, each wrapped himself in his blanket, and with his saddle for his pillow, rolled down for repose.

At about twelve o'clock of the night our watch came into camp and informed us that a bear had just entered the copse. In an instant each sprung to his feet and into the saddle. It was a still, cloudless night, and the moonlight lay in sheets on rivulet, rock, and plain. We proceeded with a cautious, noiseless step, through the moist grass of the pasture to the copse in its centre, where each one took his station, forming a cordon around the little grove. The horse was the first to discover, through the glimmering shade, the stealthful movements of his antagonist. His ears were thrown forward, his nostrils distended, his breathing became heavy and oppressed, and his large eye was fixed immovably on the dim form of the savage animal. Each rider now uncoiled his lasso from its loggerhead, and held it ready to spring from his hand, like a hooped serpent from the brake. The bear soon discovered the trap that had been laid for him; plunged from the thicket, broke through the cordon, and was leaping, with giant bounds, over the cleared plot for the dark covert of the forest beyond. A shout arose—a hot pursuit followed, and lasso after lasso fell in curving lines around the bear, till at last one looped him around the neck and brought him to a momentary stand.

As soon as bruin felt the lasso, he growled his defiant thunder, and sprung in rage at the horse. Here came in the sagacity of that noble animal. He knew, as well as his rider, that the safety of both depended on his keeping the lasso taught, and without the admonitions of rein or spur, bounded this way and that, to the front or rear, to accomplish his object, never once taking his eye from the ferocious foe, and ever in an attitude to foil his assaults. The bear, in desperation, seized the lasso in his gripping paws, and hand over hand drew it into his teeth: a moment more and he would have been within leaping distance of his victim; but the horse sprung at the instant, and, with a sudden whirl, tripped the bear and extricated the lasso. At this crowning feat the horse fairly danced with delight. A shout went up which seemed to shake the wildwood with its echoes. The bear plunged again, when the lasso slipped from its loggerhead, and bruin was instantly leaping over the field to reach his jungle. The horse, without spur or rein, dashed after him. While his rider, throwing himself over his side, and hanging there like a lampereel to a flying sturgeon, recovered his lasso, bruin was brought up again all standing, more frantic and furious than before, while the horse pranced and curveted around him like a savage in his death-dance over his doomed captive. In all this no overpowering torture was inflicted on old bruin, unless it were through his own rage,—which sometimes towers so high he drops dead at your feet. He was now lassoed to a sturdy oak, and wound so closely to its body by riata over riata, as to leave him no scope for breaking or grinding off his clankless chain; though his struggles were often terrific as those of Laocoon, in the resistless folds of the serpent.

This accomplished, the company retired again to the shanty, but in spirits too high and noisy for sleep. Day glimmered, and four of the baccaros started off for a wild bull, which they lassoed out of a roving herd, and in a few hours brought into camp, as full of fury as the bear. Bruin was now cautiously unwound, and stood front to front with his horned antagonist. We retreated on our horses to the rim of a large circle, leaving the arena to the two monarchs of the forest and field. . . . They stood motionless, as if lost in wonder and indignant astonishment at this strange encounter. Neither turned from the other his blazing eyes; while menace and defiance began to lower in the looks of each. Gathering their full strength, the terrific rush was made: the bull missed, when the bear, with one enormous bound, dashed his teeth into his back to break the spine; the bull fell, but whirled his huge horn deep into the side of his antagonist. There they lay, grappled and gored, in their convulsive struggles and death-throes. We spurred up, and with our rifles and pistols closed the tragedy; and it was time; this last scene was too full of blind rage and madness even for the wild sports of a California bearhunt. (Colton, 1850 : 214–218.)

Fights between bears and bulls probably were staged in all settlements of Spanish California.[1] In some towns a special arena was built (fig. 17) consisting of a strong wooden fence to contain the beasts and a raised platform for the women and children, most of the men staying on horseback outside the ring (Bancroft, 1888 : 433). Monterey, the capital, had an arena of adobe and stone walls. Other towns utilized the plaza, which, being surrounded by buildings, needed to be barricaded only at the street exits. Overhanging balconies provided choice vantage points for the ladies.

At Mission Santa Barbara the arena was "a large square, formed by the junction of the long corridor with a temporary fence of poles" (Robinson, 1846 : 104). Horsemen with *reatas* stood guard to keep the grizzly from clambering over the fence and venting its rage on the women and children.

According to Guadalupe Vallejo (1890 : 191), "The principal . . . fights were held at Easter and on the day of the patron saint of the Mission, which at San Jose was March 19." Having given a number of hours to prayer and religious rituals, the people were ready and eager for the noise and thrills of a holiday. Nothing satisfied their appetite for excitement better than to

---

[1] Bear-and-bull fights were held in California in the following places: Pala (San Diego Co.), 1850; San Juan Capistrano, n.d.; Santa Ysabel (Orange Co.), 1872; San Bernardino, 1864; San Gabriel, the 1840's; Los Angeles, 1854, 1857, 1866, 1868; Ventura, n.d.; Santa Barbara, n.d.; San Juan Bautista, n.d.; Castroville, 1865; Monterey, 1815, 1816, 1842, 1870, 1881; Carmel, 1827; San Jose, n.d; San Antonio (Alameda Co.), 1858; Oakland, 1854; San Francisco, 1816, 1820, 1850, 1851, 1859, 1860; Columbia, the 1850's; Mariposa, 1852; Sonora, 1851; Mokelumne, the 1850's; Brighton (Sacramento Co.), 1851; Georgetown, 1851; Nevada City, 1851; Oroville, 1857; Faggtown (i.e., Weaverville), 1859. The list is not complete.

unleash a bear against a bull. If such a fight had been arranged, the crowd, on streaming out of the church, would gather to make bets and take refreshments. Then in a high state of anticipation the people would cluster about the arena and wait for the sport to begin.

Alfred Robinson, who watched a fight at Mission Santa Barbara, said:

On this occasion everybody attended, as is customary in all their amusements, and men, women, and children took part in the discussions relative to the fight. Such exhibitions served for a topic of conversation amongst all classes for months afterwards, and the performance elicited as much applause as is usually bestowed on the triumph of some great actor in the theatres of our own country. (Robinson, 1846 : 105.)

Perhaps the most noted of these fights was that at Monterey in 1816 honoring Governor Sola. His arrival from Mexico and his inauguration were the excuse for an elaborate celebration. With the Spanish flag waving over the pine-garlanded adobes of Monterey, the celebration opened with the pomp and splendor of a military parade. There was a ceremonious reception and an elaborate state banquet. Then the governor and his party were taken to the grandstand of the arena.

As soon as they were seated, two mounted horsemen dressed in the customary brilliant array of the Spanish bull-rings made their appearance; and as they advanced strings of bells attached to the trappings of their horses kept up a jingling accompaniment to all their movements. . . . [Governor Sola] opened his eyes wide with wonder when he saw a grizzly bear, held by four mounted vaqueros each with a reata fastened to a separate leg, bound into the arena, struggling against his captors and snapping his teeth with such fury as to cause terror even in those accustomed to the sight. The governor turned with an inquiring look to the comandante, who replied that the bear was a specimen of the animals, abundant in the neighboring mountains, which often came down to regale themselves upon the cattle in the valleys. (Hittell, 1885, 1 : 637.)

A few days later the governor remarked that he had been profoundly impressed by all that had been done for him, but two things in particular ever after would stand out in his mind. One was the mimic battle between Indians at Mission San Carlos and the other was the bear-and-bull fight at Monterey.

## CAPTURING THE BEAR

The skill of the horsemen handling the grizzlies won the praise not only of the governor but of nearly everyone else who visited California in those days. Otto von Kotzebue, when at San Francisco in 1816, said that the "dragoons are so dextrous and courageous, that they are sent out into the wood for a bear, as we should order a cook to fetch in a goose. They go on horse-back, with nothing but a rope with a running-knot in their hands, which is sufficient to overpower a bear." (Kotzebue, 1821*b* : 80.)

The early Californians had two methods of capturing grizzlies for the fights. In the first, four to a dozen or more mounted vaqueros, on a moonlight night, would slaughter a bullock in the hills. Then, hidden near the carcass, they would wait for the bear. Often "the howling . . . of immense numbers of coyotes" preceded his approach, and he "usually crept along suspiciously toward the bait." (Robinson, 1846 : 103.) When he came upon the dead bullock and began to eat it, the vaqueros would break from ambush and hurl their ropes at him.

In the second method, the horsemen would go in daylight to a place where grizzlies were fond of congregating. Choosing the largest they could find, they would charge him from front and rear with whoops and yells and cries of *"Aye oso viejo, mueludo guatata!"* (Bandini, 1893 : 314.)

A contemporary account describes the reactions of the grizzly:

Seeing his precarious predicament, the bear meets the charge from the front, with such accumulating ferocity and violence, that his assailants are soon put to flight, when he shakes his ponderous head, utters a most terrific growl, and commences a hot pursuit; but soon, the *Mexican* forces, are brought to bear upon his rear; his hind-most legs are entangled in the "lasso"; and he is prostrated upon his back, uttering most piteous, growling cries. . . . a lasso is also thrown upon his neck, when the spurs are rapidly plied to the horses, which now exert every energy, every nerve, and soon, the powerful victim is distended upon the ground, in an entirely defenceless condition. (Hastings, 1845 : 97–98.)

The usual response of the grizzly was to rise on his hind legs and strike at the ropes coming at him like arrows. If he could seize one he would pull it toward himself, paw over paw, despite strenuous resistance by the horse. And since the opposite end of the rope was attached to the saddle, the safety of the vaquero depended on the speed with which a companion could lasso one of the paws and gallop off to throw the animal down. *Reatas* were sometimes greased so that the bear could not obtain a hold on the rawhide (Rojas, 1953 : 48); if he caught a horse by the tail, however, he could drag it backward.

During some of the frays, saddles were ripped from their bindings, riders were unhorsed, and steeds were toppled onto their sides. The vaqueros sought to lasso all the grizzly's legs with different ropes, for he was not in their power until they could yank him simultaneously in opposite directions.

After the bear finds himself secure and has become pretty well worried, he seats himself sullenly on the ground and lets the horse pull at the cord, stretching his leg out until the pain becomes too severe, when he will draw up his leg, horse and all, with as much apparent ease as a horse would [a] sleigh. I have been told that some of the largest bears have been known to drag two horses a considerable distance in a fit of rage, in spite of all the exertions of the horses and riders to the contrary. (Leonard, 1904 : 214–217.)

Next, "foaming with rage, gagged, and secured with a dozen ropes" (Robinson, 1846 : 103) and "beating the ground with his feet and manifesting the most intense . . . anger imaginable" (Leonard, 1904 : 214–217), the grizzly would be lashed onto a *carreta*—a two-wheeled cart—or spread-eagled across the hide of a bullock that was dragged by ropes attached to horses. During the often long and rough trip across canyons and arroyos to the arena, where the bull waited him, the bear sometimes was given water to allay his thirst and fury (Sanchez, 1928 : 18); but since this was "some trouble, and consequently contrary to the inclinations of these people," it was seldom done (Garner, 1847a : 371).

In later days, when grizzlies were found less often in the open, a chief place for capturing them near Monterey was that part of the beach where whale oil was rendered. Ramadas, or thatched huts, had been constructed to cover the vats. On moonlight nights, grizzlies came down from the hills and often tore these structures apart. When rolling in the oil-soaked sand, they might be approached close enough to be captured. (Anne B. Fisher notes.) This was the beach on which Father Ascensión had first observed grizzlies in 1602.

To keep the bear as fresh as possible for the fight, he was held captive no longer than necessary. According to Beechey (1831 : 61–63), captive grizzlies refused to eat and would exhaust themselves in their struggles for liberty. Garner (1847a : 371) said that care is always taken not to irritate them "unnecessarily, because it often happens that these ferocious animals die with rage." Robinson described the tormenting of a bear that was tied to a tree in front of Mission Santa Barbara:

> It was past noon when I rode up and dismounted to look at the poor condemned brute, who, almost exhausted with heat and rage, seemed hardly competent to the trial that awaited him. Persons were standing around, thrusting pointed sticks into his sides, till the madness of the infuriated animal knew no bounds. A sailor, rather the worse for "aguardiente," reeled up to take part in the fun, and with his recklessness and wit added infinitely to the amusement. At length an unfortunate stagger brought him within reach of Bruin's paw, who seized him by the leg and drove his teeth quite through the calf. (Robinson, 1846 : 103–104.)

## THE FIGHT

The adversaries that the grizzlies met in the arenas were not the stolid, domesticated Herefords commonly seen on the California range today but were "the lithe, thick necked Spanish bulls, sharp of horn, quick of foot, always ready for a fight, and with a charge like that of a catapult" (Kingsley, 1920 : 22). These bulls, combining weight, speed, agility, and sharpness of horn with bad temper, were exceedingly dangerous to both man and bear. Wistar (1937 : 127) called the wild Spanish bull

> the noblest game in America, with possibly the single exception of the . . . California grizzly. He knows no fear, and shrinks from no enemy, having been accustomed all his life to fighting his rivals and other formidable animals, and when surrounded by his family is always spoiling for a fight. He will come a mile for his enemy, and will as lief charge a hundred men as one.

After the animals had been brought into the arena, a hind leg of the grizzly would be attached by a "leathern cord" (Carter,

1929 : 152), about twenty yards in length, to a forefoot of the bull; this kept the antagonists close together and also discouraged the bear from climbing the barrier and rampaging among the crowd. Then the handlers withdrew, and, as was observed by Robinson (1846 : 104), the two creatures

remained sole occupants of the square. The bull roared, pawed the earth, flung his head in the air, and at every movement of his opponent seemed inclined to escape, but the lasso checked his course, and brought both of them with a sudden jerk to the ground. Bruin, careless of the scene around him, looked with indifference upon his enemy, . . . but the jerk of the lasso aroused him as if to a sense of danger, and he rose up on his hind legs, in the posture of defence.

Quite often, the bull, infuriated by the indignities to which he had been subjected and not aware of the presence of the bear, would race around the arena in an attempt to gore men over the barrier. All the while, the grizzly, eyeing his adversary, would crouch low or, more frequently, rise up on his hind feet to full height. Although he might then be stirred to a frenzy as the bull, dashing hither and yon, yanked the rope with enough force to trip the bear onto his face, he rarely initiated the attack; instead he waited for the bull to charge.

If the grizzly was reluctant to fight, as sometimes happened, he was pricked with "a nail fixed in the end of a stick" (Leonard, 1904 : 214–217) or was roped and dragged again and again against the bull until the bull was exasperated (Garner, 1847b : 187). But these inducements rarely were necessary. More often, the bull, soon realizing that its enemy was at the other end of the rope, would stop his wild dashes about the ring and look carefully at the bear; then, curving his neck, he would advance slowly as though taking aim, and finally charge straight at the target with all the speed and fury he could muster. In the face of this onslaught, the grizzly invariably held firm, although different individuals met the attack in different ways. Some, standing erect, drove a paw straight into the muzzle of the oncoming bull. Others sprang for the neck and attempted to gain a stranglehold with their powerful arms.

The usual way by which the bear countered the bull was to crouch and, as the horns smashed against his own ribs, sink his teeth into the opponent's sensitive nose, swing his arms over and behind the head, and squeeze mightily. More often than not, the impact of the charge threw the grizzly onto his back; but so long as he retained a hold on the bull's nose and embraced the neck, he had the advantage. The bull, then suffering intense pain and bellowing horribly, could not gore him and would be forced to make frantic efforts to extricate itself. "The noise was terrific and the dust rose in clouds, while the onlookers shouted as they saw that the fight was deadly and witnessed the flow of blood" (Hittell, 1885, 1 : 638). Sometimes the bear suddenly wrenched the bull's head to one side and snapped the spine (Bell, 1930 : 113), or he might try to bite off one of the forefeet. Garner (1847*b* : 187) said, "I have seen a bear get hold of a bull between the horns with his teeth and hold him there with the bull's nose on the ground for the space of ten minutes, and on being hauled off by the horsemen, again catch the bull by one of his forefeet and bite it or tear it completely off by the lower joint."

If the bull won, triumph usually came early in the struggle; only with full fresh strength could he plunge his long, curving horns into the bear's body, toss his adversary high in the air, and then gore him to death as the bear lay prostrate on the ground.

"The beginning of this mortal struggle was always in the bull's favor"; but when the grapples were prolonged, the bull would be severely torn and lacerated. And "when some deep bite or the fatigue from the combat" or hot thirst from loss of blood "forced him to thrust out his tongue, the bear never failed to seize . . . this sensitive part, and to bury his terrible claws into it; not letting go his hold, whatever struggles his adversary made. The bull, conquered, reduced to bellowing frightfully, torn in every part, fell exhausted, and bled to death." (Carter, 1929 : 152.)

Many other travelers who saw fights testified to the viciousness of the tongue-clawing technique. Wilkes (1844 : 212) even said that "the only part of the bull" the grizzly "endeavors to attack is the tongue, by seizing which he invariably proves the victor." José Arnaz, a merchant in southern California during 1840, told of a bear that

killed "three bulls, one after the other. . . . when the bull approached, the bear thrust a paw in its face, or caught it by one knee . . . In this way the bear forced the bull to lower its head, and when it bellowed caught it by the tongue. It was then necessary to separate the combatants to prevent the bear from killing the bull immediately." (Sanchez, 1928 : 18.)

Another witness described a struggle in which a bear with its entrails dragging ripped off the tongue, the ears, and much of the lower jaw of the bull.

Once a bear got hold of the bull's tongue, the battle usually soon ended in complete triumph of the grizzly. The bull was either killed outright by its opponent or, being obviously unable to continue to fight, was shot by the men in charge. One fight is recorded as having lasted for as long as two hours (Cordua, 1933 : 281), but most of them were much shorter.

Our tabulations of fight results during the Spanish period show that the grizzlies most often triumphed. Robinson (1846 : 105), however, thought that a strong bull could cope with two bears in an afternoon; Garner (1847*b* : 187) said that "an old mountain bull" was sure to be the victor; and Gibbs (1853 : 111) stated that the conflicts usually ended with "the death of both parties." Leonard (1904 : 214–217) wrote that "the bear is much the strongest, but it has no chance of avoiding the thrusts of the bull, in consequence of the smallness of the pen; but in an open field, a grizzly bear will conquer a bull in a few moments."

In contrast to these statements, Wilkes (1844 : 212) had the positive opinion that bears always won regardless of the size and temper of their adversaries. Bancroft (1886, 2 : 434) cites a record of a bear that killed three bulls. Pattie (Flint, 1930 : 304) observed a contest in which fourteen bulls were conquered by five bears. Bell (1930 : 108–113) tells of a grizzly that killed three bulls one after the other and then was overcome by a fourth.

## UNUSUAL INCIDENTS

Any man who came within reach of the contestants was in danger of his life, for the grizzly was then quite likely to forget the bull. Tinsley (1902 : 158) recorded such an incident at San Gabriel in the early 'forties:

In the excitement of the mortal fight between the beasts, a man accidentally fell over the railing to the floor of the pen below. In a second the big hulking bear dove from the bull straight at the man, striking one paw at his head. The man was literally scalped, and in a second more the grizzly had torn the man into a horrible mass.

A bear escaped early one morning while being held for a fight at Mission Dolores. It climbed into the belfry and awakened and terrified the town by banging against the bells in its efforts to get away (verbal, Harry Downie). This presumably was the episode reported in 1859 by the San Francisco *Herald* on August 18 of that year.

According to Frank Post of Big Sur, mountain lions sometimes were taken in the live traps built near Monterey to catch bears for the arena. Then a bear-and-lion fight would be arranged. Mr. Post saw such a contest at Castroville in 1865 when he was six years old, and remembered it vividly. The lion, which seemed to have no fear, leaped onto the bear's back and while clinging there and facing forward scratched the grizzly's eyes and nose with its claws. The bear repeatedly rolled over onto the ground to rid himself of his adversary; but as soon as the bear was upright, the cat would leap onto his back again. This agility finally decided the struggle in favor of the lion.

Bell (1930 : 106) mentioned a fight staged in Mexico between a grizzly and a lion that had been imported from Africa: "When, a few years ago, a Los Angeles County grizzly was sent to Monterrey, Mexico, to be pitted against the man-killing African lion 'Parnell' the great Californian handled the African king as a cat would a rat. He killed him so quickly that the big audience hardly knew how it was done."

Another bear-and-bull fight which reputedly took place in Mexico was different from any ever held in California. The bear was the "Samson" of Grizzly Adams (see chap. 9), according to Albert Evans, who had seen that bear in California.

In January, 1870, I saw that same bear in the Plaza de Toros, in the city of Vera Cruz, Mexico, dig a hole large enough to hold an elephant, take a bull which had been sent to fight him in his paws as if he were an infant, carry him to the pit, hurl him into it head foremost, slap him on the side with his tremendous paws until his breath was half knocked out of his body, and then hold him down with one paw while he deliberately buried him alive by raking the earth down upon him with the other. Samson had not a tooth to bite with at that time, they having been in the course of many years and many fights worn down to the gums. (Evans, 1873 : 65.)

Some sixteen years earlier a bear of this name had been matched against another grizzly in Oakland, according to a notice seen at that time by the Reverend William Taylor:

One day, in crossing the bay in the Oakland "steamer Clinton," I saw a man posting on the side of the wheelhouse the following bill, in large letters: "Great bear fight, in front of the American Hotel, in Oakland, between the red bear Sampson, and a large Grizzly, on Sunday, January 29, 1854. The steamer Clinton will make two extra trips for the accomodation of the public. (Taylor: 1858 : 180.)

That is the only report we have of a staged fight between two grizzlies. Apparently the Spaniards were not interested in matching bear against bear; the Americans thought up this variation.

## THE AMERICAN PERIOD

With the American period, which began with the gold rush, came a change in the attitude toward bear-and-bull fights. The fights, which had been rooted in the traditions of Spain and were an integral part of the fiestas and religious holidays of Spanish California, now were cheapened and commercialized for the benefit of the newcomers. In all sincerity a Spanish Californian could say that " 'A bull and bear fight after the sabbath services in church was indeed a happy occasion. It was a soul-refreshing sight to see the growling beasts of blood' " (Bancroft, 1888 : 433). But to the American of those days—who did not care a whit for the traditions of California, and who could not comprehend the Spaniard's emotional association of religion, violence, and blood—a bear-and-bull fight was a disgusting spectacle. He believed that anyone who attended such a fight did so because of curiosity or a depraved lust for the sordid and sensational. A newcomer to California said, "A bull and bear fight is of all exhibitions . . . the most cruel and senseless" (Marryat, 1855 : 251).

However, "dissipation and rioting was a universal indulgence, and in the absence of other kinds of amusements bull and bear fights became a very popular divertisement. Admission to these shows of animal ferocity ranged from ten to twenty-five dollars." The fight was conducted by two Mexican managers in such a way that "usually several combats could be had between a bull and bear before either was killed, which made this novel sport one of immense profit to those who owned the animals." (Buel, 1882 : 256.)

The revenue-producing aspect of these contests early received official attention. In 1853 the California legislature made provision for licensing bear-and-bull fights, imposing a tax of twenty dollars per exhibition, payable into the local county treasury (*Stats.*, 1853, chap. 127, art. 2). In 1856 an amendment provided that half the receipts go to the state (*Stats.*, 1856, chap. 130), and in 1863 (*Stats.*, 1863, chap. 194) the fee was raised to twenty-five dollars and the permit was to be issued by the State Comptroller (letter, T. H. Mugford, May 19, 1954).

At first the commercialized fights, advertised by garish posters, attracted great crowds. They were held on Sundays in most of the mining towns as well as in the booming pioneer cities of Sacramento and San Francisco. It is said that Horace Greeley was so impressed by the fights that he coined the terminology which is still used by Wall Street (Potter, 1945 : 269).

Preparations for the first fights held in Sacramento in the summer of 1851 were noted by the German traveler Carl Meyer (1938 : 233): "A long grandstand was built at the mile-long race track in Brighton, six miles from the city. The entrance price was $2.50 and enormous placards on every street corner announced for weeks ahead the glorious fight between the American gray bear 'General Scott' and the Mexican bull 'Sant Anna.'" The bear was valued at $1,500.

One man in the Mother Lode during 1851, a professional bear hunter, trapper, and fighter, said that when he was fortunate enough to take a large grizzly alive he realized a rich return. The bear was bound head and foot, placed in a large, strong cage on a wagon, and dispatched to a nearby mining town. There handbills and posters soon announced that, on the Sunday following, "the famous grizzly 'America' would fight a wild bull, &c., &c. Admission five dollars" (Marryat, 1855 : 251).

Another traveler to the gold fields, Hinton Helper (1885 : 116–117), when in San Francisco during the early 'fifties was shocked to hear the sounds of drum, fife, and clarionet on the

Sabbath. Looking out of the window, he saw a tremendous grizzly bear, caged, and drawn by four spirited horses through the various streets. Tacked to each side of the cage were large posters, which read:

FUN BREWING—GREAT ATTRACTION!
HARD FIGHTING TO BE DONE!
TWO BULLS AND ONE BEAR!

The citizens of San Francisco and vicinity are respectfully informed that at *four o'clock this afternoon, Sunday, Nov. 14th*, at *Mission Dolores*, a *rich treat* will be prepared for them, and that they will have an opportunity of enjoying a fund of the *raciest sport* of the season. TWO LARGE BULLS AND A BEAR, all *in prime condition for fighting*, and under the management of *experienced Mexicans*, will contribute to the *amusement of the audience*.

Programme—In two Acts

ACT I.

BULL AND BEAR—"HERCULES" AND "TROJAN," will be conducted into the arena, and there *chained together*, where they will fight *until one kills the other*.

JOSE IGNACIO,
PICO GOMEZ,     Managers.

The admittance price was three dollars. Approximately 1,250 people attended.

## FIGHT AT MOQUELUMNE HILL

Probably the best description of the fights held in the mining towns of the Sierra Nevada is the straightforward, objective account by J. D. Borthwick of one at Moquelumne [Mokelumne] Hill in the early 1850's (fig. 18).

At the time of my arrival in Moquelumne Hill, the town was posted all over with placards, which I had also observed stuck upon trees and rocks by the roadside as I travelled over the mountains. They were to this effect:

WAR! WAR!! WAR!!!
The celebrated Bull-killing Bear,
GENERAL SCOTT,
will fight a Bull on Sunday the 15th inst., at 2 P.M.,
at Moquelumne Hill.

The Bear will be chained with a twenty-foot chain in the middle of the arena. The Bull will be perfectly wild, young, of the Spanish breed, and the best that can be found in the country. The Bull's horns will be of their natural length, and *"not sawed off to prevent accidents."* The Bull will be quite free in the arena, and not hampered in any way whatever.

The proprietors then went on to state that they had nothing to do with the humbugging which characterised the last fight, and begged confidently to assure the public that this would be the most splendid exhibition ever seen in the country.

I had often heard of these bull-and-bear fights as popular amusements in some parts of the State . . . on Sunday the 15th, I found myself walking up towards the arena, among a crowd of miners and others of all nations, to witness the performances of the redoubted General Scott.

The ampitheatre was a roughly but strongly built wooden structure, uncovered of course; and the outer enclosure, which was of boards about ten feet high, was a hundred feet in diameter. The arena in the centre was forty feet in diameter, and enclosed by a very strong five-barred fence. From the top of this rose tiers of seats, occupying the space between the arena and the outside enclosure.

As the appointed hour drew near, the company continued to arrive till the whole place was crowded; while, to beguile the time till the business of the day should commence, two fiddlers—a white man and a gentleman of colour—performed a variety of appropriate airs.

The scene was gay and brilliant, and was one which would have made a crowded opera-house appear gloomy and dull in comparison. The shelving bank of human beings which encircled the place was like a mass of bright flowers. The most conspicuous objects were the shirts of the miners, red, white, and blue being the fashionable colours, among which appeared bronzed and bearded faces under hats of every hue; revolvers and silver-handled bowie-knives glanced in the bright sunshine, and among the crowd were numbers of gay Mexican blankets, and red and blue French bonnets, while here and there the fair sex was represented by a few Mexican women in snowy-white dresses, puffing their cigaritas in delightful anticipation of the exciting scene which was to be enacted. Over the heads of the . . . spectators was seen mountain beyond mountain . . . and on the green turf of the arena lay the great centre of attraction, the hero of the day, General Scott.

He was . . . confined in his cage, a heavy wooden box lined with iron, with open iron-bars on one side, which for the present was boarded over. From the centre of the arena a chain led into the cage, and at the end of it no doubt the bear was to be found. Beneath the scaffolding on which sat the spectators were two pens, each con-

taining a very handsome bull, showing evident signs of indignation
at his confinement. Here also was the bar, without which no place
of public amusement would be complete.

There was much excitement among the crowd as to the result
of the battle, as the bear had already killed several bulls; but an idea
prevailed that in former fights the bulls had not had fair play, being
tied by a rope to the bear, and having the tips of their horns sawed
off. But on this occasion the bull was to have every advantage which
could be given him; and he certainly had the good wishes of the
spectators, though the bear was considered such a successful and
experienced bull-fighter that the betting was all in his favour. Some
of my neighbours gave it as their opinion, that there was "nary bull
in Calaforny as could whip that bar."

At last, after a final tattoo had been beaten on a gong to make
the stragglers hurry up the hill, preparations were made for beginning the fight.

The bear made his appearance before the public in a very bear-
ish manner. His cage ran upon very small wheels, and some bolts
having been slipped connected with the face of it, it was dragged
out of the ring, when, as his chain only allowed him to come within
a foot or two of the fence, the General was rolled out upon the
ground all of a heap, and very much against his inclination, apparently, for he made violent efforts to regain his cage as it disappeared.
When he saw that was hopeless, he floundered half-way round the
ring at the length of his chain, and commenced to tear up the earth
with his fore-paws. He was a grizzly bear of pretty large size, weighing about twelve hundred pounds.

The next thing to be done was to introduce the bull. The bars
between his pen and the arena were removed, while two or three
men stood ready to put them up again as soon as he should come
out. But he did not seem to like the prospect, and was not disposed
to move till pretty sharply poked up from behind, when, making a
furious dash at the red flag which was being waved in front of the
gate, he found himself in the ring face to face with General Scott.

The General, in the meantime, had scraped a hole for himself
two or three inches deep, in which he was lying down. This, I was
told by those who had seen his performances before, was his usual
fighting attitude.

The bull was a very beautiful animal, of a dark purple colour
marked with white. His horns were regular and sharp, and his coat
was as smooth and glossy as a racer's. He stood for a moment taking
a survey of the bear, the ring, and the crowds of people; but not
liking the appearance of things in general, he wheeled round, and
made a splendid dash at the bars, which had already been put up
between him and his pen, smashing through them with as much ease

as the man in the circus leaps through a hoop of brown paper. This was only losing time, however, for he had to go in and fight . . . He was accordingly again persuaded to enter the arena, and a perfect barricade of bars and boards was erected to prevent his making another retreat. But this time he had made up his mind to fight; and after looking steadily at the bear for a few minutes as if taking aim at him, he put down his head and charged furiously at him across the arena. The bear received him crouching down as low as he could, and though one could hear the bump of the bull's head and horns upon his ribs, he was quick enough to seize the bull by the nose before he could retreat. This spirited commencement of the battle on the part of the bull was hailed with uproarious applause; and by having shown such pluck, he had gained more than ever the sympathy of the people.

In the meantime, the bear, lying on his back, held the bull's nose firmly between his teeth, and embraced him round the neck with his fore-paws, while the bull made the most of his opportunities in stamping on the bear with his hind-feet. At last the General became exasperated at such treatment, and shook the bull savagely by the nose, when a promiscuous scuffle ensued, which resulted in the bear throwing his antagonist to the ground with his fore-paws.

For this feat the bear was cheered immensely, and it was thought that, having the bull down, he would make short work of him; but . . . neither the bear's teeth nor his long claws seemed to have much effect on the hide of the bull, who soon regained his feet, and, disengaging himself, retired to the other side of the ring, while the bear again crouched down in his hole.

Neither of them seemed to be very much the worse of the encounter, excepting that the bull's nose had rather a ragged and bloody appearance; but after standing a few minutes, steadily eyeing the General, he made another rush at him. Again poor bruin's ribs resounded, but again he took the bull's nose into chancery . . . The bull, however, quickly disengaged himself, and was making off, when the General, not wishing to part with him so soon, seized his hind-foot between his teeth, and, holding on by his paws as well, was thus dragged round the ring before he quitted his hold.

This round terminated with shouts of delight from the excited spectators, and it was thought that the bull might have a chance after all. He had been severely punished, however; his nose and lips were a mass of bloody shreds, and he lay down to recover himself. But he was not allowed to rest very long, being poked up with sticks by men outside, which made him very savage. He made several feints to charge them through the bars, which, fortunately, he did not attempt, for he could certainly have gone through them as easily as he had before broken into his pen. He showed no inclination to renew the combat; but by goading him, and waving a red flag over the bear, he was eventually worked up to such a state of fury as to make another charge. The result was exactly the same as before, only that when the bull managed to get up after being thrown, the bear still had hold of the skin of his back.

In the next round both parties fought more savagely than ever, and the advantage was rather in favour of the bear: the bull seemed to be quite used up, and to have lost all chance of victory.

The conductor of the performances then mounted the barrier, and, addressing the crowd, asked them if the bull had not had fair play, which was unanimously allowed. He then stated that he knew there was not a bull in California which the General could not whip, and that for two hundred dollars he would let in the other bull, and the three should fight it out till one or all were killed.

This proposal was received with loud cheers, and two or three men going round with hats soon collected, in voluntary contribution, the required amount. The people were intensely excited and delighted with the sport, and double the sum would have been just as quickly raised to insure a continuance of the scene. A man sitting next me, who was a connoisseur in bear-fights, and passionately fond of the amusement, informed me that this was "the finest fight every fit in the country."

The second bull was equally handsome as the first, and in as good condition. On entering the arena, and looking around him, he seemed to understand the state of affairs at once. Glancing from the bear lying on the ground to the other bull standing at the opposite side of the ring, with drooping head and bloody nose, he seemed to divine at once that the bear was their common enemy, and rushed at him full tilt. The bear, as usual, pinned him by the nose; but this bull did not take such treatment so quietly as the other: struggling violently, he soon freed himself, and, wheeling around as he did so, he caught the bear on the hind-quarters and knocked him over; while the other bull, who had been quietly watching the proceedings, thought this a good opportunity to pitch in also, and rushing up, he gave the bear a dig in the ribs on the other side before he had time to recover himself. The poor General between the two did not know what to do, but struck out blindly with his fore-paws with such a suppliant pitiable look that I thought this the most disgusting part of the whole exhibition.

After another round or two with the fresh bull, it was evident that he was no match for the bear, and it was agreed to conclude the performances. The bulls were then shot to put them out of pain, and the company dispersed, all apparently satisfied that it had been a very splendid fight. . . .

I took a sketch of the General the day after the battle. He was in the middle of the now deserted arena, and was in a particularly savage humour. He seemed to consider my intrusion on his solitude as a personal insult, for he growled most savagely, and stormed about in his cage, even pulling at the iron bars in his efforts to get out. . . . I lighted my pipe, and waited till he should quiet down into an attitude, which he soon did, though very sulkily, when he saw that he could not help himself.

He did not seem to be much the worse of the battle, having but one wound, and that appeared to be only skin deep. (Borthwick, 1857 : 289–299.)

## AMUSING INCIDENTS

Borthwick stated that the bear pit at Moquelumne Hill was separated from the stands by a strong fence. Some arenas, though, had no provision for keeping the animals within bounds if the chains holding them together broke. When a grizzly got loose at Mariposa in 1852, men shot their pistols at it from all directions and killed the animal without hitting more than one person (San Francisco *Daily Alta California,* Nov. 14, 1852). At Sonora in 1851 the chains came apart, and the animals, forgetting their differences, charged the crowd. A stampede resulted, in which the bull, smashing through the arena, gored a Mexican, and the grizzly seized a white man by the leg. (*Ibid.,* Nov. 30, 1851.)

According to the San Francisco *Alta California* of July 27, 1854, a "Grand Tragi-Comedy" occurred at Iowa Hill that month:

On Saturday last a grand bear and bull fight was advertised to come off . . . At ten o'clock in the morning a crowd of about two thousand persons had assembled . . . A large amphitheatre had been erected, with ample accommodations for the spectators, underneath which they crowded in anxious expectancy, to witness the rare entertainment. The sports of the day commenced with a cock fight, after which the huge bull, Chihuahua, was ushered into the ring. The bear, a full grown animal of the grizzly species, was led from his cage, tethered by a raw-hide lariat and chain. Chihuahua surveyed his antagonist, pawed the dirt over his back, and prepared to "pitch into" bruin, who not relishing such sport, made one bound, freed himself from the thongs which bound him, and commenced ascending the seats on which set the spectators. A scrambling scene then ensued which beggars all description. Bruin succeeded in attaining the fourth tier of seats, when he either fell through or leaped to the ground, on the heads of the dense mass below. One unfortunate gentleman raised the canvas to effect his escape—bruin perceiving the opening made, darted through, overturning the man in his passage, and made for the deep canon which runs by the foot of the town. In ascending the hill, he overtook Mr. Courtney, of Mad Canon, and with one stroke of his paw almost entirely denuded him. Happily, however, he sustained no other injury than the loss of his "unmentionables." The gentleman who was upset in making his escape through the canvas, received a severe contusion, and had his head gashed to the skull, from the centre of his forehead to the crown. Meanwhile the rage of the bull having reached boiling heat, with a bound and a bellow, dashed through the crowd, overturning all in his way, and in the opposite direction from bruin, disappeared in the woods. . . . The evening wound up with a wrestling match for a hundred dollars, . . . and the whole "topped off" with a dog fight.

One peculiarity of fights that entertained the miners is that, in contrast to most of the battles of earlier times, the bull usually was the victor, as well as being the favorite of the crowd. Newspaper accounts frequently carried such statements as "Bruin had little desire to fight, finally waked up a little, and after an hour was taken away considerably gored" (*Alta California*, Sept. 9, 1851). To make the contests more nearly equal, the horns were sometimes sawed off before the bulls went into battle.

Seemingly, then, the bears used during the American period often were inferior specimens endowed with less spirit and ferocity than the giants that had fought in Spanish and Mexican days. Very likely the miners shot the larger bears with rifles and trapped mainly small ones for the fights—possibly even black bears—whereas the Spanish vaqueros had roped the biggest and toughest grizzlies they could find.

Meyer (1938:237), after observing fights in Sacramento, said that the bear was rarely the victor.

The bull was soon thought to be a too powerful opponent for the king of the California forests and he was supplanted by a—donkey. A California longear, or several of these, were brought into the arena with the bear, and it was horrible to see the bear quench his bloodthirstiness on these weak creatures. Of course some rough kicks were directed at the bear's head but sometimes the angry bear bit the donkey's leg off or bit his head off.

Alfred T. Jackson, on February 2, 1851, at Nevada City saw a big poster which said there was going to be "a grand fight betwen [*sic*] a ferocious grizzly bear and the champion fighting jackass of the State." The bear proved to be a black, which, on approaching its opponent, received "a couple of thundering kicks in the ribs." Whereupon the jackass returned to eating grass, and the bear went over the fence in two jumps and fled to the chaparral, scattering the crowd. (Canfield, 1906 : 46–48.)

So great was the degradation of the fights that even tiny burros were pitted against bears:

If the bear was a real grizzly, he always won, so far as I know, but the burro would worry him desperately for a long time. The bear would suffer terrific jolts on the jaw from the burro's heels, that would send him staggering back time and again. When the grizzly would finally get hold of his lowly but far from humble antagonist the burro would bite and hang on with its teeth like a bulldog. I must say that I always considered a match like this unfair, brutal and barbarous. (Bell, 1930 : 107–108.)

Finally, somebody with a perverted sense of humor conceived the idea of letting hundreds of city rats loose into a well-closed arena, where they tormented the grizzly to distraction by swarming over him and crawling under his fur. Viewing this loathsome spectacle, Meyer (1938 : 237) commented that bear-and-bull fights had "passed through all stages, from the heroic to the lowest and the Yankees mocked the dignity of the bear as they do that of a king."

## END OF BEAR-AND-BULL FIGHTS

Inevitably a reaction set in against bear-and-bull fights or any form of staged battle between animals. The novelty to the American was wearing off, the popularity of the fights was diminishing, the proprietors of the bull pens and arenas were beginning to lose money, and newspapers were calling for enforcement of the law. As early as 1851 the Nevada City *Journal* editorialized about the bear-and-bull fight:

It is a practice vitiating to public morals and taste, frequently productive of accident, and prohibited by municipal law—although this latter has never been enforced. The pleasure a set of civilized beings can find in witnessing the forced conflict of animals, whose instincts even prompt them to avoid each other, rather than engage in the expected fight, is of that base sort which the savage feels when he dances round the stake of his victim, and luxuriates in his agonies. (San Francisco *Daily Alta California*, Oct. 25, 1851.)

On February 24, 1852, the San Francisco *Alta California*, under the heading, "Sunday Barbarities at the Mission," called

upon the authorities to interfere and prevent further continuance of the exhibitions known as "bull fights" or "bear baits," in this city and vicinity. For the past six months these disgraceful and revolting spectacles, offered as Sunday sports, and characterized by every species of low depravity and brutality, have been permitted to go on, gathering in disgusting and sickening detail whenever presented. We now call upon the authorities for their total suppression.

These exhibitions are a disgrace to our city, to society, to our laws and to humanity. Offered as "amusements," they are a gross deception and imposition, and by the irreverent selection of the Sabbath day for their presentation they become a desecration. They are offensive to the tastes of a majority of our respectable citizens, and with their flaunting banners and bands of music paraded through our streets during the hours of religious service, they are a nuisance against which in behalf of the community we feel compelled to speak.

As public feeling was aroused, laws were enforced and new ones were passed. And so in the late 'fifties and early 'sixties, the bear-and-bull fight, which had reached a higher stage of development in Spanish California than anywhere else, began to be legislated out of existence. In Sacramento the first step was to prohibit fights on Sunday to insure better observance of the Sabbath (Thomas, 1856 : 52). In 1860 the young and booming city of Los Angeles passed an ordinance prohibiting them (Cleland, 1941 : 114)—only twenty-five years after bearbaiting and bullbaiting had been outlawed in England by act of Parliament.

Despite these California enactments, some fights are of record in later years—1868 and even 1881—but the practice finally came to an end because of public opinion and the scarcity of grizzlies.

# 7 Grizzlies and Americans

The events of history in California before the gold rush had no sharp beginnings or endings. For centuries, Indians, grizzlies, and other wild beasts had the land to themselves. Then, in the latter part of the eighteenth century, Spanish explorers and missionaries slowly entered from the south to start their enterprises. As missions and ranchos were developed in the ensuing years, grizzlies were benefited by an increased food supply in the herds of cattle (see chap. 5). Blending into this picture came the Americans and other "foreigners"—a few at first, then gradually increasing numbers, and finally a spectacular invasion following the discovery of gold. This change was destined to have a fatal effect on the grizzly population.

## EARLY AMERICAN ARRIVALS

The first Americans interested in California came by way of the sea, and their chief concerns were in products from three mammals—the pelts of sea otters, the oil of whales, and the hides and tallow of Spanish cattle. There are fascinating stories about all of these enterprises—not to be told here, although all three are related to our present subject, the grizzly.

Only a few years elapsed between the founding of the first missions and the beginning of the sea otter trade. A hint from Captain James Cook's voyage of 1776 to the northwest Pacific Coast led English and American merchants to realize the value of otter pelts, especially for the China trade. Ships of these traders first visited the California coast in 1793 and 1796, and soon the coastal waters and ports were dotted with vessels hunting sea otters and trading their pelts during the summer months. The whaling and hide and tallow enterprises developed later.

News of California and its coastal wealth stimulated early exploring navigators to visit these shores—Vancouver for the English in 1792–1794, Kotzebue for the Russians in 1816 and 1824, Beechey for the English in 1826, and La Pérouse and Du Petit-Thouars for the French in 1797 and 1837. Then the American, English, and French trappers, in the land drive for furs, began to infiltrate across the still unguarded and undefined boundaries on the east and north. These men were the first to meet grizzlies in the interior valleys and hills—Jedediah Smith in 1828, John Work in 1832–1833, and George Nidever in 1833, to mention a few. Of the many persons who sailed, rode, or walked into California, some remained to trap and hunt, others became merchants and traders, and still others became farmers. Some of the settlers married into Spanish Californian families. Bancroft (1886, 2 : 681–682) lists 146 foreign pioneers who came to California before the end of 1830.

In May, 1846, Governor Pío Pico, speaking before the Departmental Assembly in favor of annexing California to England, said:

> We find ourselves threatened by hordes of Yankee immigrants
> . . . Already have the wagons of that perfidious people scaled the
> almost inaccessible summits of the Sierra Nevada . . . and penetrated
> the fruitful valley of the Sacramento. . . . What that astonishing
> people will next undertake, I can not say; but in whatever enterprise
> they embark, they will be sure to be successful. Already these ad-
> venturous voyagers, spreading themselves far and wide . . . are
> cultivating farms, establishing vineyards, . . . [and] sawing up lum-
> ber . . . (Cronise, 1868 : 51.)

Governor Pico's prediction proved all too true. The "hordes of Yankee immigrants" that came with the gold rush were far greater than even he could have visualized.

According to Bancroft (1886, 6 : 643), crude estimates gave the population of the settled parts of California in January, 1848, as 7,500 Spanish Californians, 6,500 "foreigners," and 3,000 to 4,000 former mission Indians living in a somewhat civilized manner near towns or on ranchos.

James Marshall's discovery of gold was to change California for all concerned, men and bears. News of the event traveled slowly at first, then exploded to the Atlantic seaboard and to Europe. Soon a vast stream of immigration was flowing into California. The energetic and resourceful Americans, hard-working, hard-living, and hard-shooting, came in ever-increasing numbers. Grizzlies and Americans could not live in close proximity, and the rising tide of human population foredoomed that of the bears.[1]

## JEDEDIAH SMITH AND GRIZZLIES

The first American to record experiences with grizzlies in California was the trapper Jedediah Smith (Sullivan, 1934). He entered by way of the Mojave Desert in November, 1826, and spent some time in the coastal regions. His notes on grizzlies started while he was trapping beaver in the Sacramento Valley in 1828. At Mormon Slough, ten miles east of Stockton, on January 19 he wrote: "I saw a Grizzly Bear and shot at him but did not kill him." Next day one of his men had a similar failure (*ibid.*, p. 56). On March 7, on the Feather River (below Marysville) "the trappers . . . had killed a large Brown Bear which was in good order and were of course . . . feasting, for the hunter of the Buenaventura [Sacramento] Valley at the distance of 2000 miles from his home may enjoy and be thankful for such Blessings as heaven may throw in his way" (*ibid.*, p. 67). On the following day a companion, Harrison Rogers, killed one and wounded a second. The next morning,

Mr. Rodgers went after the wounded Bear in company with John Hanna. In a short time Hanna came running in and said they had found the Bear in a verry bad thicket. That he suddenly rose from his bed and rushed on them. Mr. Rodgers fired a moment before the bear caught him. After biting him in several places he went off, but Hanna shot him again, when he returned, caught Mr. Rodgers and gave him several additional wounds. I went out with a horse to bring him in and found him verry badly wounded being severely cut in . . . 10 or 12 different places. I washed his wounds and dressed them with plasters of soap and sugar. (*Ibid.*, p. 68.)

Mr. Rogers did not recover for two weeks or more.

On March 31 two bears were killed on the Sacramento

---

[1] The population of California increased from 92,597 in 1850 to 379,994 in 1860, 560,247 in 1870, and 1,213,398 in 1890 (U.S. Decennial Census [for 1930], 1931, 15, vol. 1 : 127).

River near the mouth of Chico Creek, and the next day another was killed and a stone arrowhead and part of an arrow were found in its "lights" (*ibid.*, p. 76). On April 7, when three miles from the Sacramento River on Mill Creek, Tehama County, Smith wrote:

In the evening we shot several Bear and they ran into thickets . . . Several of us followed one that was Badly wounded . . . We went on foot because the thicket was too close to admit a Man on horse back.

As we advanced I saw one and shot him in the head when he immediately . . . fell—Apparently dead. I went in to bring him out without reloading my gun and when I arrived within 4 yards of the place where the Bear lay the man that was following me close behind spoke and said "He is alive." I told him in answer that he was certainly dead and was observing the one I had shot so intently that I did not see one that lay close by his side which was the one the man behind me had reference to. At that moment the Bear sprang towards us with open mouth and making no pleasant noise.

Fortunately the thicket was close on the bank of the creek and the second spring I plunged head foremost into the water. The Bear ran over the man next to me and made a furious rush on the third man Joseph Lapoint. But Lapoint had . . . a Bayonet fixed on his gun and as the Bear came in he gave him a severe wound in the neck which induced him to change his course and run into another thicket close at hand. We followed him there and found another in company with him. One of them we killed and the other went off Badly wounded.

I then went on horse Back with two men to look for another that was wounded. I rode up close to the thicket in which I supposed him to be and rode around it several times halloeing but without making any discovery. I rode up for a last look when the Bear sprang for the horse. He was so close that the horse could not be got underway before he caught him by the tail. The Horse being strong and much frightened . . . [exerted] himself so powerfully that he gave the Bear no opportunity to close upon him and actually drew him 40 or 50 yards before he relinquished his hold.

The Bear did not continue the pursuit but went off and [I] . . . returned to camp to feast on the spoils and talk of the incidents of our eventful hunt. (Sullivan, 1934 : 77–78.)

Smith called the stream Grizzly Bear Creek, but later it received the more commonplace name Mill Creek.

Smith's pioneer report of encounters between men and grizzlies in 1828 in the interior valleys contains all the elements, except a human fatality, that become familiar in the many later episodes of the American period (see our chap. 8). Here are grizzlies singly or in groups taking refuge in the then dense river-bottom thickets; distant shots when the bear makes off; a wounded bear, followed into a thicket, makes a surprise attack and wounds the man; a melee in dense cover where several men and several bears tangle in battle at close range, with resulting confusion; escape by diving into a creek; a stab wound that turns a bear's attack; a bear springing and grabbing a horse; a feast on bear meat fresh-cut, -cooked, and -eaten; and finally, talk of the fight, round a campfire.

## HARMLESS ENCOUNTERS

The grizzly in his meetings with men followed no single pattern of behavior. The type of reaction depended not only upon the temperament of the individual bear but also upon the circumstances. Females with cubs had to be avoided because the mothers protected their offspring fiercely. Persons who came anywhere near a family group were likely to experience trouble. In fact, bears of any sort when startled often became aggressive, even when there was no other provocation. Some, however, on being approached, merely stared for a moment, or stood up to reconnoiter, or departed precipitately. An injured bear, of course, retaliated violently.

Duhaut-Cilly, in San Francisco in 1827, said that the Spanish Californians claimed that grizzlies "seldom attack passers-by, and that, only when one happens to be near them, or arouse their savageness by teasing them, do they make use of their terrible claws and their extraordinary strength" (Carter, 1929 : 146). Colton (1859 : 118), likewise, said that the grizzly "rarely attacks a man unless surprised or molested. The fellow never lies in wait for his victim. If the hunter invades his retreat or disputes his path he will fight . . ." And J. S. Hittell (1863 : 109–110) wrote that the grizzly "never makes the first assault unless driven by hunger or maternal anxiety. The dam will attack any man who comes near her cubs, and on this account it is dangerous to go in the early summer afoot through chaparral where bears make their home. Usually a grizzly will get out of the way when he sees or hears a man, and sometimes, but rarely, will run when wounded."

The experienced hunter George Nidever (1937 : 53–54), accustomed to the violent reactions of a wounded or aroused bear, was astonished at the indifference of an undisturbed grizzly to the presence of a Yankee youth. Nidever wrote: "I went out to hunt one morning, leaving him alone in camp. Upon returning, I found a large grizzly walking leisurely around the camp, pausing occasionally to stand up and watch the New Englander, who was seen whittling, wholly unconscious of the presence of the bear." This was in 1837 in what is now San Luis Obispo County.

A family living near the base of the mountains not far from Los Angeles had an even more astounding experience in the early days. One day while the husband was away tending his herds, the young wife went to the spring to wash clothes and was absent for about an hour, leaving her three-year-old child unattended.

When she returned what was her alarm and horror to find an immense grizzly playing pranks and cutting up rusties with the infantile Vicente, the two seeming to be on terms of the most affectionate intimacy. The old bear would lay on her back, and would hold the little fellow up in her great paws, and would toss him around and tenderly hug him, and the little don would scream with delight, so pleased he seemed to be with his new found friend. (Bell, 1927 : 262–263.)

There are several accounts of grizzlies visiting camps at night, presumably in search of food, without molesting men asleep or pretending to be asleep. Allen Kelley (1903 : 16–17) tells that when Samuel Snedden was bedded for the night in Lockwood Valley (in Monterey County?) with a campfire blazing close by, a bear attempted to reach a haunch of venison hanging well up on a branch above one of the party. To keep from attracting the bear's attention, Snedden lay quiet with half-closed eyes. When the grizzly found the meat beyond standing reach he came down on all fours, crushing a tin cup that lay within a foot of the other man's head, sniffed at Snedden's ear, and went noiselessly away.

Near Shasta in 1849 or '50, a miner who had found good

diggings for gold kept the news to himself and made solitary trips to town for supplies. Returning late one Sunday with some fresh beef, he hung the meat up to foil the thieving coyotes and made his bed below.

During the night he was waked by a tremendous pressure on his body which seemed as though he must be squeezed flat. He was not long in finding that a huge grizzly, attracted by the savory smell . . . was standing astride of him making little jumps after the meat which he could just reach but not lay hold of. After each failure he would take a seat on the Dutchman and grunt a little to himself on the tantalizing nature of the situation. At last by getting a good footing on Dutchy he made a successful grab and went off with the plunder, paying no attention what ever to the lawful proprietor. The latter lost no time in getting back to town, where . . . he told his moving tale to all who would listen, whereby his carefully concealed bonanza at once became public to all the world. (Wistar, 1937 : 187–188.)

## GRIZZLY PROWLERS

In early days, grizzlies sometimes came close to settlements. At Martinez, Contra Costa County, in 1856 one broke into the corral at a house "not ten minutes' walk from the wharf, or the center of town" (N 35), and in 1860 a grizzly and two cubs were seen but "a short distance" from the then small city of Oakland (N 59).

During the winter of 1860–61 a grizzly "paid several nocturnal visits to the people of Susanville and that vicinity." One night the animal prowled around a house. A boy who was sleeping there went to a window and threw it up. "Just then the bear reared up on his hind legs in front of the window and the boy found his face close to that of the bear. He was scared half to death, and shutting down the window, ran up stairs and locked himself into a room and stayed there all night." (V. J. Bordette, in Fairfield, 1916 : 237.) A few nights later, a man living in a cabin north of the same town heard something walking close outside. He opened the door slightly, dimly saw an animal, fired his musket, and slammed the door shut. Next morning he found a dead 800-pound grizzly not far away. (*Ibid.*)

In a dense redwood forest on Eel River near Humboldt Bay, about 1850, a small log storehouse had been built,

with roof, door, and a single window shutter of heavy "puncheons" or plank split several inches thick. It was at some distance from any other house, and among other goods contained a store of barley in bags, for the mules of packers who sometimes fitted out there for the interior. A few days before our arrival a grizzly had torn off the shutter at night, reached in and carried off a bag, and . . . [had] repeated his theft almost every night, on each occasion appropriating . . . one bag and no more. . . . [when we camped] there one night in company with an experienced "mountain man," the storekeeper . . . begged us to kill this persistent depredator . . . Having heard plenty of such panicky yarns before, and not much expecting the bear to come when he was really wanted, we nevertheless carried our blankets down and slept on the roof, which was nearly flat and about six feet high at the eaves.

In the course of the night the shutter, though firmly spiked on, was torn off with a crash, and through the dense gloom . . . under the tall redwoods, we could dimly and doubtfully make out the bear's huge bulk moving off with his plunder. . . . I rashly fired at what I took for the small of his back, hoping that if the shot missed the spine it might in ranging forward reach the heart. The bear at once dropped his booty and came for the broken shutter, which he minutely examined, walking once or twice around the small cabin and returning again to the window, all the time growling a vicious soliloquy to himself. Suddenly detecting us, . . . he stood up, placing both paws on the roof. . . . Francois . . . planted his bullet at the base of the throat at arm's length. The bear dropped back on all fours and made off. . . . we prudently remained where we were till daybreak. The morning light disclosed a bloody trail which we warily followed for quite a hundred yards, when we almost stumbled over the bear lying in an upright position on all fours, his head resting on his paws as if asleep, which is not an unusual position of the grizzly when getting ready to "pass in his checks." Approaching through the dense undergrowth on opposite sides we found him nearly cold, having been dead for hours. The last shot had done the business, the first having been merely exasperating. (Wistar, 1937 : 186–187.)

In El Dorado County during the summer of 1850 when G. W. and John Applegate were walking from Horseshoe Bar on the American River, toward Georgetown, a bear followed them through the water of Canyon Creek, and they ran to the cabin of a man named Work. The cabin was rudely built of shakes, with a window covered by greased paper. The grizzly sniffed at the door and then pawed a hole to look in. Since none of the several men present had firearms, they all speedily climbed on crossbeams. The bear withdrew but soon returned and poked through the papered window. One man seized an iron-pointed Jacob's staff and plunged it into the bear's chest. The animal disappeared, taking the staff. Next day it was found dead near by, the staff through its heart and body; it was a female with several rifle wounds. On the previous day, hunters from Georgetown had captured two cubs and wounded their mother; this was evidently the same grizzly. (Kelley, 1903 : 20–21.)

## PROPERTY DAMAGE BY GRIZZLIES

To the Spanish and Mexicans and the few "foreigners" in the coastal region of California before the gold rush the grizzlies were a rather important problem. When miners swarmed into the foothills of the Sierra Nevada and other Americans began to farm in the interior valleys, the presence of the big bears was a matter of real concern. In the first years of American occupation, grizzlies occasionally wandered into the new settlements; they were present in all the thickets and chaparral areas of the mining districts; and they were a hazard to travelers en route between the new centers of human activity.

Grizzlies preyed on practically all types of domestic animals —cattle, hogs, sheep, and horses. In Spanish days (see chap. 5) when livestock roamed the hills and valleys freely, there being practically no fences in that period, the bears inflicted great damage on the herds. Many early Americans likewise had free-ranging stock and suffered from depredations by grizzlies, but others confined their animals in home pastures and corrals.

Tired pioneers, after having crossed the desert wastes of Nevada, were confronted with a new hazard as they climbed the Sierra Nevada—grizzlies attacked the oxen. In 1849 one man found remains of three or four oxen that had been killed by bears; two of the carcasses had been dragged through the snow for some distance (Bruff, 1949 : 237, 243, 251).

In southern California during the early 1850's, according to Horace Bell (1927 : 255), "grizzly bears were more plentiful than pigs; they were, in fact so numerous in . . . Topango Malibu, La Laguna de Chico, Lopez and other places, as to make the rearing of cattle utterly impossible." In 1854, two cattle owners in the central inner Coast Ranges were so troubled by bears that they offered Grizzly Adams (see our chap. 9) $100 dollars a month and all the beef he wanted if he would remain with them and hunt for a few months (Hittell, 1860 : 325). The "big bear of Marin" plundered bullocks, calves, and pigs in that county for four or five years during the 1850's (N 58).

The menace to cattle persisted well after the bear population had been greatly reduced. On Austin Creek, Sonoma County, in 1868, a grizzly reportedly killed $5,000 worth of cattle, sheep, and other livestock (N 71). A noted bear hunter, Captain Anson Smith of Carmel Valley, was much in demand by stock owners; in 1879 he worked through the Jolon district to protect the cattle there (Anne B. Fisher notes).

Bears, grizzly or black, are proverbially fond of pork, and the big ones of California demonstrated this frequently. Between 1856 and 1868 eight or more newspaper articles told of bears killing hogs in the counties of Shasta, Marin, Contra Costa, San Joaquin, Santa Cruz, and Mariposa. One bear shot on Mount Diablo in March, 1856, had destroyed about $200 worth of livestock, and its stomach held parts of three pigs (N 34). In Santa Cruz County a grizzly took a 300-pound hog from a pen, killed it, and then covered the carcass with leaves (Welch, 1931 : 262). In 1863, J. S. Hittell wrote:

The farmers in those districts where the bears are abundant, shut up their hogs every night in corrols or pens, surrounded by very strong and high fences, which the bears frequently tear down. After having killed a hog, if any part of the carcass is left, the grizzly will return the next night and feast upon the remains, and go [on] until it becomes putrid. He prefers, however, the fresh pork if it can be had. (J. S. Hittell, 1863 : 110.)

According to Alex Kennedy of Long Valley in Lake County (interviewed in March, 1953), grizzlies actually ate stock that was alive. A bear would mash a hog down and then start eating it while the animal still squealed. The ranchers found many hogs still alive after their hams had been partly devoured the night before by grizzlies. A grizzly would also fell a cow and begin eating without first killing the victim.

An early settler in hills near Healdsburg, Sonoma County, had a number of hogs, then scarce locally and worth about $75.00 each. A "big monster" grizzly one night dispatched a fat hog but left the uneaten half under a live oak. Expecting the bear to return the next evening, the owner, Cyrus Alexander, and his hired man drove all the porkers that could be found into a pen, leaving a gap so that stragglers could enter. The men stationed themselves with guns where they could see both the pen and the partly devoured carcass. After they had waited about three hours in the dark, the hogs in the pen began to squeal and give other evidence of being disturbed. The men found the bear at the entrance, where it could attack any pig that tried to get out. Two pigs were killed and several others lacerated. As the men came up, the bear made off (Anon., 1880 : 60).

One writer (Nordhoff, 1875 : 235), in discussing Rancho Tejon, stated that "the grizzly does not usually attack sheep"; but there are several reports to the contrary, telling of severe damage in San Diego, Orange, and San Benito counties from 1872 to 1878 (N 77, N 86, N 89). Slightly earlier (1868), newspapers mentioned sheep killing in Sonoma County (N 71) and destruction of "small stock" about Round Mountain, Shasta County (N 73). A report from Kern County mentioned two hundred sheep in one flock having been killed by a single grizzly, adding that once a bear got this habit it had to be killed to prevent further destruction (Grinnell *et al.*, 1937 : 78). The grizzly, like the coyote, on occasion evidently went berserk in the presence of a flock as the startled sheep ran about in trying to avoid their enemy.

There was also concern for the well-being of the shepherd. A *tepestra* was built for him to sleep on. This was a platform about twelve feet high on stout poles set solidly in the ground, erected at the entrance to the corral where sheep were gathered for the night. The shepherd carried a gun—or kept it close by him—both night and day. (Nordhoff, 1875 : 235.)

Horses were sometimes attacked by grizzlies, although it is unlikely that many were killed on the open range. A ranch near The Geysers, Sonoma County, was reported to have lost a hundred horses to one grizzly (Brace, 1869: 247). Near Los Angeles, according to Don Francisco Gallardo,

In the river bottom below the town there was a pasture where our best horses were kept, and to protect them from wild animals a high fence was built. Yet, even over that Chapuli [one of the riding horses] could jump; and by so doing he saved my life. For an enormous grizzly . . . had got inside the fence and, hiding in the heavy willows, he struck down every horse that passed. (Coolidge, 1939 : 19.)

Near the Yuba River in the Sacramento Valley one day in November, 1848, Buffum (1850 : 46) and his companion heard "a loud braying, followed by a fierce growl . . . and in a few minutes a frightened mule, closely pursued by an enormous grizzly bear, descended the hill-side." One of the party fired a shot, the mule scampered away over the plains, and the hunters ran from the wounded bear.

A stock rancher sometimes posted a reward for the destruction of a particularly troublesome grizzly, in the hope of enlisting the aid of neighboring hunters. In 1876 the California Legislature adopted a law (*Stats.*, 1876, chap. 341) providing a bounty of ten dollars for killing a panther or a grizzly bear in the counties of Del Norte, Humboldt, Mendocino, Lake, Colusa, Placer, and San Luis Obispo. Lesser amounts were stipulated for black or cinnamon bears, coyotes, and so forth.

The claimant had to file a sworn affidavit and present the scalp with both ears as evidence. County supervisors had to pass on claims, and payment was to be from the state's general fund. The act was repealed the next year (*Stats.*, 1877, chap. 4). The peculiar pattern of counties to which the 1876 statute referred may have been the result of local political pressure or of opinions about the local populations of predators; grizzlies, however, already were on the wane.

The business costs on early ranchos included expenditures for bear control, and men were employed regularly to hunt bears, or engaged for a time when damage was severe, or paid on a per-head basis for destroying them. An account book kept by J. B. R. Cooper on the Sur Ranch in Monterey County contains the following entries (supplied by Mrs. Amelie Kneass, Jan. 23, 1954):

| July 17, 1857 | To Sandoval for killing a bear, pd. $4.00; owe him yet $20.00 | |
|---|---|---|
| July 22, 1866 | To Escobar 2 bear | $20.00 |
| September 20 | Cash on the bear act | $35.00 |
| July 24, 1867 | Jacobo Escobar killed 3 bear | $70.00 |

The development of commercial apiaries in the foothills of

southern California while grizzlies were still present on the upper chaparral-covered slopes led to raids in which the bears broke open many hives and carried others away intact, feasting on both bees and honey. Some of the last bears in the southern counties were destroyed while engaged in these depredations on farms below the San Gabriel Mountains. There is at least one report of damage to vineyards. Benjamin S. Eaton in 1865, who had upwards of 5,000 grape vines at Fair Oaks, Pasadena, was often awakened at night by his dogs barking at bears that were eating the grapes (Newmark, 1930 : 336–337).

## WOODEN TRAPS

Yankee ingenuity was manifested in many ways as the Americans took over California. The immigrants were far more clever than the native Indians and the people of the Spanish period. The miners roughhewed the equipment they needed to serve many purposes. Among other things, they contrived log-cabin traps to capture grizzlies. The idea of the "box trap" to capture an animal alive was not a new one. It can be traced definitely to 1590, when it was described in the first known book on traps, published at London in that year (Anon., 1590). In all probability this type stems from the time when man first made tools for felling and cutting up trees. The superiority of the Americans lay in the speed and skill with which, using only hand tools, they cut down forest trees and built traps large and strong enough to contain and hold a grizzly.

During the second half of the nineteenth century many of these huge traps were built in California—from the northernmost counties south at least to the San Jacinto Mountains and in both the Coast Ranges and the Sierra Nevada. Most of them were of pine; and wood-consuming insects, fungi, weather, and fire have long since erased them from the local scene. Grinnell and Swarth (1913 : 375) report that in 1908, "a few old log bear traps" could still be seen in the brushy hills bordering Thomas and Coahuila valleys" of the San Jacinto region. In 1952, we saw one grizzly trap, of redwood, in the mountains of Monterey County. This, perhaps the only surviving example, is described beyond.

In principle a box trap is an enclosure that when open, or "set," permits an animal to go in freely—usually attracted by a suitable bait placed within or attached to a release mechanism that acts to drop or close a door behind the animal. Naturalists of a century ago when seeking to capture small mammals used an inverted dish, pan, or box held up by a figure-4 trigger; when tripped, the enclosure descended over the animal. The real box trap is usually a rectangular structure with floor, top, and walls, closed at one end and having at the other end a drop door ready to fall when the animal pulls at bait on a trigger inside. Some have such a door at each end. In the West, traps for grizzlies were built thus, with one or two doors sliding in vertical guides, and were of heavy logs; hence they were called log-cabin traps. They were strongly fastened together and braced to withstand the actions of a huge beast that became enraged when confined.

The best contemporary description of a wooden bear trap is of one built by Grizzly Adams during 1853 in "eastern Washington" (probably central Montana). The tools available were axes, saws, hatchets, augers, chisels, picks, shovels, and drawing knives.

There are required, for the construction of a good trap, about seventy pieces of timber, and a large number of wooden pins. It is usually made in size about ten feet long, five wide, and five high. When the ground is selected and levelled off, two parallel trenches are dug, into which the sleepers are laid; and upon these is pinned the floor, which consists of logs placed side by side, resembling a section of corduroy road. The sides are made by placing a number of large timbers, similar to the sleepers, one above the other, and pinning them firmly together, so that solid walls of timber are formed; and upon these are pinned the top timbers, which resemble those of the bottom. Above these, along the sides, are laid what are called string-pieces, which are not only pinned down, but at the ends they are connected with the sleepers by perpendicular ties, made of limbs with crotches or hooks at their extremities. Thus bound, the body of the trap is very strong, and might be turned over repeatedly, like a box open at the ends, without displacing a timber. But, for the purpose of making it still stronger, a spot is generally selected between two trees, into which the side timbers, besides being fastened together, are also pinned, doubling the strength. Where there are no trees, posts are generally planted for the same purpose; or where there is one tree, a post is used on the other side. Strength is the great object.

The doors are made of split boards a few inches thick, pinned together crosswise in a square form of the required size. They are intended to slide up and down in grooves, made by pinning slats at the ends of the trap. When they are down, the trap is perfectly inclosed, a complete box. But the most nice and particular work is the apparatus for setting and springing the doors. Two upright, forked pieces of wood, a few feet high, are mortised into the middle of the string-pieces, one on each side; and in their forks rests a beam. Over this beam play two levers, eight or nine feet long, to the outer ends of which, by short chains or ropes, the doors are attached. The butt-ends of these levers are sharpened, to fit the notches of a small but important piece of timber, which holds them in their places. The levers lying across the beam, with the weight of the doors at their outer ends, there is a strain tending to draw their butt-ends apart; the notched stick supports this strain, and holds them in position. While the levers, and by them the doors, are held up by the notched stick, the stick itself is held by the levers,—the whole being a sort of self-sustaining, double dead-fall arrangement. The bait is attached, by a rope running up through a hole in the top of the trap, to this notched stick; so that if an animal enters the trap, and jerks at the bait, it pulls the notched stick away, and displaces the levers from the notches; and the doors, being no longer supported, of course fall, and the animal is inclosed. . . . sometimes only one door is made . . . three traps occupied our time [4 men?] for a couple of weeks; and we labored steadily . . . (Hittell, 1860 : 52–54.)

To obtain boards for the doors, when a tough log did not yield to the axe or wedges, an auger hole was bored in the tree and filled with gunpowder; the blast split the tree in two equal pieces (*ibid.*, p. 55).

From Adams' description, as conveyed to Hittell, we have ventured to make drawings of the details (p. 178). The sizes of logs used were inferred from our measurements of the Monterey County trap. The materials needed were somewhat as follows:

> 2 sleepers, 12 in. $\times$ 10 ft.
> 15 floor pieces, 8 in. $\times$ 5$^+$ ft.
> 14 side pieces, 8 in. $\times$ 10 ft.
> 15 top pieces, 8 in. $\times$ 5$^+$ ft.
> 2 stringers, 10 in. $\times$ 10 ft.
> 4 ties, 5 in. $\times$ 7(?) ft.
> 2 forked pieces
> 1 beam
> 2 levers, 8 or 9 ft. long
> 1 notched stick
> 8 door guides
> 4 tough logs split for doors
> _____
> 70 pieces in all

The trap built specifically in the region near Yosemite Valley to take the huge grizzly "Samson" (see chap. 10), later exhibited by Adams, was an even heavier structure:

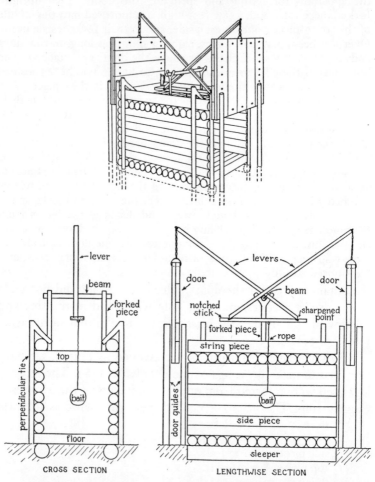

Double-doored log trap for capturing grizzlies and other large mammals. Sketches based on description by Grizzly Adams (Hittell, 1860 : 52–54). Size of logs inferred from old trap in Monterey County. *Above*—Trap as set for use. *Left*—Cross section. *Right*—Lengthwise section.

Two white pine logs, about two feet in diameter and fourteen feet long, were sunk in the ground level with the surface, and six feet apart. The upper sides were scored, to afford a flat surface; and across these were laid pine logs, about one foot in diameter, flattened at the ends on the under side, and scored on the upper. These were laid closely together and firmly pinned to the two huge sills, forming the floor of the trap. We then proceeded to build up the sides and one end with logs of the same size as those composing the floor, matching or locking the corners after the manner of a log-house, and pinning them all strongly together. They were smoothed on the inner side, and rose to the height of six logs. The cover was constructed in the same manner as the floor . . . At each end of the trap, on one side, stood two good-sized trees, and to these the trap was fastened; also, to two corresponding posts, firmly driven, on the other side. The trap was now completed all but the door. This was made of flattened logs, strongly fastened together; and was so constructed as to slide up and down, between high posts—two at each side—so set as to form grooves for it to work in. On each side of the trap, a few feet back from the door—in order to afford sufficient leverage—were erected two upright posts, rising several feet above the trap; and a beam was laid on these, securely fastened at the ends, and completing a very respectable looking gallows. Over this beam was laid a long lever, the short end of which was fastened to the door. The long end being pulled down, it raised and opened the door; and as the same end was secured by a spindle or contrivance similar to that of the box-trap, the moment the bait was disturbed, up went the long end of the lever and down went the door, completely closing the trap. . . . The trap was duly set, and baited with the half of a deer . . . I continued to change the bait every few days. . . . Sometimes with my mule I would drag a piece of bloody meat in all directions, intersecting his [the bear's] trail at different points. (*New York Weekly*, May 31, 1860.)

Four months or more passed before the big bear was trapped. After Adams moved to San Francisco he built a grizzly trap north and west of Crystal Springs, San Mateo County, in 1857 and captured a bear for his menagerie (San Francisco *Daily Alta California*, April 19, 1861).

In the Santa Lucia Mountains of Monterey County, on Seneca Creek, in 1952, we examined the moderately well preserved remains of a log-cabin trap, presumably intended to capture grizzlies (fig. 22). The condition of the structure, the amount of filling by debris, and the almost complete decay of some long-buried parts indicated that it had been constructed well before the end of the nineteenth century. The site is a small rise of land between the creek and a tributary, shaded by redwoods and alders and with hills close on two sides—a likely avenue of travel for bears.

The external dimensions of the trap, between the notched

logs, were 10 ft. 6 in. by 7 ft. 1 in.; the inside measurements were 8 ft. 8 in. by 5 ft. 9 in. The inside height in 1952 was only 34 inches, but the interior was almost filled with leaves and mold and one side of the trap seemingly had sunk with decay of the basal timber. The bottommost log (sleeper) on one side was 16 inches in diameter, and the three logs above were 11, 8, and 6 inches, respectively; those of the back measured, in turn, 12, 11, 9, and 8 inches. Where side and end logs crossed, each was notched, as in a log cabin, and the butts projected 6 to 12 inches beyond the overlap. No evidence of a log floor could be found. The doorway was about 28 inches wide and 30 inches high. Seven crosswise logs remained of the top, measuring about 5 by 2½ inches, and there were spaces between them. One pole was found, 7 ft. 5 in. long, notched on one end, that had holes matching pegs set in the outer surface of the side logs near the front; there may have been two of these "forked sticks" (as in Adams' 1853 trap) to hold the axis of the door suspension lever. No relic of a door or door guides was seen, although the fronts of logs margining the doorway were smoothed off for a distance of 10 inches. The log crossing above the doorway had 1½-inch holes, 39 inches apart; these may have received pins fastening the inner pair of door guides.

Augustín Escobar, whose portrait hangs in the Custom House at Monterey, was renowned for building traps and taking grizzlies for fights in that city (statement of Frank Post). He may have constructed the one on Seneca Creek.

A similar log trap once stood near Jamesburg, Monterey County, according to Mr. L. Cahoon (interviewed by L.P.T.). It resembled the Seneca Creek trap, was built of redwood, and had a drop door but no floor; a bear had to be removed soon after capture lest he dig out. This trap took bears for bullfights at Monterey. A captured bear was moved from the trap to a cage, which was then dragged over poles to the road and there loaded onto a *carreta* for the trip to town. W. Warner Wilson remembers log-cabin bear traps with vertical sliding doors in Lake County during the early 1900's. In the Tehachapi Mountains in 1889, Allen Kelley (1903 : 34–35) saw two traps, evidently in the Adams' style but made of oak logs. Ten years later

they were still "fit to hold anything that wears fur." Early newspapers occasionally mention log traps; one in Nevada County
in 1851 had "a sliding door, made to work with a figure 4" (N
14). Similar traps and trapping in the two decades after 1850
were mentioned in newspapers of Tehama, Plumas, Nevada,
Napa, and San Mateo counties (N 15, N 44, N 23; Welch,
1931 : 259).

In northwestern California, at Elk Camp in the Bald Hills
in 1849 or 1850, Isaac Wistar (1937 : 183) and two companions
built "a heavy crib bear trap" of as large logs as they could han-
dle. The first night it took a panther, the second a wolf [coyote?],
and the third a well-grown grizzly cub, "but his mamma in order
to extricate her baby had torn it to pieces and left only a ruin."
An early-day hog raiser near Healdsburg, when hogs were worth
about $75.00 each, was pestered by a grizzly. He built an 8- by
10-foot cabin trap, baited with an entire pig, and next morning
it contained "a monster weighing nine hundred pounds."

A log-cabin trap occasionally caught more than one bear.
Men from Iowa Hill, Placer County, on June 10, 1852, placed
a trap in a canyon.

While watching it from a distance, they observed a dozen grizzlies
around the trap, and after waiting a while, had the satisfaction of
seeing seven descend. Owing however to the fall [door] being too
light they retained only three. Two of these they have taken out,
but the larger of the three, (who will weigh about seven hundred,)
they feel a little timid in approaching. He has torn the floor of the
trap (three inch plank) to splinters. A stout cage is being built for
his Bruinic majesty. (N 26.)

Borthwick described another log device to capture live
bears, remarking that one of these animals then "was worth about
fifteen hundred dollars." Choosing a suitable site in the mountains, where bears were numerous

a species of cage is built, about twelve feet square and six feet high,
constructed of pine logs, and fastened [together] after the manner
of a log-cabin. This is suspended between two trees, six or seven
feet from the ground, and inside is hung a huge piece of beef, com-
municating by a string with a trigger, so contrived that the slightest
tug at the beef draws the trigger, and down comes the trap, which
has more the appearance of a log-cabin suspended in the air than

anything else. A regular locomotive cage, lined with iron, has also to be taken to the spot, to be kept in readiness for bruin's accommodation, for the pine-log trap would not hold him long; he would soon eat and tear his way out of it. The enterprising bear-catchers have therefore to remain in the neighborhood, and keep a sharp lookout.

Removing the bear from the trap to the cage is the most dangerous part of the business. One side of the trap is so contrived as to admit of being opened or removed, and the cage is drawn up alongside, with the door also open, when the bear has to be persuaded to step into his new abode, in which he travels down to the more populous parts of the country, to fight bulls for the amusement of the public. (Borthwick, 1857 : 299–300.)

Still another type of wooden structure for taking grizzlies was described by Isobel Meadows, a native of Carmel Valley. In 1856 bears frequently stole calves from the family ranch corral, but the men were fearful of hunting down the stock-killer directly. Instead,

They made a trap on Meadows ranch by building a great stockade of redwood stakes driven into the ground very deep—so the bear could not escape. Above the head [entrance] they would have a big V-shaped log on which was a pole—this would release the gate and shut it, when the bear touched the piece of fresh meat they would put inside the stockade. Before they set the trap they would drag freshly killed meat all around the ground so the bear would smell it—and they would drag it through the open door toward the bait. When the bear touched the meat and the great door came down, the men would climb up into nearby trees and shoot the bear. (Anne B. Fisher notes.)

## STEEL TRAPS

Craftsmanship with metals was a much later development than woodcraft; the adding of carbon in the smelting of iron to produce a tempered product, steel, is not an old practice. The trappers who explored western North America and exploited its fur resources early in the last century had traps of steel with springs. Today similar and better traps are available in various sizes for taking fur bearers and for destroying unwanted predaceous mammals. These range from small sizes (no. 0) with 4-inch jaws, for rats or weasels, through the coyote-beaver size (no. 3) with 5½-inch jaws, up to those for black bears and mountain lions (nos. 5, 50, etc.) with jaws spreading to 10 or 12 inches; indeed, there is currently available a modern grizzly trap (no. 6) with toothed jaws having a 16-inch spread and weighing more than 50 pounds.

The early Americans in California had steel traps for grizzlies, some produced by local blacksmiths and others presumably manufactured in the eastern states. The Custom House at Monterey has a trap, received in 1852, that was used in the Big Sur Valley. Its principal dimension are as follows: total length, 45 inches; springs and jaws, each 15 inches; jaw spread, 18 inches; jaws, cross section, ½ by 1⅛ inches; jaw teeth 2 inches long by ¾ inch thick; trigger pan, 6 inches in diameter; chain stock, ½ inch in diameter. The Shasta Museum has relics of a "homemade" trap with jaws that spread 22 by 22 inches; its maker and history are unknown. Alex Kennedy of Long Valley, Lake County, owns a trap (fig. 20), made about 1860 by Dan Tremper at Lower Lake, that took several grizzlies and later was used for black bears. The base piece, of 1½-inch-square wagon-axle stock, is 84 inches long. The jaws spread 20½ by 21 inches, the trigger pan is 8 inches square, and the eight teeth on the jaws are 2¾ inches long. A trap obtained by William Farschon from a sheepherder in Amador County weighed 78 pounds, was 66 inches long, and had jaws spreading to 17 by 22 inches. We have reports of other traps still in private ownership.

Only a few notes have been found on the use of steel traps for grizzlies. Martin (1887 : 36) says that they were set in places bears were frequenting, the trap being penned in to keep domestic animals away. On the Russian River, potatoes were stolen from the garden of John Cook. He set a trap, of grizzly bear dimensions, and caught the thief—an Indian, whose leg was broken and required amputation. (Hobbs, 1875 : 226–227.)

One of a party of miners near the South Yuba River during March, 1852, discovered a grizzly feeding on a horse carcass. In the hope of capturing the bear alive to sell for bullfighting a large old-fashioned gin trap was procured; it was attached to a 20-foot chain with a hook at the end, the whole weighing perhaps 100 pounds. The trap was concealed under bear tracks near the back of the carcass and fastened by long forked sticks driven deep into the earth. The chain was linked round a tree, and both trap and chain were covered with leaves and withered grass. Next morning the trap was gone. The bear had pulled the stakes "as though they were but skewers, and released the linked chain by gnawing a root fully six inches in thickness." Shortly the bear was discovered upslope amid chaparral, the trap on one forepaw and the chain hook deeply buried in a scrub-oak crotch; half the chain was round the tree butt. The bear lay exhausted from his struggles. The animal had wound and unwound himself, forming a ring at the limit of the chain, within which saplings were uprooted and stripped, the brush was flattened, and the earth torn up "to such an extent that it seemed impossible for any animal to do so much mischief in so short a time."

When the men appeared, the bear rose on his hind legs, lifted the huge trap with little effort, growled furiously, and sprang at them; but the chain tumbled him flat. A vaquero was sought to lasso the grizzly, whereupon the men tied his hind feet to a forepaw. When the trap was removed from the bear's paw he sprang and again dispersed his would-be captors. Two men tried to gag him. The animal sprang downhill, crushing one man (who was disfigured for life in consequence) and breaking the lasso, then fell dead. (Spurr, 1881 : 145–152.)

## OTHER METHODS OF TAKING GRIZZLIES

One unique device reportedly used at Mariposa in the winter of 1851–52 was made of a sapling pole having the butt fixed to the base of a large tree so as to extend out horizontally a few feet above the ground. To the springy tip a heavy weight—perhaps a rock—was tied. In the ground immediately below the weight a can was sunk containing bait. A bear would push the weight aside to get at the bait, and the sapling would swing the stone back and hit the bear in the head. (Perlot, 1897 : 171–172.)

Still another special means of dealing with a grizzly was tried by two Sierran miners who had neither firearms nor the energy to build a log trap. They

cut a sapling about thirty feet long, fastened the butt end between two stumps, fixed a long knife with an eighteen inch blade in the other end, and then bent the pole by means of a "Spanish windlass," so that the knife would range in the path where the bear usually came; the spring was fastened by a "figure-four" trigger, with a piece of beef attached. Some time in the night the trap "went off," the conspirators treed, and the bear left overwhelmed with astonishment, and never came back again. By some accident the knife missed the animal, but the boys thought he got a powerful "lick" from the pole. (Sacramento *Age,* in San Francisco *Evening Bulletin,* June 6, 1857.)

In 1852 Oliver Allen, the inventor of a whaling gun, was farming near the mouth of Tuolumne River and used his device for killing grizzlies. His method was described as follows:

Mr. A. loads his gun in the same manner as for shooting whales, with a harpoon and line attached. The gun is then secured to a tree, and the end of the line attached to the harpoon is tied to a broken limb or some movable object, and a bait is attached to a string pending from the gun in such a manner that the instant it is touched the gun discharges its contents into whatever is before it. The gun is so arranged that it cannot be approached except in a direct line with the muzzle, and sure destruction awaits whatever dare touch the bait, either a piece of fresh meat or a salmon. Mr. Allen has recently killed several bears in this manner . . . (N 22.)

A whaling gun was also used successfully near Half Moon Bay, San Mateo County, in the late 1860's to kill a bear that had been destroying cattle (Welch, 1931 : 261).

A set gun, consisting of a double-barreled shotgun loaded with balls and having release strings tied to the triggers, was placed near the carcass of a hog that had been killed by a grizzly. This was in Round Valley, Contra Costa County, on January 12, 1856. The gun went off during the night, and next morning *two* grizzlies were dead at the site, a single ball having passed through both bears. The animals weighed about three hundred pounds each. (N 33.) An eighty-pound grizzly was killed by a spring gun on San Lorenzo River, Santa Cruz County, in 1866 (N 67).

Albert Kennedy of Clear Lake Oaks, Lake County, has a 32-caliber "pistol" (fig. 19) once used as a "set gun." Pistol and bait were lashed to a tree, and a string connected the trigger and bait. When a grizzly in reaching for the bait pulled the string, the pistol discharged six bullets directly into the bear's face.

Colin Preston, an early hunter in the San Luis Obispo region, who killed many grizzlies with his gun, described still another method.

We make large and dangerous bears drunk, when they have cubs in February, and are too savage. The bear goes to and from his den or cover—usually a hollow among rocks—by certain paths, called "beats." A bear will use the same beat for years, going by night on one beat, and in the day taking another, more circuitous. You will often find a tree fallen across the beat, or you fell one, and wait till the savage has examined the new barricade, and finding that it is not a trap is willing to climb over it. Then you make a hole in it with an axe, large enough to contain a gallon of rum and molasses. Bears are greedy of sweets. In countries where there is wild honey they will overturn all obstacles to get at it. Of sugar and molasses, and sweet fruits, strawberries, mulberries, and the like, they are passionately fond. The bear reaches the log; he pauses over the hole full of sweet liquor; examines it, tastes of it, drinks all at a draught, and is drunk. And what a drunkenness is that! The brute rolls and staggers, rises and even bounds from the earth, exhausts his enormous strength in immense gambols, and falls at last stupefied and helpless, an easy prey to the hunter. We have killed many in this way, but it is treacherous, and I do not like it. (Anon., 1857 : 819.)

Poison was another means of destroying grizzlies. Jonathan Wright and Captain Anson Smith kept bees in Carmel Valley during the 1870's and were much bothered by the bears. They built a large stockade and baited it with strychnine-poisoned honey. When the bears became ill from the poison, the apiarists shot them. (Mrs. Bolts, daughter of Wright, Anne B. Fisher notes.) In 1882 or 1883, W. S. Tevis visited this place, nine miles from Carmel, and saw seventeen grizzly skulls in a shed; these bears may have been killed by this means. One of the skulls is now in the possession of the junior author (see App. A). Another account dealing with damage to apiaries in Monterey County states that honey or sugar with strychnine was put out each night and gathered up the next morning for fear bees would take the poisoned sweets and people might later be poisoned. At a later date, tallow with poison was suspended over a trail and also smeared on the ground and twigs. (Grinnell *et al.*, 1937 : 88.)

Writing in 1876, Henshaw (p. 308) said "a supply of strychnine is part of the outfit of every shepherd, and by means of this the number of Bears is each year diminished, till in many sections where formerly they were very abundant they have entirely disappeared." This poison was also used to protect cattle from grizzlies.

We infer that the role of strychnine in reducing the grizzly population was far more important than these few brief notes would indicate. Because poisoned carcasses were a hazard to dogs of ranchers it is likely that users of strychnine did not publicize their activities.

## FLESH, OIL, AND HIDES OF GRIZZLIES

Grizzlies in early days were taken for various reasons, as we have already shown. Some were caught alive for use in bear-and-bull fights (see chap. 6) and some to exhibit as captives (chap. 10). The majority were slain by men in self-protection or in protection of their livestock. Still others, however, were killed so that their flesh could be used as food, their fat as oil, or their hides as rugs or bedding. In pioneer situations the world over, man depends on wild animals to supply some of his important needs, and his vigorous outdoor life entails a heavy demand for food of high energy value. This was true of explorers, miners, farmers, and even city folks in early California. They found "by-products" from the then common grizzly convenient and useful. There were market hunters who made a business of seeking various kinds of wild game, including bears, to satisfy the current demand. Grizzlies were rarely hunted for their meat before the gold rush, but the large influx of energetic Americans in that period resulted in a great increase in "meat hunting."

At the mining camp of El Dorado in 1849, the meat of one grizzly, of slightly more than 1,100 pounds, sold at $1.25 per pound, netting the hunter about $1,300 (Anon., 1856 : 106). In Sacramento in 1850 there were sales of grizzly meat on February 2 and 7 at $1.00 per pound (N 29, N 5); and on March 11 a hunter operating about 30 miles west of that city brought in a wagonload of meat—about 500 pounds of venison that sold at 20 cents per pound and a female grizzly with three cubs, totaling 470 pounds, that brought 50 cents per pound (N 10). In June, 1850, the Fulton Market in San Francisco sold grizzly meat at 50 cents per pound (N 6). Borthwick (1857 : 181) wrote that a carcass fetched $100 or more according to its weight; in another place he mentions bear meat as being worth $2.00 per pound (*ibid.*, p. 287).

During 1850, five hunters killed a female grizzly with two cubs and packed them into Sonora, Tuolumne County; a butcher cut up the carcasses, and soon a line of men was waiting to buy grizzly at $1.00 per pound. The hunters received $500 for their efforts. (Kelley, 1903 : 91–92.) A man named Jim Lyons, on the South Fork of the Tuolumne River during 1855, was called "Grizzly Jim" because he supplied meat of grizzly, deer, and other animals to men then digging ditches for water to supply the mines (San Francisco *Call*, April 2, 1882). In San Mateo County during the 1850's, according to Chase Littlejohn, vaqueros would rope a grizzly, bring it to town for an exhibit, then kill the bear and ship its carcass and hide to San Francisco (Ellsworth, 1931 : 88). A market hunter of San Diego shipped bear meat in winter by boat to San Francisco (Grinnell, 1938 : 80). These are only a few of the many references to market hunting of grizzlies for the mining camps and cities. We conclude that this practice was rather extensive and continued so long as there was a fair supply of the bears. The shooting of other wild game for market sale—deer, elk, antelope, ducks, geese—continued for years after grizzly flesh had become a rarity. By about 1886 not only were grizzlies scarce but their flesh was evidently less relished—the carcass of an old female, weighing 642 pounds when dressed, brought the hunter only 10 cents per pound at a market in Santa Cruz (Welch, 1931 : 262).

In camp, travelers and hunters relished grizzly meat because it was rich with fat. Farnham (1849 : 381) wrote that "a steak cut from the haunch of the grisly bear, and roasted on a stick by a camp fire, is by no means despicable fare . . ." At Clear Lake in 1847 a hungry traveler, after enduring an enforced fast, came on a hunter's camp with "an abundance of fat grisly bear meat, and the most delicious and tender deer meat. The camp looked like a butcher shop. The pot filled with bear flesh was boiled again and again." (Bryant, 1858 : 354.) Borthwick (1857 : 181) sampled grizzly meat in the Mother Lode camps, and thought it was too rich. He said it should be well spiced, because in cooking it became saturated with fat.

The method of preparing and enjoying a fine game dinner outdoors without cooking utensils was described by Bartlett (1854, 2 : 37–38), who was in the Napa Valley during March, 1852:

First, a number of sticks are cut about two feet in length, the size of one's finger, divested of their bark, and sharpened at one end. These correspond to the spits in civilized roast-ovens. The meat is now cut up into pieces about three quarters of an inch in thickness and half of the size of one's hand. Through these the sharpened stick is thrust, and its lower end planted in the ground before the fire. As our fare consisted of venison and bear's meat, successive layers of each were put upon the sticks, the fat of the latter, as it dripped down, basting and furnishing an excellent gravy to the former. In fifteen minutes, with occasional turning, the dinner was pronounced ready to be served up.

Being . . . [without] a table, we seated ourselves on the grass, beneath the wide-spreading boughs of a tree, and a few yards from the fire, in order to be near the kitchen, and to have our meats and coffee warm. Before each person was stuck in the ground a stick of the roasted meat. A bag of hard bread (pilot bread), some sugar, salt, and pepper, were placed near, and each man was provided with a tin cup filled with coffee. . . . we fell to, and never was a feast more heartily appreciated. . . . those who were not satisfied with one stick of meat, found another ready at the fire when the first was gone.

Hunters were sometimes more critical in the midst of plenty. From a big grizzly killed in the San Bernardino Mountains—at Bear Lake—in 1858, "the liver and tenderloin steaks with the broiled ribs made us a sumptuous repast." The meat was very fat. (N 53.) If the hunter was afoot and unable to take much from his kill, he hastily hacked off a few steaks from the bear's thigh. The flesh of grizzly cubs was reported as being "delicious —fat and tender" (N 81).

"Roast grizzly and bear steaks were always prominent features of the bills-of-fare in mining camp restaurants, selling at a dollar a share, payable in gold dust" (Herrick, 1946 : 179). This was a modest charge in view of the prices paid hunters for meat —50 cents to $2.00 per pound—and the size of the portions—the hard-working miners had hearty appetites not readily satisfied by a portion the size of the modern beef filet.

A bill of fare at one miner's hotel in Mariposa during Janu-

ary, 1850 (Chamberlain, 1936 : 41–42) included among others
the following items (italics ours):

| | |
|---|---|
| Beef, Mexican (prime cut) | $ 1.50 |
| " , Up along | 1.00 |
| Beef, Plain | 1.00 |
| " , with one potato (fair size) | 1.25 |
| Beef, Tame, from the States | 1.50 |
| Hash, Low Grade | .75 |
| Hash, 18 Carets | 1.00 |
| Codfish Balls, per pair | .75 |
| *Grizzly, Roast* | 1.00 |
| " *, Fried* | .75 |
| Jackass Rabbit (whole) | 1.00 |
| Square Meal, with Desert | 3.00 |

There were many favorable reports on grizzly meat when
it was served in city restaurants. Franklin Buck, in Sacramento
during February, 1850, said of a grizzly-steak dinner: "It was
the finest meal I have ever eaten, so tender and juicy" (White,
1930 : 65). Bruff (1949 : 469) in the same city, wrote *"Christmas
Day* what a contrast with the last!—I dined . . . at the Fremont
House. Grizzly bear steak, venison pie, fine vegetables, and delightful mince pie, garnished with wine."

In San Francisco, during 1849, Bayard Taylor (1850, 2 : 62)
became eloquent about the rich supplies of game that winter:

Fat elks and splendid black-tailed does hung at the doors of all the
butcher-shops, and wild geese, duck, and brant, were brought into
the city by the wagon load. "Grizzly bear steak," became a choice
dish at the eating-houses; I had the satisfaction one night of eating
a slice of one that had weighed eleven hundred pounds. The flesh
was of a bright red color, very solid, sweet, and nutritious; its flavor
was preferable to that of the best pork.

J. S. Hittell (1863 : 111), who probably had frequent opportunity to eat grizzly in the same city, said, "The meat of the
young grizzly resembles pork in texture and taste, exceeding it
in juiciness and greasiness; but the meat of the old he-bear is
extremely strong, and to delicate stomachs it is nauseating." So
also is the flesh of an old he-pig today!

In 1846 William Garner (1847*b* : 187) said that the feet and
hams were good eating, however prepared, but not other parts.
Another early diner wrote that bear meat was "satable, but very

devoid of flavor, and I think the grizzly indulges in too much gymnastic exercise to qualify him for the table of the epicure. He figures on the bill-of-fare at all California Restaurants . . ." (Marryat, 1855 : 121.)

Refrigeration was scant at best during early times, but the demand for meat was so heavy that much game was probably eaten soon after being killed. That taken in winter could be kept and hung for several days and its texture thereby improved; but in hot w; :ther there could be little delay. Dried meat was therefore a widely used staple. It could be carried and chewed as the traveler became hungry, or soaked and stewed when he made camp. Grizzly Adams (see our chap. 9) mentioned eating this food himself and providing it for his animal captives. James Hobbs of San Diego County stated that "the meat, dried, meant more profit, since it was wanted by the miners;—and . . . dried bear meat brought twice the price that deer meat did . . ." (Grinnell, 1938 : 80). An early hunter of the Sonoma region is said to have cured some bear bacon (Hobbs, 1875 : 227).

A special delicacy, noted by several writers, was "bear paws." Bidwell (1897 : 83), in Colusa County during 1844, shot a grizzly but took "the only part fit to eat—the foot." And Revere (1849 : 143) speaks of cutting off the forepaws in the Russian River Valley. "They are excellent eating—being very tender and gelatinous . . ." Some army officers at Fort Tejon in 1857 shot two grizzlies and took the paws; in the words of one participant:

never did I taste anything more exquisitely delicious than the bear's paws; they were placed in a hole about a foot deep, beneath the camp fire, and remained there till morning, when they were taken out, skinned, and served up. Each foot weighed about 2 pounds, the bottom of the foot is one mass of delicious marrow—only more delicate. (Los Angeles *Star*, Nov. 21, 1857; Giffen and Woodward, 1942 : 84).

The modern bakery "bear paws" may reflect a memory of the pioneer delicacy.

According to Herrick (1946 : 179), "the head baked in a rough earthern oven filled with heated stones, was once considered an especially delectable tidbit."

The grizzly of California when food was plentiful stored fat both under the skin and amid the internal organs. This served the bear as extra insulation in cold weather and also for bodily energy when food became scarce; it also tided females along when nursing young cubs in the den. The fat on grizzlies was a boon to early settlers in California. It improved the flavor of bear flesh, but more important, it could be tried out to become an oil. A bear killed in 1845 by Clyman (1926 : 262) was a "noble animal yeelding more than three Hundred pounds of oil." Two 300-pound grizzlies in Contra Costa County during January, 1856, afforded about seventeen gallons of oil (N 33); and the fat of a "big bear" tried out by Tom Lucas filled five oil cans (letter, Arthur Carter, Feb. 21, 1939, MVZ files).

Bear fat or oil was used not only for cooking but also as a pomade for the hair. In a famous early school in Sacramento which H. H. Granys attended, the teacher used oil from grizzly bears on her hair, and the children were distressed to find that she used the same oil for frying their flapjacks (verbal, Mrs. C. G. Murphy, of Sonoma).

At least one early writer said that the miners did not use bear grease for pomade (Borthwick, 1857 : 181); but a reminiscent account of early days states "that is what the young sports used to put on their hair in the good old times to make it lie down flat when they went to a party in the parlor and had hot biscuit and pumpkin pie, and how it did smell when it got good and warm" (Arthur Carter, *loc. cit.*).

Hides of grizzlies were useful as rugs on floors and sometimes served as bedding or as a ground mat under camp beds. In San Luis Obispo County during 1837, Nidever (1937 : 51, 52) was able to sell one grizzly hide to a trading vessel because "it was much larger than any bullock's hide they took on board" but otherwise found no demand for bear pelts. They were seldom saved except for use as gifts. During the gold rush, however, bearskins became a more useful commodity. In 1850 a skin sold for $35 at Sacramento (N 3) and an early hunter sold a hide to a Digger Indian for $15.00 (Isbell, 1948 : 8). J. W. Audubon, who was in California at this time, wrote that the grizzly pelt "makes a first rate bed under the thin and worn blanket of the [gold] digger" (Audubon and Bachman, 1856, 3 : 149).

One novelty from the grizzly was a pair of grizzly bear boots made by skinning out the hind feet and lower legs. A pair (fig. 21), obtained by General Joseph Hooker in the 1850's at Agua Caliente, Sonoma County, has survived in the Jansen family. In the intervening century the boots at various times lay out in the sun, were soaked in water, and were used in play by children. About 1903 they were worn in a masquerade at the Vallejo home in Sonoma. The present dimensions are: height, 14 inches; length of foot including claws, 12 inches, without claws, 10 inches; width of foot, 6 inches. The "track" therefore is now 10 by 6 inches. The hairs are a uniform brown which presumably was once dark brown or black; they are now sparse, and broken at the tips. The Department of Anthropology in the American Museum of Natural History, New York City, has a similar pair from southeastern Alaska.

# 8 Grizzly Hunting in California

The largest stock of information about grizzlies in California has come from narratives of hunting experiences. Hunters are prone to relate their exploits in the fullest detail, and accounts of "How I killed the bear" were as wordy in the middle 1800's of California as they are in books and magazines describing similar episodes in other regions today. The larger the game, the more spectacular and lengthy the account; few words are put down about squirrels or rabbits, but grizzly hunting seems to require every detail. Only a few of the more significant and representative accounts can be included here.

When grizzlies were abundant in the lowlands and foothills a hunter could find the animals with little effort, and often the bear found the hunter rather than the reverse.

## THE DURABLE GRIZZLY

The tenacity of life in grizzlies that had received a multitude of rifle balls—or one in a particularly vital spot—was remarked by many contemporary writers. The power of early-day muskets was appreciably less than that of later rifles with grooved barrels, cylindrical bullets, and more powerful charges of powder. The ball from a muzzle-loader, fired at some distance, often had only enough energy to pierce the heavy skin and lodge in the thick body fat outside the muscles; such wounds were merely aggravating. Adams (Hittell, 1860 : 161) said of one bear, "several balls had struck her in the sides, but had not gone through the fat."

"The grizzly is very tenacious of life, and he is seldom immediately killed by a single bullet. His thick, wiry hair, tough skin,

194

heavy coats of fat when in good condition, and large bones, go far to protect his vital organs; but he often seems to preserve all his strength and activity for an hour or more after having been shot through the lungs and liver with large rifle-balls." (J. S. Hittell, 1863 : 109.) Near Livermore in 1854 one lived half an hour after a ball from a 5-inch Colt revolver had passed completely through the heart (N 27). Another reportedly "turned and showed fight" after his skull was split with an axe, "scattering his brains on the ground" (N 43).

In the somewhat lurid tale of "How Old Pinto Died," Allen Kelley (1903 : 171–191) wrote that the bear was finally killed by a 45-70-450 bullet entering at the "butt" of the ear and passing through the base of the brain. The previous evening one shot "had nearly destroyed a lung." In all there were eleven bullet holes, but only two or three bullets had lodged, "the others having passed through, making large, ragged wounds and tearing the internal organs all to pieces." Of another bear, which had been fired at repeatedly by members of a party, Vachell (1901 : 251–252) stated that "when we skinned him, we found that he had been shot through the heart, through the lungs, through the head, and through the loins!"

The grizzly of California was "big game," as is his counterpart today in Canada and Alaska. It is remarkable that so many grizzlies were killed in early days when rifles lacked the power now deemed essential when hunting the larger animals.

## FAMOUS GRIZZLY HUNTERS

Several of the men who came between 1828 and 1848 to hunt otters or to farm are of record as having slain a good many grizzlies. Indeed, the titles "great bear hunter" and "greatest bear hunter" occur frequently in the chronicles of that period. Since more than a century has passed, we shall not attempt to evaluate these claims but merely mention a few of the more outstanding hunters. George Nidever wrote (1937 : 53): "I think I must have killed, on this coast, at different times, upwards of 200 grizzlies," and Colin Preston (Anon., 1857 : 819) said that he had destroyed more than that number in a single year. William Gordon who settled on Cache Creek, in what is now Yolo County, during

1842, was said to be "one of the most successful bear-hunters in California, having killed nearly fifty in one year" (Gilbert, 1879 : 31). William B. Elliott of North Carolina, who came to George C. Yount's rancho in Napa Valley in 1845 and by 1855 was settled at Upper Lake, Lake County, "became noted as a great bear hunter, and with his boys . . . probably killed more grizzlies than any other man in the state" (Anon., 1881 : 198). Yount himself was another famed bear hunter; a bas-relief of a bear adorns his gravestone.

There is word of at least three bear hunters in the central Sierra Nevada near the Merced River. Thomas R. Scull was "well known as the 'Bear Hunter of the Mercedes'—a name which he . . . earned by many acts of fearless courage, evinced in numerous and victorious encounters with that formidable animal." But he disappeared from the record as early as 1851 when he sailed from San Francisco (*Daily Alta California*, Nov. 15, 1851). His role was taken up the next year in the same region by Grizzly Adams, whose famous exploits are detailed in chapters 9 and 10. From about 1865 on, Jim Duncan and his companion David Brown were of repute on the South Fork of the Merced at Wawona. Duncan is said to have killed forty-nine bears in nine years, keeping record by notches on a timber of his cabin at Crescent Lake. (Muir, 1898 : 619–620.) Some of these, however, may not have been grizzlies.

## GRIZZLY HUNTING

The single-shot muzzle-loading rifle was the principal weapon used, but a few hunters had only a shotgun. Many of them also carried a pistol or two. Almost every pioneer had a substantial hunting knife but in emergency might pick up the butcher knife from camp. Most hunting was evidently on foot, although some persons used horses. They would dismount to stalk or shoot, because without the repeating gun of later years it was of utmost importance to aim and fire the one rifle ball as accurately as possible. Many men sought and fought grizzlies singly; but there was advantage in having companions, because more than one bear might be stirred up (fig. 23). Some hunters used dogs, others did not. Before shooting from the ground, the hunter usually selected a nearby tree as a retreat should the bear be only wounded and charge. Occasionally a specially built platform or a tree was used as vantage point and a freshly killed carcass might be dragged over ground in the vicinity to attract bears. Some doughty souls crawled into the chaparral shelters of the grizzlies or entered dens directly but these were particularly hazardous ventures.

The best account of grizzly-hunting procedure is that by George Nidever, who was famous and highly successful in this activity. It was written as of 1837 in what is now San Luis Obispo County.

. . . If I discovered them feeding I rode as near as possible with my horse, keeping always to leeward; then dismounting, I tied him with a small cord which he could easily break if the bear got after him. I then approached the bear cautiously and under cover if practicable, taking advantage of the ground, timber, &c. I would get as near as 40 or 50 yds. and under no circumstances did I shoot at a longer distance than 100 yds. I never shot at the head, as unless a ball could be put fairly in the eye or ear, the constant motion of the head when feeding and its shape make the glancing off of the bullet almost certain. The breast is a good place to aim at if the bear is facing you, but I prefer the side, just back of the fore shoulder. The aim should be at a spot well back of the shoulder, as a bear's heart lies much farther back than that of a deer or buffalo. Here a shot is almost sure to drop a bear dead in his tracks; sometimes they will run a few yards, but very rarely do they move from where they are shot. One objection to shooting them in the breast is the danger of striking some of the larger bones against which the bullet flattens. On the side the ribs are the only bones and they are easily broken or even pierced by an ordinary sized bullet. If I approached them through timber, I always looked about me before firing and selected a tree that I could readily climb in case of need. When feeding, they are easily approached, by keeping to leeward, as, although their scent is extraordinary, they keep no watch; relying no doubt upon their poweress [prowess].

When a she bear with cubs is found, it is best to shoot the old one first and the young ones can be shot at leisure, as they will not leave the mother. I never shot at a bear unless I could see him plainly and was within good range, and in this way I *never* missed one. In a thicket a bear has one most decidedly at a disadvantage, and under no circumstances should they be followed, however severely wounded. If a person is cool, a good shot, and above all prudent, there is but little danger in hunting bears. Most of those I shot . . . I killed almost as easily as I could a squirrel. A person who is not a good shot, cool, and cautious, and . . . [lacks] a good rifle, has no business to hunt bears. . . . we had only muzzle loaders and although, by carrying two guns, we had a reserve shot, our chances were poor if a wounded bear got after us. It will sometimes happen that, in passing through bushes, one comes suddenly upon a bear and then of course his chances are poor. When a grizzly is lying in a thicket and hears a noise, it keeps perfectly quiet until the animal or person is within reach. (Nidever, 1937 : 51–52.)

Another hunter, Isaac Wistar, who shot grizzlies about 1850 in several parts of California, had somewhat different advice as to the best procedures. In his reminiscences he stated that

. . . with the muzzle loaders of those days having an extreme safe range of less than a hundred yards, one must kill at first fire at the peril of his life, because if the bear showed fight, as is not uncommon, it was rarely possible for a single hunter to reload in time for a second shot. Hence few hunters cared when alone to molest the animal except under peculiarly favorable circumstances. When a bear has his nose to the ground drinking, digging or feeling for acorns, as is often the case, a square right-angled shot at the top of his head will reach the brain and is the surest of all shots. [Here Wistar differs from Nidever.] Even when he is coming for a hunter, unless excited by wounds and rage, he is given to rearing on his hindquarters for a better view or smell, and in that act often shows a whitey brown spot at or below the base of the throat, which is a fair mark for a heart shot. There are men reckless enough to risk a side shot at the point of the shoulder, but for a solitary hunter with a short range muzzle loader all these shots are uncertain and dangerous except the first. With a companion of course the case is different, as the bear nearly always loses some valuable time whenever his attention is freshly attracted. (Wistar, 1937 : 183–184.)

It is worthy of note that several skulls of grizzlies in the U. S. National Museum contain bullet holes; some hunters obviously held the same opinion as Wistar.

The method of loading a rifle used for grizzlies has been described by the daughter of David Hopper, who began ranching in 1857. She writes:

he sure killed an awful lot of them, and it's a wonder he hadn't been eaten up by them as he had to hunt with an old muzzle loading rifle and only had one shot. I can remember seeing him load the old gun many times. He had a leather pouch that he carried over his shoulder and it had a small horn that just held the proper amount of powder (kept his powder in a bottle). Then he had little pieces of domestic [cloth] a little more than an inch square and he would lay one of those little pieces on the muzzle of his gun, lay the bullet on that and hammer it down with his ram-rod. When he was in a great hurry to shoot again he would . . . [load] with a naked ball, would not take time to use the patchin, as he called it, but the rifle shot much more accurately when it was loaded properly. He also carried a large pistol in his belt and a hatchet. (Letter, Ida J. Cox to Rollo E. Darby, Aug. 17, 1946.)

Representative of many individual accounts of bear hunts but of especial historical interest is that of W. H. Eddy, a member of the ill-fated Donner party. On November 13, 1846, faint from lack of food, he came on a large grizzly track, and was eager to find the maker.

At the distance of about ninety yards he saw the bear, with its head to the ground, engaged in digging roots. The beast was in a small skirt of prairie [? near Donner Lake, Nevada County], and Mr. Eddy, taking advantage of a large fir-tree . . . kept himself in concealment. Having put in his mouth the only bullet that was not in his gun, so that he might quickly reload in case of an emergency, he deliberately fired. The bear immediately reared upon its hind feet, and seeing the smoke from Mr. Eddy's gun, ran fiercely toward him with open jaws. By the time the gun was reloaded, the bear had reached the tree, and, with a fierce growl, pursued Mr. Eddy round it, who, running swifter than the animal, came up with it in the rear, and disabled it by a shot in the shoulder, so that it was no longer able to pursue him. He then dispatched the bear by knocking it on the head with a club. Upon examination, he found that the first shot had pierced the heart. (Thornton, 1855 : 124–125.)

More spectacular and with more serious result was the encounter of James St. Clair Willburn in Hetten Valley, Trinity County, during October, 1857. The twenty-six-year-old school principal had turned from hunting gold to the surer and more profitable business of supplying fresh and dried meat to those who mined. With two Indian helpers he was dressing several nice bucks when a grizzly, attracted by the scent of meat and blood, came directly toward them. Willburn aimed and fired, but the ball merely creased the bear between the ears and lodged high on its shoulder. The animal charged, crazed with pain, and the Indians took to trees. The hunter, thinking he could finish off the bear, fired with his 5-shot, cap-and-ball Navy pistol. Despite Willburn's accurate marksmanship, the bear came on, reared on its haunches, snapping and growling, and seized the man in his "arms." The young hunter stood poised with his hunting knife, using his left arm to shield his face. The bear "brought its huge jaws down on the left arm of Willburn crushing the bones like so many matches. As the bear continued to chew on his arm with snarls of rage, Willburn . . . plunged the heavy hunting knife into the animal's heart. The grizzly dropped dead at the hunter's feet . . . Willburn saved his life, altho he lost his arm . . . " (J. J. Jackson, 1945.) A considerably different account of Willburn's grizzly battle is presented by Allen Kelley (1903 : 201).

Another episode concerned a hunter who became involved with a she-bear and cubs. Usually this was very serious business, and it was often fatal for the hunter; but here it had a ludicrous and rather unique ending. Fred Stacer

was walking along down on one side of a steep descending ridge or backbone, and suddenly came upon two young grizzlies, and shot one of them dead. Hastily reloading his rifle he took after the other, which ran along the mountain side in a horizontal line, which soon brought it and also its pursuer to the backbone or summit of the ridge. The cub had from the first set up a terrific squalling, and it so happened that the old she bear had been on the opposite side of the ridge when her first cub was killed, and followed in the direction taken by the frightened young survivor. The result was that the old she bear, Fred Stacer and the cub all met on a converged line. When the old bear saw Fred she ran back a few paces, stopped, looked at him for a moment, and then commenced to walk deliberately toward him. Fred knew he could hit her directly in the eye, so he quietly awaited her approach until she got within ten feet of him, when he pulled away, and lo! for the first time his gun missed fire. He had forgotten to put a cap on the tube. As quick as a flash the old bear sprang upon him, and the two commenced to roll down the steep mountain side, Fred struggling to escape, and the bear plying teeth and toenail as best she could. The further they went the more rapid became their motion, and finally the two plunged over a perpendicular, rocky precipice, more than fifty feet high, and lodged in the top of a live oak tree that grew at the bottom. Fortunately when they struck the tough but yielding branches of the tree Fred was on top, and lodged, and held on for dear life, while the bear went crashing through to the bottom . . . (Bell, 1927 : 264–265.)

While prospecting for gold in the San Gabriel Mountains of southern California, a miner named Hobbs saw two grizzly cubs and shot at one, breaking its shoulder; the cub's outcry promptly brought the mother. The miner, not having time to reload, climbed rapidly on a high rock and from that vantage point fired at the female with his revolver, wounding her badly. She could not reach Hobbs, but retained enough vigor and perseverance to keep him "treed" for the night. The next morning he was able to kill the female and the remaining cub. (Hobbs, 1875 : 333.)

William Kelly (1851 : 99–101) met a grizzly in the hills near Cottonwood, Shasta County, probably in 1849. After tying his mule, he walked up a gully and saw the grizzly seated and feeding on manzanita. His rifle ball, intended for a side shot to the heart, glanced along the ribs and shattered shoulder bones. The bear staggered, then came toward him on all fours. Kelly bolted for his mule, leaped astride, and applied his big Mexican spurs, but he had only the mule's lariat in hand. The unguided mount raced under a tree; Kelly was stunned by a branch and wiped from the saddle. He scrambled aloft, just as the bear arrived, then with a knife severed the tendons of a forepaw as the beast reached for him. The bear "instantly fell, with a dreadful souse and horrific growl, the blood spouting up as if impelled from a jet; he arose again somewhat tardily, and limping around the tree with upturned eyes, kept tearing off the bark with his tusks." A revolver shot behind the head finished the animal.

Dr. H. W. Nelson, who practiced medicine in Placer County (and later in Sacramento), is stated to have killed a grizzly with a shotgun. He was hunting quail in the hills and in climbing a narrow ravine choked with chaparral heard some men on the hill above shout that a wounded bear was running down the ravine. Escape being impossible, the physician backed slightly into the chaparral, and when the bear's snout was within three feet he discharged both barrels. The fine shot entered the nasal passages and smashed the front of the bear's skull. The doctor was knocked down by the momentum of the dead grizzly but he was only bruised. (Kelley, 1903 : 198–199.)

Isaac Wistar (1937 : 184) tells of a hunter in the Russian River region of Sonoma County who hunted alone except for a small, noisy terrier. "That wary but irrepressible animal had been taught to remain at heel till his master fired, and then make straight for the enemy's rear where he kept up such an insulting and alarming snarling and snapping as to tempt the bear to waste precious time by attending to his case first, giving the hunter time to reload and get in a second shot." The man had the reputation of having killed many grizzlies.

A grizzly on Round Mountain, Shasta County, that had been killing small livestock in 1868 was made the object of a hunt by men in the neighborhood. For this a "pack of dogs" was used. When the bear was descried, men and dogs followed for several miles until it took refuge in a cave. Several dogs entered to continue the attack, and two were killed; but the grizzly was eventually downed by a volley of shots. (N 73.)

## HUNTING PARTIES

Grizzlies were a constant menace to early ranchers, and neighbors often joined in a "community bear hunt" to rid themselves of an objectionable and dangerous competitor. James Clyman describes one of these affairs:

generally two or three men go in company and when the bear is discovered they all aproach in good rifle distance one firing one at a time in slow succession when if their balls take a good impression it so confuses the animal that he is kept continually fighting the ball holes ... so that he has no time to attact the hunters untill it is late—

one which we had the Luck to kill was seen passing to his lair in the morning after sun rise  two men attacted him and gave him five shots at a vital part of his body when he made his Escape to an allmost impenatrable thicket  in an hour after three of us well mounted followed him more than a mile whare we found him badly wounded and in good disposition for a fight

I however had the luck to get a shot at him takeing him close behind the shoulder when he broke back for a desperate [dense ?] thicket  several guns ware fired at him on his retreat but he made his Lair and defied all our methods to draw him out again untill one man at the risk of himself and horse ventured in to the thicket

cutting open a retreat with his butcher knife   at length the bear
charged on him   the other man standing on an Eminence shot at
him as he passed an open aperture through the brush and had the
luck to shoot him in the head   on butchering him we found nine
balls had taken good effect but owing to the greate thickness of the
fat on his side only one had passed in to his lungs (Clyman, 1926 :
261–262).

Nathan Coombs narrowly escaped death in an encounter
with a grizzly bear. In the fall of 1843, a grizzly had been prowling in the vicinity of William Gordon's place, on Cache Creek
in what is now Yolo County, and had eluded the rancher.

It was therefore decided to have a general hunt in which all the men
were to participate. . . . they started out—some five or six strong—
one day, all mounted, and commenced beating the brush along the
creek for their game. Mr. Coombs was riding a partially broken
colt, and . . . forced it into the thicket where the bear was concealed, and, bruin, disturbed in his lair, made straight for the venture-
some horseman. The colt being frightened, stood as if paralyzed
until too late, when it turned, but received a blow from the bear
that knocked it down, and with it the rider, who was seized before
he could rise, and the flesh torn from the bone of his arm. At this
critical moment, when it seemed that nothing could save the unfortunate man's life, there came out of the bushes like a flash of light
another form, that, cleaving the air with one sweeping bound, alighted
squarely upon the back of the enraged grizzly, then leaping to the
ground seized, as a vice, the animal's haunch. This new combatant
. . . forced the bear to defend itself from the attack in the rear . . .
Who has not seen a dog . . . undertake to catch its own tail? He
who has can imagine the bear attempting to get hold of "Tinker,"
his assailant. The dog hung to the grizzly's posterior like grim death,
shaking whenever he could get his feet upon the ground, growling
with overflowing wrath, and in the rapidly-revolving combat was
apparently being transformed into an overgrown bear's tail that the
owner was furiously intent upon laying hold of if he could ever turn
around fast enough to catch up with it. One of the hunters spurred his
horse on to the field of action between the combatants and the prostrate man, who, springing to his feet, made a successful retreat out
of harm's way. A few well directed bullets put an end to the bear
and relieved "Tinker," the hero of the contest, from his perilous
position. (Gilbert, 1879 : 31.)

A "community hunt" of 1850 by five hunters from Sonora,
Tuolumne County, was described many years later to Allen

Kelley by one of the party, William Thurman. The men separated, two going above and three below a large chaparral patch in the hills near the new town. A female and two half-grown (yearling?) cubs soon appeared, and the hunters hit two. The bears ran back into the brush, whereupon the three lower hunters fired. The noise of the running animals stirred up others. Shortly, five more grizzlies broke from cover in the chaparral, and then three more. Eleven bears were now in sight of Thurman. Though all but the original three ran away, it took the party of hunters a long time to dispatch the mother bear and her cubs. The battle lasted from eleven o'clock until three, and the female received more than thirty bullets before she was killed. (Kelley, 1903 : 88–91.)

Tucked away in a prosaic *Handbook and Directory of Napa, Lake, Sonoma and Mendocino Counties* (Paulson, 1874 : 203–204) is one of the liveliest descriptions of a "party hunt" in all the California literature. North of Cahto, Mendocino County, a rancher had been losing sheep. A party was therefore formed both to aid the owner and have some sport. After a tedious ride for several hours on a narrow trail, the party stopped for lunch.

We sought out a favorable spot on a little knoll, divested ourselves of coats and vests, placing them together with our arms, like inexperienced and badly-schooled hunters, under a tree about 20 yards distant. . . . We had just seated ourselves . . . [when] we beheld the largest gentleman of the party, who had just absented himself from our midst, closely pursued by a large Grizzly, which had already succeeded in removing the most necessary portion of his nether garment. Our ally, the Rancher . . . seized a heavy stone with which he dealt the bear a powerful blow, diverting the attack to himself. The entire party now scattered . . . to avoid the bear and reach the weapons. Meanwhile . . . his Bearship was making rare sport of our domestic arrangements. At length one of the party reached the guns . . . and fired. [The bear] . . . with an angry growl and a slight limp, denoting the trifling nature of the wound, . . . started in the direction of the shot.

The rancher ran for his gun; "the bear, smarting from his wound, and growling ferociously, turned in his direction," when the man tripped and was precipitated to the ground. . . . Paralyzed with horror, each stood rooted to the ground, unable to move. Before he could possibly arise, the bear would be upon him, and one blow from that powerful paw would end his existence or lacerate him to such an extent as to render him a cripple for life. But just as the bear is about to close with him, a man without coat, vest or hat, and showing signs of having himself been roughly handled, rushes forward . . . Hesitating not a moment, he throws himself before the prostrate form, and ere the bear can clasp him, buries his knife in the animal's heart. With two or three rapid plunges of his keen knife he finishes the encounter. Both go down together; the hero and our rescued leader are stained with the blood of the victim. . . . Mr. W. explained to us that he had acquired this mode of finishing his game on the occasion of hunting the boar in Europe.

The bear was skinned, the camp kit reassembled, and soon the hunters "partook of a hearty meal" of bear steak. Although Mr. W. had secured a magnificent bear skin, his wearing apparel was defective in one essential point so that he would be obliged to remain at Cahto until a messenger could be sent to Ukiah to procure from his valise a garment, for none sufficiently ample could be obtained here. We tried to patch and mend and even called to our aid the services of a good tailoress . . . but to no avail; the cloth was literally torn to shreds and could hold no stitches. (Paulson, 1874 : 204.)

While in the Russian River Valley during the late 1840's Revere participated in a social bear hunt in which an Indian, a Spaniard, and an American used their respective weapons coöperatively to kill a grizzly.

Stringing his bow with the rapidity of lightning, old Colorado shot two arrows up to the feather into the recumbent and unsuspecting foe, and hastily retreated, while I discharged the contents of my holster pistols into the monster as he made towards me. . . . The rest of the party soon came up, and an active, well-mounted, half-breed vaquero, named Hidalgo, whirled his riata with a whiz around the head of the enraged animal, and he was soon despatched, after a sharp and unbearable conflict with the whole party. (Revere, 1849 : 142–143.)

## SHOOTING FROM A TREE OR A PLATFORM

Some persons, instead of going out to hunt bears, arranged for the bears to visit them. For greater safety they erected a platform, or dug a pit, and then decoyed the bear with bait. According to one account, when a rancher discovered where a grizzly had "buried a steer, hog, or sheep," he would "construct a platform high up on a large tree . . . or dig a pit if no tree is near, and on the platform or in the pit await the bear's return at night." These platforms were used often both in the Sierra Nevada and the Coast Range. (Evans, 1889 : 60.)

A variation of this procedure was for the hunter to "fasten the entrails of a calf, or deer, to the end of a lasso, tie the free end of the lasso to the saddle, and ride across the country several miles, drawing it after him." He would then "ride over the bear's 'beats,' or paths [and] bring the trail, finally, to the foot of an oak tree" with an open space around it, where the bear could be seen as he approached from a distance. The hunter then fastened "the offal to the lower branch, just within reach— perhaps five feet from the ground" and took his perch for the night. (Anon., 1857 : 820.)

In the San Luis Obispo region, at a place where bears had been destroying pumpkins, according to Colin Preston, two men "erected a wooden stage with a platform eight or ten feet high, in the middle of a field, with a wall or wickerwork of brush for an ambuscade, and from this point they watched [for] the bears. . . . by moonlight they saw a large bear enter the field and come toward the platform." One man "fired upon the bear; who instantly ran to the staging and overthrew it, tumbling our two hunters upon the ground. They escaped over the fence a good deal bruised and not a little frightened." (*Ibid.*, p. 819.)

At El Dorado during the autumn of 1853, J. H. Bachman, a "used-up miner" who was attempting a garden, became discouraged by frost and bears. He sat in a tree all one night (October 19), hoping to kill a bear, but none came. Then when he was en route home in the morning one chased him up another tree! Later (on November 12), when he and a companion were sitting at night in trees, three bears came. The men put three balls into one but were unable to kill it, because of distance and uncertain aim. Meanwhile they built a bear trap and took two, which they killed with a shotgun loaded with balls. At the time Bachman wrote there were "any number of Grizzlys in the Valley." (Bachman, 1943 : 78–79.)

## WOUNDS AND DEATH FROM GRIZZLIES

"If you play with a bear, you must take the bear's play" was a common saying; but the full force and significance of this was appreciated only by those who tussled with California grizzlies (Evans, 1873 : 62).

The accounts of persons variously injured, deformed, or scarred for life, lead to the conclusion that the California population of the middle 1800's must have included a good many people who were severely blemished or maimed as the result of encounters with grizzlies. In those days, the setting of fractures was still a rather primitive art. Probing for bone fragments or other debris without the use of anesthetics must have been painful to the patient. There was no repair or skin-grafting comparable with the skillful surgery of today, and antisepsis was not practiced regularly—"laudable pus" undoubtedly exuded from many grizzly bear wounds. Hospitals were all but nonexistent, and persons injured away from medical facilities had to be patched, bandaged, and poulticed by the self-taught first aid of rugged pioneer friends working with whatever commonplace knives, needles, cloths, and chemicals might be at hand in an emergency.

The wounded grizzly was a fearsome beast (fig. 24). Quiet or placid as he might be or seem when undisturbed, the sting of a rifle ball changed him into a raging monster that feared nothing, and he charged his assailant forthwith. Only exceptionally did he turn and run. Occasionally, if standing upright, he closed to the attack without dropping to the ground. More often he advanced on all fours, and in spite of his seemingly cumbersome build and gait it was only a matter of seconds until he was upon the man. After the first shot, a lone hunter seldom had time to reload unless the bear came from a distance; and at long range the shot rarely incapacitated the beast at once, even when he was hit vitally. Injured bears were highly vocal, and one reporter said they stamped the ground more heavily than usual as they came on. When the bear reached the hunter he sometimes reared to standing position when both jaws and claws were ready to inflict deep and serious wounds. The hunter might be bitten and clawed at the same instant. Some victims were scalped by a rake of the heavy claws; others had their arms, chest, or back scored deeply. A bite of the massive jaws could penetrate parts of the human skull, mince a hand or arm or thigh. If the hunter was knocked down he was vulnerable in all parts; and the bear continued to bite, chew, or claw so long as the man showed any movement or other sign of life. Only the heaviest of buckskin clothing helped to save its wearer; the claws would slip over the smooth surface and the teeth could not take good hold. Other garments were of scant protection, and many hunters were variously disrobed in encounters with grizzlies. If the bear sat or fell on a person, the victim was all but crushed by the weight.

As the Americans invaded grizzly domain and sought to kill the big animals, many were successful, but many were not —some were injured but recovered, some were maimed or disfigured for life, and a number were killed outright or died of their wounds. J. S. Newberry (1857 : 47–48), who traveled widely in California during 1855, observed of the grizzlies that "large numbers are annually killed by the hunters, and . . . not a few of the hunters are killed annually by the bears." J. S. Hittell (1863 : 109) at San Francisco in 1863 wrote that "about half a dozen men, on the average, are killed yearly in California by grizzly bears, and as many more are cruelly mutilated." This was a perspective after fifteen years of the gold rush. Later, in southern California, Vital C. Rèche said: "probably three-fourths of the recorded killings and woundings of human beings by bears was done by the mother bears," and that he "knew personally of six men thus killed in the Temecula Mountain region [Riverside and San Diego counties] in one ten-year period." He stated that "when the repeating rifle (Winchester) came in, there was much more careless, therefore dangerous shooting than in the previous era of the single-shot rifle, when a man knew that he had just one chance." (Grinnell, 1938 : 79.)

A place in northwestern Kern County was dubbed Grizzly Gulch because it was infested by the bears; mining had to be discontinued until they were exterminated, and several prospectors reportedly were crippled or killed (Hoover *et al.*, 1948 : 98).

We have fully thirty reports from early newspapers, 1850–1882, of persons injured by grizzlies in various parts of California, and many more are to be found in early books. The newspapers in about the same period record twelve persons killed by grizzlies, and at least as many more are mentioned in other literature. Early California had no adequate statistics of morbidity and mortality, else there would be a far more extensive record of persons who suffered from the bears. Yet with all these episodes, "deaths of violence" from grizzlies were probably far fewer in proportion to the population than those from automobiles of the present day. California was perhaps safer in the days when grizzlies abounded than in recent times. Human fatalities occurred almost throughout grizzly territory—in the Coast Ranges, the Great Valley, and the Sierra Nevada—and in practically all months of the year. Some attacks were unprovoked, the bear perhaps being frightened by the sudden appearance of a person; others were retaliations for injuries inflicted by the hunter. Both young men and old, green and experienced, were among the victims.

Near Bodega, Sonoma County, a young man armed only with pistols, followed a mortally wounded bear into a thicket and was literally torn to pieces (Revere, 1849 : 258). In southern California, Andy Sublette, a guide and "mountain man" with experience in Colorado, the Mexican War, and the gold rush, made grizzly killing his favorite sport. He is of record as having killed a big one near "Cauenga" in October, 1852, and being "shockingly bitten and mutilated" by a wounded bear near the seacoast in the following May. In spite of this, he soon had two cubs as pets. In December, 1853, he went hunting at Elizabeth Lake and shot and killed a grizzly; but in doing so he was so severely wounded that he died. (Hafen, 1953.) Isaac Slover, a durable trapper-hunter who had been a member of the Pattie party on the Colorado River in 1827, survived to hunt and fish in the San Bernardino Mountains at the age of seventy-seven and then was killed four years later by a bear on the north slope of Mount San Antonio (Beattie and Beattie, 1939 : 104, 105). In early days an unknown man was killed by a grizzly in what later became Strawberry Canyon on the University of California campus at Berkeley (Carleton, 1938 : 49).

The first victim in California of whom we have record was

Peter Lebec (Le Bec, Le Becque, Lebeck?), killed and buried October 17, 1837, on the site that became Fort Tejon in Kern County. The grave was under an oak tree. One surface of the trunk had been hewn flat and bore the epitaph: "PETER LEBECK killed by a X bear Oct^r. 17 1837" (R. F. Wood, 1954). Blake (1857 : 47) viewed the marker in 1853 and wrote, "it is a durable monument." Edgar (1893 : 26–27) camped at the tree in 1857 and was told by local Indians that

many [20] years previously some trappers were passing through the cañada, when seeing so many bears [it was the season of acorn abundance], one of the party went off by himself in pursuit of a large grizzly and shot it under that tree, and supposing that he had killed it, went up to it, when it caught and killed him, and his companions buried him under the tree, upon which they cut his epitaph.

On April 5, 1936, the Native Sons and Daughters of the El Tejon and Bakersfield parlors marked the site with an appropriate granite monument. Peter Lebec's origin is uncertain but his name is perpetuated in the place Lebec on the Ridge Route (U.S. 99) a few miles south of Fort Tejon.

Perhaps the quickest fatal encounter on record was that of a man named Boggs in the Russian River country during early days. Starting with two companions, he left them

to hunt down the opposite side of a narrow bottom closely fringed with large timber, having a close undergrowth. Before long, hearing an excited shout from the other side, we ran quickly in that direction, and scarcely three minutes could have elapsed before we reached the body of Boggs, dead and terribly mangled, his entire left side having been torn off by a blow of the bear's paw. His gun, broken but undischarged, lay near. . . . it appeared that B. had approached within twelve feet of a well-used, but now empty lair, from which the bear had sprung upon him without warning, and struck him down before he had time to fire. . . . the bear had made off so quickly with her cubs, that we did not get a glimpse of her. . . . (Wistar, 1937 : 184–185.)

One of the most notable cases in which a man was severely injured by a grizzly and then survived for many years is that of John W. Searles. There is at least one contemporary newspaper report of the event, and there are three in books (Kelley, 1903; Ingersoll, 1904; Chickering, 1938). These differ in respect to the exact locality and some of the hunter's activities before the attack but agree substantially on the manner in which the bear performed and the damage inflicted. The incident occurred on March 15, 1870, probably in the mountains of Kern County where Searles was on a general hunt with companions. He heard a bear and was seeking to locate it when the beast reared up, its nose not two feet away. Searles could not back away because of the dense brush. He pointed his gun toward the bear's jaw and fired. The bear pitched to its forefeet, gasping and pawing at its eyes where the flame of the cartridge had burned the hair —but it was not seriously hurt. Three times Searles tried to fire, but the misfit cartridges failed him. The bear rose, open-mouthed, and the hunter jammed the rifle into its jaws. The animal brushed it aside, felled the man, and holding

one foot on his breast bit off his lower jaw. The next bite was in the throat, severing the windpipe and laying bare the artery as well as the jugular vein, and then it grabbed the flesh of the shoulder, laying bare the bones and cutting a blood vessel, from which the blood spurted up so that Searles, lying there, saw it stream in a curve above his face.

As the bear pulled this mouthful of flesh clear of the bones its foot slipped and Searles rolled over. His coat was all in a hump on his back, and the bear bit into that once and then went away. . . .

[Searles] was as near dead as ever a live man was, but part of his discomfort saved him. It was turning cold rapidly [the end of a late winter day in the mountains] and the wet clothing began to freeze, and this sealed up the torn blood vessels. (Ingersoll, 1904 : 372.)

Then, "with his lower jaw dangling about his throat in shreds and his left arm useless," in frightful pain Searles walked and crawled to his horse, mounted, and rode four miles to camp. Three days later he was in the Los Angeles Hospital, and after surgery he was up in three weeks. While he was in the hospital a surgeon proposed drilling into his undamaged upper teeth to provide anchorage for his lower jaw. The "invalid" kicked the surgeon halfway across the hospital room! Later, the old Spencer rifle dented by the grizzly's teeth and a two-ounce bottle with twenty-two fragments of teeth and bone from his facial wounds became his chief treasures. Under his beard the jawbones seemed half missing, to the touch, and he could not turn his head readily. Yet this indomitable hunter and explorer, for whom Searles Lake is named, was later superintendent and chief owner of the San Bernardino Borax Mining Company works. He survived twenty-seven years, dying on October 8, 1897, at age sixty-nine. (*Ibid.*, 372–374.)

Accounts of grizzly attacks often emphasize the fact that the animal sought the victim's head or face, as in the Searles episode. Victor (1872 : 301) wrote: "this trick of the grizzly—striking a man on the head, or 'boxing his ears'—is a dangerous one. It is not . . . rare . . . to find men . . . who have had their skulls broken by the blow of its immense paw." But injuries were not confined to the head. Sometimes an arm was severely lacerated or practically bitten off, great gashes in the body were made by the huge foreclaws or teeth, or a thigh was slashed. A man named McCoy at Tomales Bay in 1846 unfortunately walked near a female with cubs. She knocked him down, bit him in the legs and back, and tore out some cords (tendons) in one thigh (Nidever, 1937 : 61–62).

Colonel William Butts and a party tried to destroy a huge grizzly in San Luis Obispo County on March 29, 1853. Several shots hit the bear, which took refuge in a brushy arroyo, and Butts followed. The grizzly sprang too soon for the man to use his rifle, and bore him to earth. The Colonel finally killed the bear with a knife. When he was found by his companions, the bones of his face were so crushed that he was thenceforth disfigured; there were several fractures in his left arm and right leg; some ribs were crushed; and his body and legs were literally cut into strips—yet he survived. (Bell, 1927 : 258–259.)

In Placer County during 1851, a man named Wright who had had no experience with grizzlies shot a bear in the side with a small caliber rifle. The animal came on, struck the man into the snow, bit at his face and arms, and gave a stroke of its paw that almost tore off his scalp. A piece of skull some three inches in diameter was broken out "and lifted from the brain as cleanly as if done by the surgeon's trephine." Wright survived, but the "hole in his head" incapacitated him for work. The local miners took up a collection and sent him home to Boston. (Kelley,

1903 : 24–26.) Grizzly Adams (see our chap. 9) had similar injuries.

Charles McKiernan, a lumber merchant of San Jose, shot a bear in the chest with an ounce ball, but the animal retained enough vitality to close for an attack. McKiernan swung his rifle against the bear's head and broke the wooden stock; as he stooped to recover the barrel his head came close to that of the grizzly. The beast clamped its jaws over the man's head; one canine tooth entered the left eye socket and another sank into the skull. At this McKiernan fell, and the bear slid down slope. In spite of his injury, the man regained his horse and rode to camp. He was patched up by a surgeon and, though he had lost an eye, later went to prospect on the desert. (Kelley, 1903 : 75–79.)

Another man was bitten through the neck and soon died; still another had the whole lower jaw torn away and died from loss of blood (*ibid.*, pp. 22–24).

The most effective measure when attacked was to lie absolutely quiet, feigning death. "If the man lies still, with his face down, the bear will usually content himself with biting . . . for a while about the arms and legs, and will then go off a few steps and watch. . . . the bear will believe him dead, and will soon . . . go away. But let the man move, and the bear is upon him again; let him fight, and he will be in imminent danger of being torn to pieces." (J. S. Hittell, 1863 : 109.)

Various persons saved themselves by this ruse, though some were injured in spite of it. There was another hazard, or at least an inconvenience. "The man who has the courage and nerve to lie still as if dead, and never cringe when he is lifted by the bear's teeth, stands a chance of being buried under a pile of loose leaves and rubbish" (Evans, 1889 : 59–60)—just like any bit of bear provender covered to discourage competitors!

One day in 1850, three men in Humboldt County encountered five grizzlies but wounded only one, and it escaped. The next day they met eight. One of the men "treed" at once; the others advanced to about a hundred yards from the nearest bear. Rifle shots downed two bears but did not kill them. Both hunters took to trees, but a wounded grizzly grabbed one by the ankle and pulled him to the ground. Soon the two bears seized the victim, one on an ankle, the other on a shoulder. His hip was dislocated and he received many flesh wounds. His clothes gave way, and soon only part of his coat and shirt remained in place. After the bears had divested him of clothing they seemed unwilling to take hold of his body. One bear went entirely away; the other moved off a hundred yards, sat, and watched the man as he lay perfectly still. After several minutes he moved, and the bear came pell-mell, roaring at every jump. "She placed her nose violently" against the man's side, "raised her head and gave vent to two or three of the most frightful, hideous and unearthly yells . . ." Then she moved off to watch. The man, in spite of his injured leg, climbed into a buckeye tree, and the bear immediately took position below. His companion aimed but did not shoot. Soon the bear left. One of the wounded bears died, and the men subsisted on it for twelve days. (L. K. Wood, 1932 : 59–61.)

Although grizzlies tore and mangled many victims, curiously they did not eat them. At least, only one report of a grizzly eating human flesh has come to our attention. In the mountains about four miles west of Sierraville, Samuel Berry went out on a Sunday early in December, 1874, to tend some traps. He was in shirt sleeves, carried only a hatchet and bowie knife, and was on snowshoes. When he did not return, search parties went out —two, ten, and then about forty men; on Wednesday his body was found. The evidence indicated that while he was walking along a large grizzly had sprung out from a hollow tree. The man had discarded his snowshoes and had run only about ten rods when the grizzly caught and killed him. He apparently had not "made much if any fight. The head and breast had been eaten away." A report of this casualty was published promptly in the Downieville *Messenger* and was reprinted in the San Francisco *Alta California* on December 13, 1874 (N 85). Allen Kelley (1903 : 103–110), writing possibly of the same episode, says that it occurred in the record snowfall of 1889–1890, that the man was a mail carrier on skis, and that the attack was by a female with two cubs. It is possible, however, that the evidence was not properly interpreted; the animal that ate part of the body may have been a coyote.

Men have at times frightened away bears by using fire. Once when some campers, overtaken by night, made a fire and were preparing to lie down, a huge grizzly arrived. Being short of ammunition, each took a firebrand and walked slowly toward the bear; the animal moved sulkily "around the fire, sideways, snorting and showing his teeth, and at last went off." (Bruff, 1949 : 199.) Another and later story is of a man in Kern County during 1896 who shot and injured a grizzly. The grizzly bit the man, and a second adult bear made a snap at his ribs and bit his coat. Her teeth set off a box of matches in one pocket. The man fainted, and when he came to a few minutes later the bears were leaving. (Tinsley, 1902 : 161.) He was one of the last persons injured in California by a grizzly.

A few people frightened grizzlies away with noise. When Frémont's party was camped on the headwaters of the Salinas River early in March, 1847, there was a stampede among the horses after midnight. The cause of the alarm was not Indians but "gray bears"—grizzlies. The first impulse was to attack the bears, but Don Jesús gave a succession of loud halloos, and the bears retreated. (Colton, 1850 : 381.) Another traveler was in the mountains on his way to Mission San Fernando when his horse gave out and he had to proceed on foot. A grizzly appeared in the path. "Afraid to fire," the man "unslung his tin prospecting pan, and drawing his ramrod, commenced a clatter on the pan, which soon drove the grizzly off." (Kip, 1921 : 32.) The San Bernardino *Guardian* of June 6, 1874, told of a man named Bayley who became lost on a mountain trail, his gun not loaded and without ammunition. He came in sight of several animals which he at first thought were hogs but soon discovered were bears with cubs. As they advanced, he picked up two stones, struck them together and shouted, long and loud. The bears paused, then turned and bolted for the brush. (Ingersoll, 1904 : 370.)

The philosophy of the California pioneer regarding the hazard of grizzlies was stated most pointedly by one of them:

The risk was great, to be sure. I knew several gentlemen in California who had been horribly mutilated by these ferocious animals. One had the side of his face torn off; another had one of his arms "chawed up," as he expressed it; a third had suffered paralysis from a bite in the spine; a fourth had received eighteen wounds in a fight with one bear; and I knew of various cases in which men had been otherwise crippled for life or killed on the spot. Hence the peculiar charm of a fight with a grizzly! If you kill your bear, it is a triumph worthy enjoying; if you get killed yourself, some of the newspapers will give you a friendly notice; if you get crippled for life, you carry about you a patent of courage which may be useful in case you go into politics . . . Besides, it has its effect upon the ladies. A "chawed up" man is very much admired all over the world. (Anon., 1861 : 602–603.)

# 9 Grizzly Adams

## THE MAN

During the middle of the nineteenth century many persons in California—padres, vaqueros, hunters, trappers, miners, farmers, and others—met grizzlies in various circumstances. Some suffered injury or death from the big animals and, in turn, many hunters accounted for fifty, a hundred, or even more bears.

Among all these people, one man, known as James Capen Adams,[1] or Grizzly Adams, is preëminent in the literature on California grizzlies because of the extent and variety of his recorded experiences with these bears (fig. 25). His story is extensively documented in a book (Hittell, 1860), several newspaper accounts, a series of weekly magazine articles, and several dime novels. Adams was unique because he made a regular business of capturing and dealing in live animals—grizzlies, mountain lions, smaller carnivores, deer, elk, and others. A chronology of his activities derived from the literary sources indicated hereafter (see pp. 234–238) is as follows:

1812—Born October 22 at Medway, Mass.

1849, *autumn*—Arrived in California via Mexico (Chihuahua).

---

[1] There is some confusion regarding the identity of "Grizzly Adams." According to Farquhar (1947), he was not James Capen Adams but most probably John Adams (born Oct. 22, 1812, at Medway, Massachusetts; died Oct. 28 or 25, 1860; buried at Charlton, Mass.). Family records show that John had a brother James Capen Adams. In two newspaper articles by Theodore H. Hittell (San Francisco *Daily Evening Bulletin,* Oct. 11 and 17, 1856), the hunter designated himself as William. The book by Hittell (1860) uses the appellation James Capen Adams but contains several references to a brother William, who reportedly visited the grizzly hunter at his Sierran camp and was a partner in the shipment of captive animals from the Pacific Coast; the family records, however, give no indication of a brother of that name.

217

1849–1852—Variously occupied in mining, raising livestock (near Stockton), and trading.

1852, *autumn*—Established camp in central Sierra Nevada near Yosemite to hunt and trap.

1853, *spring to autumn*—On trip to eastern part of Washington Territory,[2] collected pelts and trapped live animals, including the bear "Lady Washington," taken as a yearling; transported and sold the animals and skins in Portland; returned to winter at his camp in the Sierra Nevada.

1854, *spring to autumn*—Hunted southeast of Yosemite Valley and captured the bear "Ben Franklin" as a cub; crossed the Sierra Nevada (in April) and the Great Basin to hunt and trap in the Rocky Mountains; returned over the Sierra Nevada, after the first snows, and reoccupied his camp.

*November*—Built a trap for the big grizzly.

1854–1855, *winter*—Captured the bear "Samson." Moved camp down to Merced River.

1855, *spring*—Took his collection of animals to Corral Hollow, eastern Alameda County. Hunted near Corral Hollow; returned briefly to Sierran camp, where he received a scalp wound from a grizzly and Ben Franklin was injured; trip down inner Coast Range to Tulare Lake and to mines in Kern County, hunting and trading; to Tejon Pass; episode with jaguar.

*Autumn*—Returned up west side to San Jose and San Francisco; Charles Nahl evidently portrayed one of Adams' bears at this time. Hunted in Russian River Valley, 1855 or later.

1856—In San Francisco at least from September 23 on; established Mountaineer Museum at 143 Clay Street; first interviewed (October) by Theodore H. Hittell (San Francisco *Daily Evening Bulletin*); California Menagerie removed to California Exchange, Clay and Kearny streets, and name changed to Pacific Museum.

1856–1859—Maintained museum; Ben Franklin died January 18, 1858; Adams interviewed by Hittell July, 1857, to December, 1859; sketches by Nahl for Hittell's book probably made during this period.

1860—Adams departed January 7 from San Francisco on clipper ship *Golden Fleece*, with menagerie; three and a half months on voyage round Cape Horn to New York; ex-

---

[2] Probably Montana east of Flathead Lake, since Adams tells of shooting a number of buffalo and in 1853 these animals did not range west of the continental divide (J. A. Allen, 1876 : 122 ff., map).

hibited animals with P. T. Barnum for about six weeks; series of articles about Adams in the *New York Weekly* beginning May 31; Adams on tour with menagerie in Connecticut and Massachusetts, probably in July and August; Hittell's book *The Adventures of James Capen Adams, Mountaineer and Grizzly Bear Hunter, of California* published in August or earlier in San Francisco and by September in Boston. Adams died at Neponset, Mass., October 28 (or 25) and was buried at Charlton.

Adams was trained as a shoemaker, but upon attaining his majority broke away to hunt wild animals in Maine, New Hampshire, and Vermont for a traveling show. Injured by a tiger in the exhibit, he returned to shoemaking in Boston for about fifteen years. He invested in a cargo of boots and shoes but lost it all in a fire at St. Louis, Missouri.

In 1849 he started west and reached Los Angeles in the autumn. In California his fortunes rose and fell three times. Then his great interest in wild animals reasserted itself, and in the fall of 1852 he took to the mountains with two oxen and an old wagon, two rifles and a pistol, several bowie knives, and some tools and clothing.

Adams was an enterprising Yankee, tireless, courageous, and resourceful, a keen shot and possessed of much manual skill. Like other early mountaineers he traveled widely and far on foot or by wagon when roads were few and poor. He fought and killed many grizzlies singlehanded, often following his rifle shot with a hand-to-paw encounter in which his knife served for the final coup. He made log traps (see chap. 7) to catch other bears and built numerous cages to contain and transport them. He did a rather large business in the sale and export of pelts and live captives, including bears, mountain lions, elk, deer, and lesser beasts and birds; and he reared and trained two grizzlies to be his constant companions and aides, even teaching them to serve as pack animals on occasion; for display he rode them "bear back" and taught them to wrestle (see chap 10).

Grizzly Adams, as seen in San Francisco by his chronicler, Theodore H. Hittell (1911 : xxii), was

a man a little over medium size, muscular and wiry, with sharp features and penetrating eyes. He was apparently about fifty years of age [actually forty-four years in 1856]; but his hair was very gray and his beard very white. He was dressed in coat and pantaloons of buckskin, fringed at the edges and along the seams of arms and legs. On his head he wore a cap of deerskin, ornamented with a fox-tail, and on his feet buckskin moccasins. An excellent likeness of him, as well as his favorite bear, is presented in the illustration, drawn from life by Charles Nahl, entitled "Adams and Ben Franklin" (fig. 25).

The activities of Adams were too extensive to be recounted here; they are set forth in detail in Hittell's book. Two of his major adventures will serve to illustrate his reckless courage and his ability in organizing a large expedition.

In the spring of 1855, Adams, while hunting near his camp in the Sierras and accompanied by his dog "Rambler" and trained bear "Ben Franklin," had a terrific encounter with a she-grizzly in which both he and Ben were severely injured (*New York Weekly*, July 12, 1860). While he was walking along a bear trail through shoulder-high chaparral, a large female rose on her hind legs close by him. She jumped and with one paw knocked the gun aside before he could fire, then with the other struck Adams and peeled the scalp down over his eyes. This felled the hunter face downward, whereupon the grizzly put a paw on his neck and began to tear at his shoulders and back. The bear was old, her teeth blunted, and the hunter's coat and shirt were of leather; these circumstances lessened the damage, yet heavy scars resulted from the injuries. Adams sang out "St'boy!" to Rambler and Ben, who were close behind him; the dog attacked the grizzly's rear and Ben her front. Adams sprang up, grabbed his gun and went up a small tree, laid back his scalp, and wiped the blood from his eyes. The she-grizzly had put Ben down and was "chewing and tearing his head and neck fearfully." The hunter let out a "bear screech" and the grizzly rose. He shot her in the heart and she fell backward "like a log of wood." Ben, finding himself free, went bounding for camp, "continuing his yells at every jump." When the grizzly showed signs of recovery, Adams descended and closed on her with his knife; she was so badly wounded that a half dozen strokes finished her. Back at camp, Adams, aided by his Indian boys, first bandaged Ben; then he trimmed the edges of his own scalp, washed the wounds, and poulticed his head with a decoction of "snake root and blood root." He repeated the washing and bandaging every few days, but he was confined to his cabin for a long time.

Hittell wrote (1911 : 371) that "Ben always afterward carried the scars of the conflict upon his face . . . As for Adams himself, when his scalp-wound healed, it left a depression about the size of a silver dollar near the top of his forehead, which looked as if the skull underneath had been removed." By the time Adams visited Barnum in New York five years later, he had received still other head injuries from his "grizzly students."

None of Adams' many other encounters with bears was quite so serious as the one just described. The man was intrepid, fearless, and seemingly indifferent to the risks involved in attaining his objectives.

The magnitude of Adams' collecting operations in "eastern Washington" in 1853 is indicated by a description of his caravan as it started in the autumn for Portland:

there were five horses packed with buffalo robes, of which we had about thirty-five; next, four horses packed with bear skins, and several large bear skulls; then, two packed with deer skins; two with antelope skins; one with fox and other small skins; seven with dried meat for the use of the animals on the journey, and, in part, on their intended voyage; one with boxes containing the young bear cubs last caught; two with boxes containing wolves, untamed; a mule with [live] foxes and fishers in baskets; and a mule with tools, blankets, and camp luggage. Almost all the horses besides the seven specially devoted to the purpose carried more or less dried meat,—even those we rode. But the most remarkable portion of the train consisted of the animals which we drove along in a small herd; these were six bears, four wolves, four deer, four antelopes, two elks, and the Indian dog. (Hittell, 1860 : 169–170.)

Besides Adams, who handled the "herd" of wild animals on foot, the party included six local Indians as guides and helpers, three other white men, and Adams' two Indian assistants from California—Tuolumne and Stanislaus. In all there were 36 horses, 2 mules, and 12 persons. He had bargained with the local Indian chief, Kennasket, for 30 horses and 6 Indians; rent for their services was 2 sacks of dried meat per horse and 6 fawns and 1 elk for the 6 men. The party had hunted antelope, buffalo, and deer to obtain the meat to pay for the services and to use as food on the journey. Saddles had been made "of boughs fastened together with wooden pins and a few nails, and covered with elk-skin," which likewise served as leather for pack harnesses. The party also "prepared and arranged the packs; marking and numbering them; so that our camp, in a short time, resembled a sort of bazaar, where the packages of a caravan are displayed."

## THE MENAGERIE

After his principal outdoor experiences, Adams took his collection of live animals to San Francisco and became a showman. His exhibit of large mammals and birds proved to be a major interest in the pioneer period of that city. The "museum" lasted for about three and a half years, according to a series of advertisements and short news notes in the San Francisco *Daily Evening Bulletin* from the autumn of 1856 on. The first advertisement, on September 23, 1856, reads as follows:

Great Show of Animals.
M O U N T A I N E E R   M U S E U M ,
at 142 Clay Street.

The Proprietor has just returned from the mountains with the
LARGEST COLLECTION OF WILD ANIMALS
ever exhibited on the Pacific Coast.
SAMPSON—The largest Grizzly Bear ever caught,
weighing 1,510 pounds.
LADY WASHINGTON—With her cub, weighing 1,000 pounds.
BENJAMIN FRANKLIN—King of the Forest.
Two young WHITE BEARS from the Rocky Mountains.
ELK, LION, TIGER, PANTHER, DEER—
and numberless small animals.
Open from 10 A.M. to 12 P.M.
From 2 to 5 P.M., and from 7 to 10 P.M.

A word picture of the museum is provided by Hittell (1911: ix–x):

Descending the stairway, I found a remarkable spectacle. The basement was a large one but with a low ceiling, and dark and dingy in appearance. In the middle, chained to the floor, were two large grizzly bears, . . . Benjamin Franklin and Lady Washington. They were pacing restlessly in circles some ten feet in diameter, their chains being about five feet long, and occasionally rearing up, rattling their irons, and reversing their direction. Not far off on one side, likewise fastened with chains, were seven other bears, several of them young grizzlies, three or four black bears, and one a cinnamon. Near the front of the apartment was an open stall, in which were haltered two large elks. Further back was a row of cages, containing cougars and other California animals. There were also a few eagles and other birds. At the rear, in a very large iron cage, was the monster grizzly Samson. He was an immense creature weighing some three-quarters of a ton; and from his look and actions, as well as from the care taken to rail him off from spectators, it was evident that he was not to be approached to closely. . . .

John Dean Caton (Wright, 1909 : 44) said he first looked upon Hittell's account of Adams as either an entertaining romance or much embellished. When in San Francisco during the 1880's, however, he met reliable persons who had known Adams well and had seen him passing through the city followed by a troop of these monstrous beasts which were entirely docile and oblivious to the crowds of children and yelping dogs attending his passage along the streets.

After slightly more than two months, the "California Menagerie" was moved to the California Exchange at Clay and Kearny streets and on December 8 was reopened as the "Pacific Museum." By January 16, 1857, a man named Wirsen was exhibiting snakes in the museum, and Adams had added monkeys and provided "a fine brass band" that played every evening. The *Bulletin* of January 27 reports that "a very curious bird has lately been procured . . . called the 'Race Bird,' and it is said to excel a horse in running. It is about the size of a pullet, very spry and a native of California."[3]

Additions in mid-March included a sea leopard (leopard seal?) in a reservoir, a "Golden Bear," and "Mr. Overton's Steam Wagon." The size of the daily advertisement was reduced, but the admission was raised to 50 cents, and "a great number of persons attended." The next month a baboon and a

---

[3] Obviously the road-runner (*Geococcyx californianus*).

South American vulture were added, together with other animals not specified. At the end of April three grizzly cubs about
four weeks old were received from the mountains; then a young
sea lion and a gopher joined the exhibit. More bears came in July.

The big grizzlies, though, continued to be the main attraction. In those days, stories about the ferocity of bears then
roaming the ranching and mining country of California were
commonplace, and doubtless the San Franciscans were eager to
take advantage of the opportunity to see grizzlies for themselves.
News items such as the following helped to whet their interest:

> The grizzly bear stories, which appear from day to day in the
> newspapers, naturally create some curiosity among reasonable be-
> ings to see what sort of animal a grizzly is. The creature can be
> seen to perfection at the Pacific Museum. Considerable noise has
> been made lately about killing a bear weighing 1,100 pounds some
> place up-country. Adams has one weighing 1,500 pounds. (San
> Francisco *Bulletin*, July 23, 1857.)
>
> Samson, the big bear at the Pacific Museum, weighs 1,500
> pounds. The soles of his enormous feet are very sensitive, and if
> touched even with a feather, while he is lying down, he draws back
> his legs, his eyes become red and green, and his growl resembles
> distant thunder (*ibid.*, April 23, 1857).
>
> Adams says, that common bears may be intimidated and fright-
> ened; but a real grizzly will fight till his eye-lids will no longer
> wag. He walks the monarch of the mountains. He is something like
> a man, in respect that he will take his tipple, whenever he can get
> a chance; and will get as drunk as a lord, or a sow. Adams is going
> to treat his bears, some night soon, to a steaming bowl of whiskey
> punch, to amuse and instruct his visitors, just as the old Spartans
> used to make their slaves "tight" as a warning to freemen. Adams
> is a public benefactor, and his bears are the tools by which he works.
> It is a pleasure to praise the chap (*ibid.*, Oct. 9, 1857).

J. S. Newberry (1857 : 48), a zoölogist on the Pacific Railroad surveys during the mid-1850's, evidently visited Adams'
menagerie, because he wrote:

> These grizzlies were under perfect control, and were knocked
> about entirely without ceremony by the showman, yet unresent-
> ingly, and he would even go so far as to ride upon their back. He
> used to . . . [get] up a wrestling match between the bears, when
> they would tumble one another about with considerable spirit, yet
> usually very good naturedly. The reward which more than any

other stimulated them to effort was tobacco, of which they seemed very fond. If undisturbed, however, they were very lethargic, lying the whole day through, each rolled up into a huge ball of fur, nearly as high as the animal when standing. (Fig. 26.)

During 1857 there were about one hundred short news notes about the museum and the hunter in the *Bulletin*. We are led to infer that the museum proprietor had a constant admirer and advocate in Hittell, who was then on the newspaper's staff, and that Hittell acted as an unofficial public-relations officer for Adams, keeping him constantly before the public by printing such items as:

Adams and his bears will growl and tumble about as usual to-night at the Pacific Museum.

The attendance at the Pacific Museum continues to be large. The reason of this is the fact, that Adams and his collection of California wild animals are no humbug, and because the visitor, at every new visit, finds, at the Museum, some new subject of instruction or amusement.

The wonders of the Pacific Museum are spread forth like a bouquet for those who desire to enjoy them. The most intelligent enjoy the most.

Adams, the Pacific Museum man, has a large baboon and several grizzly bears tied near together. It is a sight to see them growl, snap, spit and splurt at each other.

. . . visitors to the city, by going there, will see one of the institutions of San Francisco.

While the exhibit was in San Francisco, "Adams lived among his animals. He continued to wear buckskin; and when seen on the street, it was almost always in his mountaineer garb. He slept on a buffalo robe or bear-skin, in one corner of his exhibition room or in a small adjoining chamber. . . . he was not a business man and did not save money; so that about the beginning of 1860 . . . he was substantially as poor in purse as when he first came to San Francisco. (Hittell, 1911 : xii.)

Ben Franklin died on January 18, 1858, and was the subject of an extensive unsigned article (possibly by Hittell), perhaps the only bear to have a flowery funeral eulogy in print.

DEATH OF A DISTINGUISHED NATIVE CALIFORNIAN.

Ben Franklin, the grizzly bear, the favorite of the Museum man Adams, the companion for the last three or four years of his various expeditions in the mountains and his sojourns in the cities and towns of California, departed from this mortal existence on Sunday evening, at 10 o'clock. The noble brute, which was captured at the head waters of the Merced river in 1854, had been raised by his master from a cub, and during his life manifested the most indubitable indications of remarkable sagacity and affection. He was ever tame and gentle, and although possessed of the size and strength of a giant among brutes, was in disposition peaceful; rough, it is true, in his playfulness, but always well-disposed. He accompanied his master on hunting expeditions to the Rocky Mountains and through various portions of California, and on two occasions saved his life in long and desperate struggles with savage animals in the wilds. He frequently carried his master's pack, provisions and weapons; frequently shared his blanket and fed from the same loaf.

During the last two years Ben was the "star" animal in Adams' wonderful collection in this city and has spent his time in exhibiting his frame, rearing upon his colossal legs and shaking his immense chains, for the amusement and edification of spectators and the profit of his master. Many of our readers have often seen him tramping at the length of his chain in the midst of the other animals of the Pacific Museum. One of his eyes was observed to be injured and several scars were to be seen about his head and neck; but they were honorable wounds . . . They were all received in the service of his human friend, protector and master. As might be supposed, his loss has been severely felt by Adams. . . . The old hunter would willingly have lost all the balance of his collection to have saved Ben.

Ben's sickness was some internal disorder, which could not well be understood, and for which no remedy was known. Had any cure been possible, no cost would have been spared to procure it. After an indisposition, however, of about a week's duration, during which time the animal refused to eat or drink, he sank on Sunday evening . . . and died without a struggle. His shaggy skin was cut off yesterday, and all that can be kept of him, his form and outward appearance, will be preserved. His sagacious expression, however, and his great affectionate spirit are gone forever. Alas, poor brute! (San Francisco *Evening Bulletin*, Jan. 19, 1858.)

By August, 1858, a second story had been added to the Pacific Museum for a large exhibit of wax works, and early in May, 1859, an amphitheater and circus ring were established in the second story for equestrian entertainment. The strain on Adams' finances was evidently too great, for on May 11 Abel Guy and James A. McDougall filed suit for $1,000 against him for rent due since January 1 at $250 per month. The "attachment was issued and levied . . . upon the bears, sea lions, elks, panthers, nondescript and other wonders." The advertisement of the amphitheater appeared for only five days (May 9–13, 1859), after which the *Bulletin* printed no further items about the man and his museum.

Adams left the California scene on January 7, 1860, when he took passage with his menagerie on the *Golden Fleece*, sailing for New York. Fortunately, the prime showman of California met immediately upon his arrival the arch showman of New York, Phineas T. Barnum, a chronicler of note who has provided, in successive editions of his autobiography, vivid accounts of himself and those who participated in his lifelong years of public exhibitions. In 1860, after Barnum had reëstablished himself as proprietor of the "American Museum," he wrote:

Among the first to give me a call, with attractions sure to prove a success, was James C. Adams, of hard-earned, grizzly bear fame. This extraordinary man was eminently what is called "a character." "Grizzly Adams" . . . was brave, and with his bravery there was enough of the romantic in his nature to make him a real hero. For many years a hunter and trapper in the Rocky and Sierra Nevada Mountains, he acquired a recklessness, which, added to his natural invincible courage, rendered him one of the most striking men of the age, and he was emphatically a man of pluck. A month after I had re-purchased the Museum, he arrived in New York with his famous collection of California animals, captured by himself, consisting of twenty or thirty immense grizzly bears, at the head of which stood "Old Sampson," together with several wolves, half a dozen different species of California bears, California lions, tigers, buffalo, elk, and "Old Neptune," the great sea-lion from the Pacific.

Old Adams had trained all these monsters so that with him they were as docile as kittens [fig. 26], though many of the most ferocious among them would attack a stranger without hesitation, if he came within their grasp. In fact the training of these animals was no fool's play, as Old Adams learned to his cost, for the terrific blows which he received from time to time, while teaching them "docility," finally cost him his life.

Adams called on me immediately on his arrival in New York.

He was dressed in his hunter's suit of buckskin, trimmed with the skins and bordered with the hanging tails of small Rocky Mountain animals; his cap consisting of the skin of a wolf's head and shoulders, from which depended several tails, and under which appeared his stiff, bushy, gray hair and his long, white, grizzly beard; in fact Old Adams was quite as much of a show as his beasts. They had come around Cape Horn on the clipper ship "Golden Fleece," and a sea voyage of three and a half months had probably not added much to the beauty or neat appearance of the old bear-hunter. During our conversation, Grizzly Adams took off his cap, and showed me the top of his head. His skull was literally broken in. It had on various occasions been struck by the fearful paws of his grizzly students; and the last blow, from the bear called "General Fremont," had laid open his brain so that its workings were plainly visible. I remarked that I thought it was a dangerous wound and might possibly prove fatal.

"Yes," replied Adams, "that will fix me out. It had nearly healed; but old Fremont opened it for me, for the third or fourth time, before I left California, and he did his business so thoroughly, I'm a used-up man. However I reckon I may live six months or a year yet." This was spoken as coolly as if he had been talking about the life of a dog. . . . I had purchased, a week previously, one-half interest in his California menagerie, from a man who had come by way of the Isthmus from California, and who claimed to own an equal interest with Adams in the show. Adams declared that the man had only advanced him some money, and did not possess the right to sell half of the concern. However, the man held a bill of sale for half of the "California Menagerie," and old Adams finally consented to accept me as an equal partner in the speculation, saying that he guessed I could do the managing part, and he would show up the animals. I obtained a canvas tent, and erecting it on the present site of Wallack's Theatre, Adams there opened his novel California Menagerie [fig. 27]. On the morning of opening, a band of music preceded a procession of animal cages down Broadway and up the Bowery, old Adams dressed in his hunting costume, heading the line, with a platform wagon on which were placed three immense grizzly bears, two of which he held by chains, while he was mounted on the back of the largest grizzly, which stood in the centre and was not secured in any manner whatever. This was the bear known as "General Fremont," and so docile had he become, that Adams said he had used him as a pack-bear to carry his cooking and hunting apparatus through the mountains for six months, and had ridden him hundreds of miles. But apparently docile as were many of these animals, there was not one among them that would not occasionally give Adams a sly blow or a sly bite when a good chance offered; hence old Adams was but a wreck of his former self, and expressed pretty nearly the truth when he said:

"Mr. Barnum, I am not the man I was five years ago. Then I felt able to stand the hug of any grizzly living, and was always glad to encounter, single handed, any sort of an animal that dared present himself. But I have been beaten to a jelly, torn almost limb from limb, and nearly chawed up and spit out by these treacherous grizzly bears. However, I am good for a few months yet, and by that time I hope we shall gain enough to make my old woman comfortable, for I have been absent from her some years."

His wife came from Massachusetts to New York and nursed him. Dr. Johns dressed his wounds every day, and not only told Adams he could never recover, but assured his friends, that probably a very few weeks would lay him in his grave. . . . Among the thousands who saw him dressed in his grotesque hunter's suit, and witnessed the seeming vigor with which he performed the savage monsters, beating and whipping them into apparently the most perfect docility, probably not one suspected that this rough, fierce looking, powerful demi-savage, as he appeared to be, was suffering intense pain from his broken skull and fevered system, and that nothing kept him from stretching himself on his death-bed but his most indomitable and extraordinary will. . . .

After the exhibition on Thirteenth Street and Broadway had been open six weeks, the doctor insisted that Adams should sell out his share in the animals and settle up all his worldly affairs, for he assured him that he was growing weaker every day, and his earthly existence must soon terminate. "I shall live a good deal longer than you doctors think for," replied Adams doggedly; and then, seeming after all to realize the truth of the doctor's assertion, he turned to me and said: "Well, Mr. Barnum, you must buy me out." He named his price for his half of the "show," and I accepted his offer. We had arranged to exhibit the bears in Connecticut and Massachusetts during the summer, in connection with a circus, and Adams insisted that I should hire him to travel for the season and exhibit the bears in their curious performances. He offered to go for $60 per week and travelling expenses of himself and wife. I replied that I would gladly engage him as long as he could stand it, but I advised him to give up business and go to his home in Massachusetts; "for," I remarked, "you are growing weaker every day, and at best cannot stand it more than a fortnight."

"What will you give me extra if I will travel and exhibit the bears every day for ten weeks?" added old Adams, eagerly.

"Five hundred dollars," I replied, with a laugh.

"Done!" exclaimed Adams, "I will do it, so draw up an agreement to that effect at once. But mind you, draw it payable to my wife, for I may be too weak to attend to business after the ten weeks are up, and if I perform my part of the contract, I want her to get the $500 without any trouble."

I drew up a contract to pay him $60 per week for his services, and if he continued to exhibit the bears for ten consecutive weeks I was then to hand him, or his wife, $500 extra. . . .

The "show" started off in a few days, and at the end of a fortnight I met it at Hartford, Connecticut.

"Well," said I, "Adams, you seem to stand it pretty well. I hope you and your wife are comfortable?"

"Yes," he replied, with a laugh; "and you may as well try to be comfortable, too, for your $500 is a goner."

"All right," I replied, "I hope you will grow better every day."

But I saw by his pale face and other indications that he was rapidly failing. In three weeks more, I met him again at New Bedford, Massachusetts. It seemed to me, then, that he could not live a week, for his eyes were glassy and his hands trembled, but his pluck was as great as ever.

"This hot weather is pretty bad for me," he said, "but my ten weeks are half expired, and I am good for your $500, and, probably, a month or two longer."

This was said with as much bravado as if he was offering to bet upon a horse-race. I offered to pay him half of the $500 if he would give up and go home; but he peremptorily declined making any compromise whatever. I met him the ninth week in Boston. He had failed considerably since I last saw him, but he still continued to exhibit the bears although he was too weak to lead them in, and he chuckled over his almost certain triumph. . . . I remained with him until the tenth week was finished, and handed him his $500. He took it with a leer of satisfaction, and remarked, that he was sorry I was a teetotaler, for he would like to stand treat!

Just before the menagerie left New York, I had paid $150 for a new hunting suit, made of beaver skins, similar to the one which Adams had worn. This I intended for Herr Driesbach, the animal tamer, who was engaged by me to take the place of Adams, whenever he should be compelled to give up. Adams, on starting from New York, asked me to loan this new dress to him to perform in once in a while in a fair day, where he had a large audience, for his own costume was considerably soiled. I did so, and now when I handed him his $500 he remarked:

"Mr. Barnum, I suppose you are going to give me this new hunting dress?" . . .

"Oh no," I replied, "I got that for your successor who will exhibit the bears to-morrow; besides, you have no possible use for it." . . .

A new idea evidently struck him, for, with a brightened look of satisfaction, he said:

"Now, Barnum, you have made a good thing out of the California menagerie, and so have I; but you will make a heap more. So if you won't give me this new hunter's dress, just draw a little writing, and sign it, saying that I may wear it until I have done with it."

Of course, I knew that in a few days at longest, he would be "done" with this world altogether, and, to gratify him, I cheerfully drew and signed the paper.

"All right, my dear fellow; the longer you live the better I shall like it."

We parted, and he went to Neponset, a small town near Boston, where his wife and daughter lived. He took at once to his bed, and never rose from it again. . . . The fifth day after arriving home, the physician told him he could not live until the next morning. He received the announcement in perfect calmness, and with the most apparent indifference; then, turning to his wife, with a smile he requested her to have him buried in the new hunting suit. "For," said he, "Barnum agreed to let me have it until I have done with it, and I was determined to fix his flint this time. He shall never see that dress again." His wife assured him that his request should be complied with. He then sent for the clergyman and they spent several hours in communing together.

Adams, who, rough and untutored, had nevertheless, a natural eloquence, and often put his thoughts in good language, said to the clergyman, that though he had told some pretty big stories about his bears he had always endeavored to do the straight thing between man and man. "I have attended preaching every day, Sundays and all," said he, "for the last six years. Sometimes an old grizzly gave me the sermon, sometimes it was a panther; often it was the thunder and lightning, the tempest, or the hurricane on the peaks of the Sierra Nevada, or in the gorges of the Rocky Mountains; but whatever preached to me, it always taught me the majesty of the Creator, and revealed to me the undying and unchanging love of our kind Father in heaven. Although I am a pretty rough customer, . . . I fancy my heart is in about the right place, and look with confidence for that rest which I so much need, and which I have never enjoyed upon earth." . . . In another hour his spirit had taken its flight. . . . Almost his last words were: "Won't Barnum open his eyes when he finds I have humbugged him by being buried in his

new hunting dress?" That dress was indeed the shroud in which he
was entombed. (Barnum, 1875 : 529–542.)

In the graveyard at Charlton, Massachusetts (Farquhar,
1947 : 37), there is a stone carved with unmistakable effigies of
a tame bear and a hunter and inscribed:

JOHN ADAMS

DIED

Oct. 25, 1860

Aged 48 yrs.

And silent now the hunter lays
Sleep on, brave tenant of the wild
Great Nature owns her simple child
And Nature's God to whome alone
The secret of the heart is known
In silence whispers that his work is done

## SIERRAN HEADQUARTERS

The location of Adams' camp in the central Sierra Nevada
during 1852–1855 is of interest, because of the extensive activities of the hunter there and because it was near this site that
his big bear "Samson" was taken—a notable feature of Adams'
menagerie and the subject of Charles Nahl's drawing that became the trademark of the *Overland Monthly*. Adams did not
keep a diary, so far as we know, and had to depend on memory
later when recounting his exploits; there is one passing mention
of a diary (*New York Weekly*, July 12, 1860) but no reference
to such a document by Hittell.

In the fall of 1852 Adams "chose out a little valley, on a
northern branch of the Merced River, forty or fifty miles northwest of the famous Yo-Semite . . ." (Hittell, 1860 : 14).[4] He
mentions the giant rocks, the scanty soil, the forest of giant
evergreen trees, ash, oak, and other deciduous trees, many val-

---

[4] For reasons unknown, Hittell in the 1911 edition of his book changed his
statement to read: "on a northern branch of the Merced River, *twenty or thirty
miles northeast* [italics ours] of the famous Yo-Semite . . ." (1911 : 6). This
would have placed the camp near Tuolumne Meadows, well removed from the
Merced River and other places mentiond by Adams, and in a totally different
sort of plant cover. Elsewhere in the 1911 edition mention of the Merced River
as the site of the camp was usually changed to read Tuolumne.

leys affording abundant herbage and beautiful meadows, the canyons crowded with chaparral including "manzanita, juniper, laurel, whortleberry, and . . . many vines and weeds" (*ibid.*, p. 14).

Samson was trapped in the general vicinity of the camp "in the midst of a heavy white [sugar] pine forest, the trees ranging from one foot to four feet in diameter" (*New York Weekly*, May 31, 1860). This would mean that it was in the Transition Life Zone of biologists, where there is a good deal of snow, as mentioned by Adams.

The two editions of Hittell's book differ somewhat in the location of the camp; the earlier (pp. 186, 296), says that Adams returned in 1853 and 1854 to his old camping ground at the "headwaters of the Merced River," whereas the later (pp. 178, 288) substitutes "headwaters of the Tuolumne and Mercedes Rivers."

Adams frequently referred to visiting "Howard's Ranch," where he kept equipment, supplies, and live specimens and where he obtained draft animals. The location is not specified in the Hittell account, but according to the *New York Weekly* (July 19, 1860), the ranch was "in the San Joaquin Valley, some ten or twelve miles from Mariposa . . . On my way, I stopped at a Spanish settlement called Arnitus" (Hornitos, Mariposa County). A recent book (Collins, 1949) shows on a map (end papers) a Howard's ranch south of the Merced River against the first foothills, a place that agrees with Adams' statement quoted above. A crossing on the Merced River known early as Howard's and also as Belt's Ferry is the site of the town of Merced Falls. These places are all within reasonable distance of the "little valley" chosen by Adams in 1852 for his mountain camp. In 1854 (see Wheat, 1942 : map no. 257, State of California) a road from Mariposa led through Mount Ophir, Quartzburg, Howard's Ferry (Merced River), Snelling's, Dickinson's Ferry (Tuolumne River), Heath and Emory's Ferry (Stanislaus River), and on to Stockton.

In addition, Adams mentions visiting the town of Sonora on several occasions. Once, returning from Sonora, he killed a fat buck within a mile of Strawberry Ranch, late in the afternoon (Hittell, 1860 : 191), and next day reached his camp. He also mentions hunting at Bell's Meadows "four or five miles from camp." There are two places of similar name today near the road from Sonora to Sonora Pass: Strawberry and Belle Meadow (U.S.G.S. Dardanelles Quadrangle). These are in the Stanislaus and Tuolumne drainages, respectively; but the hunter never mentions the Stanislaus and often names the Merced in speaking of his headquarters (*ibid.*, pp. 189, 220).

After capturing Samson, it being midwinter, Adams moved his camp "to the mouth of the Merced River," taking with him all the "camp fixtures, stock, and animals, with the exception of Samson . . ." (*ibid.*, p. 304). The 1911 edition (p. 296) reads "to the Merced River below Yosemite." From this camp he made a trip to Hornitos. In the spring of 1854, while in his mountain camp, he was joined by a Mr. Solon of Sonora. The two men, traveling southeastward with a horse and mules, after a "three day trip" reached the brink of the Yosemite Valley (p. 197).

Reviewing all the evidence, we think that Adams' principal exploits took place between the Merced and Tuolumne rivers, to the west and a little north of Yosemite Valley. His base camp may have been in Mariposa County near Pilot Peak, between Bower Cave and "Big Grizzly Flat" (? named for Samson), which is about three miles south of Hazel Green (U.S.G.S. Yosemite Quadrangle, surveyed 1891–1896).

## THE LITERARY RECORD

The principal volume dealing extensively with grizzlies in California is that by Theodore H. Hittell entitled *The Adventures of James Capen Adams, Mountaineer and Grizzly Bear Hunter, of California.*[5] The book appeared in 1860 in two editions, one at San Francisco published by Towne and Bacon, the other at Boston by Crosby, Nichols, Lee and Company. The San Francisco book was reprinted in 1861 and again in 1867. A new edition was issued in 1911 by Charles Scribners' Sons; it differs from the earlier volume mainly in containing an introduc-

---

[5] The original manuscript was in the Sutro Branch, California State Library until 1924 and then was returned to the Hittell family. It was not found in the effects of Carlos Hittell; Mrs. Franklin Hittell thought that the family may have disposed of it (letter, Helen M. Bruner, Librarian, Sutro Branch, to California State Library, Sept. 4, 1947).

tion and postscript by Hittell that give some information about Adams and how the book came to be written. There are also some textual changes.[6]

In the fall of 1856, Hittell, then a young reporter on the San Francisco *Daily Evening Bulletin*, found Adams with his newly established "Mountaineer Museum" of wild animals in a basement on Clay Street. Hittell became fascinated by Adams and his exhibit and wrote four articles about the hunter and his experiences (*Bulletin*, Oct. 10, 11, and 17, 1856; Oct. 31, 1857). In the years 1857 to 1859 he had a further series of interviews with Adams, who gave at length his experiences in hunting, capturing, and taming animals in the West, especially in California. In the winter of 1859–60 Hittell wrote the book from the notes he had taken, making the account a narrative in the first person by Adams. The volume was illustrated by twelve excellent woodcuts from originals by Charles Nahl.

Of the two 1860 editions of the book, Farquhar (1947 : 39–40) concludes that the San Francisco edition was printed from type, and that stereotype plates, made in San Francisco, were shipped to Boston for the other edition. The San Francisco volume was mentioned in the *Hesperian* (vol. 4, no. 6, p. 288) in August, 1860, and was advertised in *Hutching's California Magazine* (vol. 5, no. 2) of the same month (Anon., 1860b). A copy of the Boston printing is inscribed "September 29, 1860."

The subject matter pertaining to bears and to Adams in Hittell's book is probably more reliable than the New York parts of the literary record mentioned hereafter. Hittell lived in California; he doubtless had and used opportunities to learn from other local people about grizzlies, and his visits and conversations with Adams extended over a long period—the autumn of 1856 and from July, 1857, until December, 1859. By contrast, the subject matter for the magazine articles (1860–1861) must have been obtained during Adams' rather brief stay in New York, about two or three months at the most.

Besides the book and newspaper articles there were fully one hundred short notes in the San Francisco *Bulletin* between

---

[6] Comparison of the text in the two editions shows differences in the location of Adams' Sierran camp (see "Sierran Headquarters," above).

1856 and 1859 about the man and his museum. In 1860, Adams and his menagerie moved to New York (see below). Soon after his arrival, the *New York Weekly*,[7] "a journal of useful knowledge, romance and amusement," then in its fifteenth volume, began a series of articles about Adams. The first, in the issue for May 31, 1860, was preceded by the following editorial:

There are few who do not feel interested in . . . a good bear story; and the taste for narratives of this kind has often led to the manufacture of bogus incidents . . . We have therefore resolved to give a series of genuine, matter-of-fact sketches, prepared by one whose own experience makes him every way qualified for the task. We have engaged Mr. JAMES C. ADAMS, the celebrated California hunter, and "wild man," who has succeeded in capturing animals enough himself for a good menagerie, to relate his experiences and adventures in the forest during the period he was engaged in getting up his California menagerie, now on exhibition in Thirteenth Street, between Broadway and the Bowery. They will appear in a series of short [!] sketches, which will be replete with fact, thrilling incident and instruction. . . .

The series was headed "Wild Sports in the Far West, by James C. Adams" and comprised fifteen articles, as follows:

1. How I captured the grizzly bear, Samson. 2 woodcuts. (Vol. 15, no. 27, p. 8, May 31, 1860.)
2. How I captured the grizzly bear, "Funny Joe." 2 figs. (June 7.)
3. A dead companion and a hunter's funeral. 2 figs. (June 14.)
4. A deer hunt and its consequences. 2 figs. (June 21.)
5. Chased by a grizzly—Can a man run as fast as a bear? 2 figs. (June 28.)
[6]. Exploring a cave—Unpleasant meeting with a California tiger. 2 figs. (July 5.)
7. How I was scalped by a she grizzly. 2 figs. (July 12.)
8. A true heroine—A huge he-grizzly in the kitchen—How the wife and two children escape and kill the grizzly. 2 figs. (July 19.)
9. A night with an old "brave"—His terrible encounter with a wounded grizzly. 1 fig. (July 26.)

---

[7] Francis P. Farquhar found four of these articles in the Coe Library, Yale University, and kindly told us of them; we then located the file of the *Weekly* in the Library of Congress and from it obtained the titles and later a microfilm of the fifteen articles.

[10]. A snowstorm in the mountains—Five days under a snow-drift! 1 fig. (Aug. 2.)

11. Hunting sea lions. 1 fig. (Aug. 9.) [The terminal paragraph an adieu by Adams.]

12. An attack by gold fever—The cure—How I captured two golden bears by approaching Nature on her blind side. 1 fig. (Sept. 27.)

13. I start for Correll [Corral] Hollow—Fever and ague attacks me on the road, and is put to flight by the Spanish remedy —The horrors of a night followed by a beautiful morning. 1 fig. (Oct. 4.)

14. My strange companion—Pursuing a grizzly—Returns the compliment—A jump for life. 1 fig. (Vol. 16, no. 3, Dec. 13, 1860.)

[15]. How I made several additions to my menagerie, and how I came to be a Showman, with after reflections. (Vol. 16, no. 14, Feb. 28, 1861.)

The editorial says that Adams was engaged "to relate his experiences," and we infer that someone else transcribed his accounts, although Adams is more than once listed among other "authors" contributing to the *New York Weekly* during the period when the articles were appearing. Their publication continued after his death. They may have been produced in the same manner as Hittell's book. When Adams left San Francisco (on January 7) Hittell was either still at work on his manuscript, or his book at most was in early stages of production. It seems unlikely that Adams could have had proofs of the volume that was to appear in San Francisco during the summer of 1860. We incline to the belief that Adams had a ghost writer. The wood-cuts are *de novo*, without signature; that of Adams and his pet (? Lady Washington) could have been sketched from life, but the illustration of Samson's trap and cage obviously was the re-sult of collaboration between Adams and his artist (fig. 31).

Finally, there are at least four pamphlets about Adams, all now scarce, about which little is known (fig. 28).

Life / of / J. C. Adams, / known as / [figure of Adams and one of his pet bears; see *New York Weekly*, above] / Old Adams, / Old Grizzly Adams, / containing a truthful account of his / bear hunts, fights with grizzly bears, / hairbreadth escapes, / in the Rocky and Nevada Mountains, and the / wilds of the Pacific Coast. / — / New York, 1860. / — / Price, Ten Cents. [53 pp. 16.3 cm.]

The copyright, for 1860, is by J. C. Adams. Page one is headed: "The / hair-breadth escapes / and / adventures of 'Grizzly Adams,' / in / catching and conquering the wild / animals included in his Cali-/ fornia menagerie. / — / written by himself." We have seen the copy of this pamphlet in the Huntington Library. The text is somewhat parallel in subject matter to the Hittell volume but with differences in phrasing, emphasis, and some details. Farquhar (1947 : 9) says it was issued in connection with the [New England] tour of the California Menagerie under Barnum.

Adams, J. F. C. [called Bruin.] 1874. Old Grizzly, the beartamer. New York: Beadle and Adams. 102 pp. 16 cm. Beadle's pocket novels.

Substantially the same text, condensed and with a new title, appeared in 1882 as *Old Grizzly and His Pets* (15 pp., 29 cm., Beadle's half dime library). This was designated as the "fifth edition" and as Vol. X, No. 247, of this series published by Beadle and Adams.

Adams, J. F. C. [called Bruin] 1882. The lost hunters; or the underground camp. New York. 15 pp. 29 cm. Beadle's half dime library. Prospectors in southern California.

The last two (three) titles are from Cowan and Cowan (1933 : 3).

# 10 *Captive Grizzlies—Some Famous*

Many kinds of wild mammals are indifferent subjects for exhibition because they usually sleep or remain quiet during the day, except when being fed. Bears, like monkeys, are strikingly different from other animals; as inmates of zoölogical gardens or as street or circus performers they have been favorites of mankind for centuries. This is because they display a relatively high degree of native intelligence; they have "inquiring minds" that make them responsive in the presence of human visitors; their mannerisms are often droll, indeed manlike in many respects; and they can be trained to perform in various ways. The European brown bear has long been a prime object for exhibition because of these traits, and its relative the California grizzly was similarly used during the years while it was still common after the American occupation.

Some of the captive bears had been taken as cubs, others after they had become older and free-ranging. Most of them were confined in cages or tethered by a chain or rope in public places. When not molested they evidently behaved well, being tamer and more docile or friendly than many other kinds of "wild beasts." But if taunted by unthinking visitors, they would become ill-tempered and sometimes injure their tormentors. In addition to the usual grizzlies on display during those early days, there were a few of more distinctive character about which records are available.

## MISCELLANEOUS CAPTIVES

As early as 1847, Lyman (1924 : 218) saw a partly tamed grizzly tied by a rope to a tree south of San Jose. One exhibited in Leidesdorff Street in San Francisco in 1853 was said to have "become so accustomed to his keeper as to take small pieces of meat from his fingers quite civilly and gently" (N 23).

Dr. J. S. Newberry, a scientist of the Pacific Railroad surveys, wrote (1857 : 48–49):

> A very beautiful bear, eighteen months old, and weighing nearly five hundred pounds, was confined by a long chain near a slaughter house, in the environs of San Francisco [in 1855]. . . . He had always been well fed, was very fat, and, for a bear, very good natured. Every day some one of the butchers would have a wrestling match with him, into the sport of which he would enter with great zest, yet never evincing anything like ferocity; indeed, to all mankind, he had been, so far, entirely harmless. Not so, however, toward the pigs. For pork he seemed to have a special fondness, and he exhausted all his bearish cunning to draw within the circle, of which his chain was the radius, the vagrant shoats which ranged around the slaughter house. He would leave his food half eaten or untasted, that it might attract the pigs, while he, retreating under the cart to which he was chained, watched their motions with all the silent cunning of a cat. Woe to the unlucky pig which, drawn by the bait, came within that magic circle! he ceased to grow old from that hour.
>
> Like most bears, he was also very fond of sweets, of which a poor laborer, living in the vicinity, had . . . proof. Sometimes the bear would break his slender chain, and range about the place at his own free will, doing no harm, but sometimes frightening people, until he was caught and tied up again. One day, when the bear was at liberty, this poor laborer was passing, just at evening, with . . . a sack of sugar, which he carried on his shoulder . . . He heard a step behind him, to which he paid no attention till he felt violent hands laid on his sack of sugar. Turning round, . . . [he found] himself face to face with a large bear. Of course he was frightened, dropped his bag and ran. When afterwards he returned, having gained courage and assistance, . . . the sack was empty.

A few years later, Lola Montez, an actress of early days, at her cottage in Grass Valley had various animal pets, including "two well-grown grizzlies chained at the front door" (Ayers, 1922 : 98).

Tom Gardner, who had a roadhouse at Lathrop, caught grizzlies in Corral Hollow and exhibited them in a cage. One day a visitor teased a bear and the animal grabbed the man's hat. Gardner went into the cage, took the hat, slapped the bear with it, and walked out. The bear, unaccustomed to such treatment from his master, retired to a corner and whined. (Story from Henry Miller, told by Frank Latta, April 20, 1953.) Again at Lathrop, in 1878, an "uncommonly large" grizzly was caged at the Central Pacific depot where passengers stopped to take their meals. The bear was still there in 1880, being fed peanuts, cakes, and popcorn by children. A few months later, this bear, which had long been in captivity, seized a little girl and tore off her arm; the injury was so severe that she soon died. (Buel, 1887 : 714.)

At Monterey in 1879, trained grizzlies were kept by an elderly man who exhibited them, behind the Pacific House; one bear reportedly played a flute (Mrs. Milly Birks, interview by Mrs. Anne B. Fisher, July 16, 1940).

Grizzlies were of interest to people in other parts of the world also, and some were "exported." The famous hunter Adams not only transported a menagerie including several big bears from San Francisco to New York by sailing vessel in 1860 (see chap. 9), but he successfully shipped live grizzlies and other animals in 1853 from Portland, Oregon, to Boston, Massachusetts (Hittell, 1860 : 184–185).

There is an earlier record of three cubs sent from California to London, which were in the garden of the Zoological Society in Regent's Park before September 7, 1850. The *Illustrated London News* of that date and the issue of October 12, 1850 (Anon., 1850a, 1850b) give somewhat differing details of their capture and transport, but the story is substantially as follows. About the first of June a large female was killed in the Sierra Nevada; in her den were three cubs "so young they could hardly see" (a late birth). A Judge Pacton of the gold region took them to San Francisco, thence to the Isthmus, and by the mail packet *Avon* to Southampton. Across the Isthmus the box containing the bears was carried on men's shoulders. On the ship the cubs "were very docile . . . and were allowed to run loose about the deck. One of them particularly was so tame that it would play and roll about the ship with the boys on board. They grew amazingly after they left Chagres." The present authors agree with the *News*, which said: "we can only marvel at the success attained in this venture in respect to a supply of food appropriate to the needs of young grislys that could not have been very old when captured."

## "SAMSON," THE "HUGE GRIZZLY" OF ADAMS

The capture of cubs, such as the "London trio," was rather commonplace in early California. Much rarer was the taking of large, adult grizzlies, such as "Samson," the "1500-pound grizzly" (fig. 29) trapped by Grizzly Adams in the winter of 1854–1855. Indeed, according to Adams, this was one of the major events in his short but spectacular career (see Hittell, 1860: 300–308; Adams, in *New York Weekly*, May 31, 1860). A trap (fig. 31) was built, not far from Adams' camp "between the waters of the South Fork of the Tuolumne and Mercedes Rivers." Four months after the first baiting of the trap, Adams was awakened one night "as if by the shock of an earthquake." The "unearthly bellowing" of the big bear echoed "heavily through the pines, like the mutterings of thunder." With pine torches and accompanied by his two Indian boys, he hastened to the trap and found the huge animal "taking chips out of the white-pine logs faster than I could have done it with an axe." To distract the bear's attention, Adams thrust a torch in its eyes and then assailed the beast with coals, firebrands, and red-hot irons. After the first day, Adams bedded on top of the cage, sleeping when the bear was quiet. Eight days passed before Samson calmed down; in this time "the bear did not eat ten pounds of meat." Melting snow for drinking water was placed in a wooden trough passed through a hole in the side of the trap.

Leaving his boys and a hunter to watch and feed the bear, Adams took off for Stockton—or Sonora—and obtained an iron cage. To take the cage to the trap, he hired a teamster with a wagon and two yoke of oxen. When the cage and trap were placed together and the doors opened, the bear would not move. Prodding and using firebrands were of no avail. A chain was passed through the cage and into the trap, where Adams, by working from the top, after a day of effort, finally slipped a loop over the bear's head and one leg and shoulder. With oxen attached to the chain, "by dint of burning, punching, and pulling, we snaked him into his new cage, and shut the door . . ." The bear had lost weight during Adams' absence of about two months, because there was a shortage of game for its food. The animal again was enraged, and four days passed before the trip down from the mountains began. Samson was first held at Corral Hollow, where Adams often assembled the animals he captured. Later, the big bear became an important part of the Pacific Menagerie in San Francisco (see chap. 9).

When he was in San Francisco, one observer (Evans, 1873 : 64) reported that

while a museum [Adams'] was being moved from one part of San Francisco to another, Old Samson—who chewed up "Grizzly Adams" once upon a time and rendered him beautiful for life—got out of his cage and took possession of the lower part of the city. A crowd of excited men and boys were soon at his heels, endeavoring to corral him, but for a long time without success. At length, tired of picking up damaged fruit from the gutters, upsetting ash-barrels and swill-barrels, and frightening all the women and children on the street out of their several senses, he took refuge in a livery stable, where he was speedily surrounded and cornered.

In addition to the grizzlies restrained by chains or in cages there were others that lived with people, literally as household companions.

## BOB, THE KENT FAMILY GRIZZLY

When Albert E. Kent of Chicago was on a hunting trip in Mendocino County during 1868, a female grizzly was shot and her two cubs were brought into camp in the saddlebags of a horseman. One cub was so mean and temperamental that it was put to death at an early age. The other, named "Bob," proved more amiable and was taken as a pet to the family home in Chicago.[1] There he was kept in the family stable, presided over by

---

[1] Mills (1919 : 211–216, 221–226) tells of three grizzlies that lived with men in Wyoming. One, a female taken as a cub, played with the men and dogs at a sawmill, usually ate outdoors but sometimes in the mess hall, rode the saw carriage, and also rode with a teamster. She was active the first winter, hibernated the second, and left to live in the wild in her third summer. The other two, of opposite sex, caught as cubs, lived with a ranch family for four years. With five black bears they occupied a room next to the dining room, eating at a special "bear" bench set at the ranch table. After the male had injured a visitor he was deported two hundred miles but returned eight hours before his captors.

the Negro coachman, Daniel. Neither the coachman nor the horses were afraid of him. In time the grizzly became too large for a household pet and was given to Lincoln Zoo. There Albert Kent often visited him and sometimes entered his cage. (Letters, William Kent, Jr., grandson of Albert, to Robert E. Miller, Sept. 25, Oct. 26, 1953.) The two photographs of "Bob" (fig. 30) are the only ones known to exist of a yearling California grizzly.

## GRIZZLY ADAMS' TRAINED COMPANIONS

The notable California hunter Grizzly Adams actually trained two bears to be his constant hunting companions and aides. The story about them is scattered throughout the book by Hittell (1860; 1911) but is brought together here to show more effectively the amazing ways in which these two grizzlies served their master.

In eastern Washington during the summer of 1853 Adams killed a female accompanied by two yearling cubs. Attempts to capture the cubs on foot proving unsuccessful, Adams obtained horses from a neighborhood tribe of Indians and with his companions watched through the night until the yearlings came to drink at a spring. The men then pursued the bears with *reatas* and captured them both, Adams taking the young female after a long chase in open country. This was "Lady Washington." After muzzling and binding her with cords, they took her to camp on a crude Indian cart and chained her to a tree. Adams tried kindly approaches without success, then beat her vigorously until she was exhausted. Soon she permitted him to touch and feed her, and she became so tame that he put her on a longer chain. In time she could be led about camp with a lariat; later she learned to follow Adams. Before long she was taken on a hunting trip and, although wearing a chain, assisted Adams in routing a wild grizzly. With some effort she was taught to carry small burdens. In the autumn she was among the animals that went on foot to Portland, and thence accompanied Adams and his party back "home" to his mountain camp in California. There, "the Lady," Adams, and a small Indian dog wintered together. Having killed a buck at Bell's Meadows, four or five miles from camp, Adams lashed half of the carcass on her back. She stood quietly while he did so, then tried to disengage the load, using her front claws and teeth and then rolling on her back. Persuaded by a stout stick applied by her master, she finally bore the load to camp. Thereafter she transported his equipment and blankets on various occasions. Sometimes she hauled sleds improvised to drag game over the snow to camp.

Toward the spring of 1854, Adams with Lady Washington and a Mr. Solon of Sonora went through Yosemite Valley to the headwaters of the Merced River (? Little Yosemite) and there discovered the den of a female grizzly. The mountaineer killed the parent and crawled into the den. He found two cubs, both males, "which could not have been over a week old, as their eyes, which open in eight or ten days, were still closed." The two hunters named the cubs "General Jackson" and "Ben Franklin," respectively. A greyhound in the party, having recently produced a litter of pups, became foster mother to the grizzly cubs. To prevent injury to the dog, the cubs were provided with buckskin mittens. They suckled for three or four weeks and then were given bruised meat.

By this time Lady Washington was a regular pack animal; with a saddle of green hide resembling a Mexican *aparejo* she would carry loads of up to two hundred pounds. A mountain sheep of about seventy-five pounds and a deer were her burden on one trip back to the camp. For the return toward Adams' mountain headquarters she carried, for a time, five live mountain lion kittens.

In April, 1854, Adams started across the Sierra Nevada with a wagon drawn by oxen and mules; Lady Washington was chained to the hinder axletree while Ben Franklin and a greyhound pup, "Rambler," rode in the wagon. Crossing Nevada, they were occasionally let out to exercise and chase rabbits and squirrels. By the time the party left the Humboldt Mountains, Ben and Rambler were permitted to run during the whole day. The rough ground bruised the bear's feet, but he was fretful if restrained in the wagon. Adams therefore made moccasins with soles of elk hide and uppers of buckskin, sewed with thongs, and bound them tightly to Ben's feet. They remained on for a couple of weeks, until his feet were healed. Later, on several occasions, both tame bears were provided with this kind of footwear. The party reached Salt Lake City on July 3, and shortly went farther into the Rockies.

At one camp Lady Washington was visited several nights by a Rocky Mountain grizzly. The next year, when Adams "& Co." were at Corral Hollow, she gave birth to a cub that grew to be "Fremont," a bear later exhibited in Adams' menagerie. According to Adams, this animal "gave unmistakable evidences, in the form of the body and in the color of the hair, of having the blood of the Rocky Mountain bear in its composition; and I rejoiced that I had exercised so much forbearance toward the Lady's lover, the previous summer east of Salt Lake."

Leaving the Lady and her cub, Adams went hunting with Ben and Rambler at the old Sierra campsite. When they were passing through a chaparral thicket, a huge female grizzly, with three cubs, sprang at Adams, knocked his gun away, and threw him to the ground. Ben attacked the female at the throat, and Rambler seized her thigh; this gave Adams a chance to use his gun and follow with his knife. The female was killed, but both the hunter and Ben carried scars of the fight to the end of their days.

Upon returning to Corral Hollow, Adams organized for a trip to Kern River. As usual the Lady was chained to the wagon axle while Ben and Rambler ran free. At one place, south of Pacheco Pass, the vehicle toppled sideways over a steep decline. The wagon and its contents, the mules, and Lady Washington all were damaged and bruised. The bear had been pitched over with such violence that her nose plowed a furrow in the ground. Adams said ". . . she seemed frightened and snuffed and snorted, and her hair stood on end in great agitation; but I went up and patted her head, and in a few minutes she appeared to be pacified, and licked my hand, as if she understood the affair was only an accident and entirely unintentional." Such was the degree to which Adams had tamed and won the confidence of the big she-bear.

While in the valleys of the inner Coast Range, both Ben and Rambler participated in hunting antelope, but the bear was always outdistanced by the prey. Ben attended his master closely when Adams began skinning a carcass. He was so well trained, that he never presumed to touch anything until I gave it to him; but he had a way of grumbling for food, when hungry, that was irresistible. . . . His perquisites were generally the entrails of the game, of which he was remarkably fond; but as he now had to wait until they were removed, his impatience at last assumed such a pitch, that he got excited, and grumbled more than ordinarily. I resolved to try him a little, and placed the food in such a way as to tempt him; but the faithful fellow continued true to his training, and the meat remained inviolate. Seeing this, I threw his portion to him, and he ate until I almost thought he would burst . . . (Hittell, 1860 : 333.)

When the hunting party turned out on the hot, sandy, waterless plain of the San Joaquin Valley, Ben's feet became blistered by the hot sand; Adams bound pieces of cloth about the bear's paws, but they did no good. The hunter then made a shelter of blankets and boards from the wagon for Ben and Rambler, while he went on in search of water. After dark, Adams returned, and the bear, which had not moved in the interval, drank several quarts of water but still could not travel. Finally, the wagon had to be brought back; Adams and his companion had a task in putting Ben aboard, because he then weighed in the neighborhood of four hundred pounds. They camped at a watering place for two or three days, and Adams made moccasins for Ben. After pouring bear oil into the moccasins as a salve, he bound them tightly in place and put a muzzle on the animal so that it would not tear them off. They served well, and soon the party headed for Tulare Lake.

Upon recovery, Ben assisted in further hunts for wild grizzlies. When a female with cubs was encountered, Adams would try to kill her, and Ben would grab a cub. Sometimes he was too vigorous and fatally injured it, but at other times he held on until his master could secure the cub and add it to his collection.

Adams describes an experience in the Tehachapi Mountains, not far from Fort Tejon, with a jaguar (an animal present in southeastern California a century ago; see Grinnell, 1933 : 114). He built a trap of cottonwood logs in an effort to capture the beast, and he used horses, mules, and Lady Washington to take the logs to the trap site. The grizzly dragged two, about 10 feet long and 6 to 8 inches thick, tied to her saddle by one end, the farther end dragging on the ground.

When Hittell (1911 : x–xi) first visited Adams' menagerie in San Francisco, he found both Lady Washington and Ben Franklin among the several grizzlies present.

Adams seemed to have perfect control over them. He placed his hands upon their jaws and even in their mouths, to show their teeth. He made them rear on their hind legs and walk erect, growl when he ordered them to talk, and perform various tricks. He put them to boxing and wrestling, sometimes with himself, sometimes with each other; and they went through the performance with good nature and great apparent enjoyment of the sport.

One thing especially noteworthy, in addition to the docility of the huge beasts, was the fact that the hair was worn off their backs. Upon my asking the reason, the hunter answered that it was caused by pack-saddles [this is the only allusion to Ben having carried a pack]. . . . Adams said he would show how the bears would carry burdens; and, after loosing Ben Franklin and jumping upon his back, he rode several times around the apartment. He next threw a bag of grain on the animal's back, and the bear carried it as if used to the task.

Dr. Walter K. Fisher (A. B. Fisher, 1945 : 309–310) wrote that Hittell told him in later years of having seen Adams thus ride bear-back in the menagerie in San Francisco.

The foregoing account is the only one we have found describing captive grizzlies that were trained as aides to a hunter afield in early California. One bear, Lady Washington, was with Adams from the summer of 1853 well through the days of his menagerie in San Francisco. Ben Franklin, taken as a small cub in the spring of 1854, lived in captivity until January, 1858. Later, in New York, Adams exhibited another bear trained to ride—"General Fremont" (Barnum, 1875 : 534), Lady Washington's cub of 1855.

These grizzlies and their owner-trainer were remarkable. The bears obviously had a high degree of native intelligence and were amenable to schooling, like their big brown relatives in Europe; and Adams was a keen, energetic, resourceful, and persevering custodian and teacher of the bears. They really worked for him. Together they made a team that entertained all who saw them whether they were traveling through the countryside or exhibiting in cities. They are gone from the California scene and nothing like them will be seen again. They were unique in that brief, surging, and unique period—the gold rush. As to the accuracy of this account, we place reliance in the fact that Theodore Hittell, a "cub" reporter of the 1850's and reporter on Adams, lived to become Judge Hittell, an outstanding and respected jurist and historian of the state.

## "MONARCH," THE LAST CALIFORNIA CAPTIVE

During mission days and for a time after statehood it was possible to obtain a captive grizzly with no great effort. The Californians took grizzlies almost at will for their bear-and-bull fights. The *State Journal* of Sacramento, reporting the sheriff's sale of a local "Mountain Museum" in that city quoted a wild grizzly at $15.50 and a trained one at $20.50 (San Francisco *Daily Evening Bulletin*, Jan. 22, 1858). But as the years went on, the grizzlies were hunted out and killed. Those remaining become more reclusive, and captives were harder to come by. The end of the grizzly era was approaching. The story of the bear that proved to be the last living example of his kind on display became somewhat of an epic.

This is the tale of "Monarch," a rather large male, caught about October, 1889, and caged on public exhibition in San Francisco until his death in 1911. His capture, his behavior while on display, and his one mating are matters of record, although somewhat glossed by the exuberance of newspaper reporting. Finally, Monarch has survived as a museum specimen for more than forty years; he has the distinction of being the only known example of a California grizzly mounted for exhibit (fig. 33).

Allen Kelley, a newspaper reporter with some hunting experience, was commissioned in 1889 by the editor of a San Francisco daily (the *Examiner*) to catch a grizzly. The editor "wanted to present to the city a good specimen of the big California bear, partly because he believed the species was almost extinct, and mainly because the exploit would be unique in journalism and attract attention to his paper." Of the subsequent record, Kelley in his book *Bears I Have Met—and Others* (1903) wrote: "The newspaper's account of the capture of 'Monarch' [although signed by Kelley] was elaborated to suit the exigencies of enterprising journalism, picturesque features were introduced where the editorial judgment dictated, and mere facts, such as the name of the county in which the bear was caught, . . . were distorted beyond recognition." He incidentally commented that "more than one-fourth of Joaquin Miller's 'True Bear Stories' [1900] consist of that newspaper yarn, copied verbatim and without amendment, revision or verification."

According to Kelley's account, he spent about five months without success, first in the mountains of Ventura County near Santa Paula and then in the region of Tehachapi and above Antelope Valley. The latter area was tenanted by a locally famed grizzly called "Pinto," but the bear avoided Kelley's traps. Finally, Kelley learned that a group of Mexicans had taken a bear on Gleason Mountain, in the San Gabriel Mountains, Los Angeles County, and after some dickering he purchased the animal. Kelley's own story,[2] somewhat abbreviated, is as follows:

The bear made furious efforts to escape from the trap. He bit and tore at the logs, hurled his great bulk against the sides, and tried to enlarge every chink that admitted light. Only by unremitting attention with a sharpened stake was he prevented from breaking out. For a full week he raged and refused to touch food that was thrown to him. Then he became exhausted, and the task of securing and removing him from the trap began.

First it was necessary to chain one foreleg. That task lasted from eight in the morning until six o'clock. Much time was wasted in trying to work with the chain between two side logs of the trap. Whenever the bear stepped into the loop on the floor and the chain was drawn tight around his foreleg just above the foot, he pulled it off easily with the other paw. Finally, the chain was let down between the roof logs, the bear stepped into the loop, and the chain was drawn upward, bringing the loop well up toward the shoulder.

After one of the bear's legs was well anchored, it was com-

---

[2] In September, 1899, before Kelley had written his book, Ernest Thompson Seton questioned him in regard to the authenticity of the Monarch account. Later Seton wrote (letter, 1928): "As nearly as I can make out, Kelley bought the captured bear from a Mexican who captured him. It is safe to say that many adventures ascribed to this bear belonged to various and different bears" (Grinnell *et al.*, 1937 : 90). Seton visited Monarch in San Francisco on March 18, 1905, and sketched the bear's cubs. The Kelley account is used here because it gives the most details on the procedures in securing and transporting a trapped grizzly.

paratively easy to introduce chains and ropes between the side logs and secure his other legs. He fought furiously during the whole operation, chewed the chains until he splintered his canine teeth to the stubs, spattered the floor of the trap with bloody froth, and did not give up while he could move limb or jaw.

The next operation was to gag the bear so he could not bite. The trap door was raised, and a billet of wood about 18 inches long was held within his reach, and he promptly seized it. A cord fastened to the stick was quickly wound round his jaws, with turns round the stick on each side, and passed back of his ears and round his neck like a bridle. In this way his jaws were firmly bound to the stick so he could not move them, yet his mouth was open for breathing.

Then one man held the bear's head down by pressing his whole weight on the ends of the gag, and another entered the trap to put a chain collar round the grizzly's neck. This he secured in place with a light chain that was attached to the back of the collar, and, passing it under the bear's armpits and up to his throat, he again made it fast, like a martingale. The collar passed through a swiveled ring on the end of a heavy iron chain. A stout rope was fastened round the bear's loins, and to this another strong chain was attached. Then the gag was removed and the grizzly was ready for his journey down the mountain.

In the morning he was hauled out of the trap and bound to a rough skeleton sled made from a forked limb, very much like the contrivance called by lumbermen a "go-devil." It was difficult to find horses to haul the bear; two teams were terrified, but a third proved tractable. The trip down to the nearest wagon road required four days.

At night the bear was released from the "go-devil" and chained to a tree (fig. 32). So long as the campfire was bright, he would lie still, watching it attentively; but when the fire grew low he would pace restlessly to and fro and tug at the chains, stopping now and then to seize the tree to which he was anchored and test its strength by shaking it with his forepaws. Every morning there was a struggle in tying him to the sled. He became expert in dodging ropes and in seizing them when the loops fell over his legs, so that considerable skill was required to lasso his paws and stretch him out. In the beginning of these contests the grizzly uttered angry growls, but soon became silent and fought with dogged persistency, watching every movement of his foes with alert attention and wasting no energy in aimless struggles. He soon learned to keep his hind feet well under him, and his body close to the ground, leaving only his head and forelegs to be defended from the ropes. He was so adroit and quick in the use of his paws that a dozen men could not get a rope on him while he remained in that posture of defense. But when two or three men grasped the chain round his body and suddenly threw him on his back, so that four legs were in the air at once, the *reatas* flew, and he was soon secured.

Monarch was pretty well worn out when the wagon road was reached. A few days of rest and quiet followed, while a cage was being built. The remainder of the journey to San Francisco, by wagon and railroad, was in a box of inch-and-a-half Oregon pine that had an iron grating at one end. The box was not strong enough to have held him for five minutes had he attacked it as he did the trap, but the chain round his neck seemed to be a "moral influence," and he behaved admirably during the remainder of the trip. (Kelley, 1903 : 39–43.)

The story of Monarch's capture and of his arrival in San Francisco was told in the San Francisco *Examiner* of Sunday, November 3, 1889, in a nine-column illustrated article signed by Allen Kelley. From this we have obtained a few further details. The cage containing the bear was shipped by rail (the place of shipment is not named) via Mojave, and arrived in San Franccisco late in October. From the freight depot at Third and Townsend streets the cage was trucked to Woodward's Gardens, then a popular amusement park on Mission Street. When the travel cage was backed against the exhibit cage and the doors were opened, Monarch refused to change his quarters. Four men on the bear's chain effected the transfer. At first a day-and-night watch was kept; within a few days the bear partly demolished his quarters. A new and stronger cage was made: one of the compartments was lined "with heavy iron of the toughest quality" and strengthened "with bars and angle iron." The *Examiner* editorialized about the bear and other matters in the following words:

The people of San Francisco want a zoological garden in Golden Gate Park . . . The EXAMINER starts the collection by obtaining for the proposed garden a magnificent specimen of the California grizzly, a peculiarly appropriate gift, because the grizzly is emblematic of the State and has a place upon the coat-of-arms of California. . . .

It is many years since a bear of this species has been seen in San Francisco, and the capture of a good specimen was a very difficult thing to achieve. Never before was a reporter sent on so unique an errand . . . and when the managing editor informed reporter Allen Kelly that he was to catch a live grizzly he evinced no surprise, but packed his gripsack, took a check for expenses and went into the wilderness. After an absence of five months he returned with a half-ton specimen of *ursus ferox* and an interesting account of the long hunt and final capture, which the thousands of readers of the EXAMINER can read today.

The value of a zoological collection has been indicated already by the eagerness evinced by artists to make studies of the "Monarch." The grizzly appears in everything emblematic of California, but as the artists have had no authentic model they have introduced the black, the European brown and other varieties of bear, all of which are totally unlike the real grizzly. The EXAMINER has afforded them the opportunity to correct their mistakes and study the characteristics of the California bear. . . .

On the Tuesday following there was a three-column article with sketches of Monarch and six other kinds of bears:

Monarch, the EXAMINER'S big grizzly, received many visitors yesterday, but, having been up all night trying the strength of his new house, he declined to stand up and paid but little attention to the crowd. His chain had been fastened to the bars of his cage with three half hitches and a knot, and the knot was held in place by a piece of wire. During the night he removed the wire, untied all the knots and half-hitches and hauled the chain inside, where nobody could meddle with it. Having the chain all to himself, Monarch was indifferent to his visitors and lazily stretched himself on his back. . . .

He had a good appetite yesterday and got away with a leg of lamb and a lot of bread and apples. He ate a little too heartily and had symptoms of feverishness. To-day he will not get so much food. The best time to see him is when he eats, because he lies down at all other times during the day. He has breakfast at 10 A.M., lunch at 1 P.M. and dinner at 3 P.M.

Monarch still looks travel-worn and thin, but he is brightening up, and when the abrasions of his skin, made by ropes and chains, are healed up and his hair grows again on the bare spots he will be more presentable. His broken teeth trouble him some and it will be some time before he will feel as well as he did before he was caught. . . .

Monarch has a big, intelligent-looking head and a kindly eye, and is not disposed to quarrel with visitors, but he objects to any meddling with his chain, and will not submit to insults. It was necessary yesterday to keep a watchman between the cage and the crowd to prevent people from throwing things at the bear and stirring him up. Monarch is getting along very well . . . but he has had a rough experience, is worn out with fighting and worry, is sore in body and spirit and needs rest. It is a difficult thing to keep alive in captivity a wild bear of his age, and undue excitement might throw him into a fatal fever. If Superintendent Ohnimus succeeds in his efforts to cure the Monarch of his bruises and put him into good condition, he will deserve great credit, and visitors are requested not to make the task more difficult by worrying the captive. No other zoological garden in the world has a California grizzly, and it would be a great loss to the menagerie to be established in the Park if the Monarch should die. (San Francisco *Examiner,* Nov. 5, 1889.)

Then on Saturday the newspaper arranged Monarch's "day at home" in Woodward's Gardens, following the social custom of ladies in those years. At the head of page 1 of the Friday and Saturday issues two coupons were printed:

---

### FREE TICKET.

Cut this out and it will admit any child to WOODWARD'S GARDENS tomorrow (Saturday). Go and see the EXAMINER'S GREAT GRIZZLY "MONARCH".

### FOR CHILDREN ONLY

---

The paper of Sunday, November 10, 1889, said:

Children from the bay region thronged Woodward's Garden on Saturday to see Monarch. By 9 in the morning at least a thousand had passed in the Mission Street gate. The grizzly slept but at 2 o'clock as a keeper approached with food the bear aroused. The food included raw beef, apples, biscuits and other articles. After eating the bear curled up again in the sawdust and slept. An estimated 20,000 persons attended during the day.

Meanwhile, several local artists visited Woodward's Gardens to portray the bear. Among these, Sculptor Rupert Schmidt made a model of Monarch. His comments were printed in the *Examiner* of November 5:

> . . . I have modeled many bears, but never one like this. You see in this design some figures of bears [showing a wax model of decorative capitals]. These were intended to be grizzlies, but you see they have the Roman nose, which is characteristic of the black bear. No other bear that I ever saw had the broad forehead and strong, straight nose of the grizzly. He has a magnificent head . . . I have inquired for grizzlies in zoological gardens all over the world, but never found one before."

Some laymen questioned whether Monarch was really a grizzly, as in the following letter to the editor of the *Examiner:*

> Sir: I read in Sunday's paper about Monarch . . . and this morning I went to Woodward's to see him. I was surprised to see a big black animal in a cage marked as Monarch. Is not a grizzly bear always gray or grizzled? I never heard of a black grizzly bear, though I spent several years in Colorado, where grizzlies were plentiful, and killed several. All were silver tips. Is Monarch really a grizzly?
>
> Thomas E. Powers.

To settle the question definitely, Professor Walter E. Bryant of the California Academy of Sciences, long a student of California mammals, was asked by the *Examiner* for an expert opinion. He examined Monarch very carefully and then studied the other bears at the gardens. When he had completed his investigation and stood once more before Monarch's cage he was asked:

"Well, what is he?"

"He is a true grizzly bear," answered Professor Bryant, and he added, "a mighty big one, too."

"I never before saw one of the animals with as dark a coat as his," he continued, "but that is nothing. The bear is a true grizzly, and has all the characteristics of one. As far as his color is concerned, grizzlies are of all colors; there is almost as much variety in that regard among bears as among dogs . . ." (San Francisco *Examiner*, Nov. 5, 1889.)

According to Kelley (1903 : 44–45), Monarch passed three or four years in a steel cell [presumably at Woodward's Gardens] before he was taken to Golden Gate Park. He soon accepted his confinement; but whenever a keeper approached to clean his cage, Monarch would not permit him to remove any of the shavings used for his bedding unless a fresh supply was in sight. Otherwise he would gather all the bedding in a pile, lie on it, and guard it jealously. When the fresh sack was placed in view he would move elsewhere and permit removal of the old pile.

Little more was recorded in print, so far as we have learned, except for a matrimonial experience. The San Francisco *Chronicle* (Feb. 10, 1903) reported the arrival of a mate for Monarch, a young female "silver-tipped grizzly" from Idaho, "much smaller than the ordinary species." For some time the big male had been "seemingly suffering from an extreme case of ennui. In the center of his pit, near two large rocks he dug a hole, and in this he would lie all day, his head buried in his front paws, which he placed over two rocks. The [Park] Commissioners feared that he was grieving, and that he might die of a broken heart. They decided to get him a mate . . ."

The box containing the female was put down, and she was released into a nearby pit. Then "old Monarch began to tear things up." He loosened a few square yards of earth and tried in other ways to attract her attention. She backed away, the full length of her pit, and was "ferocious," and "as cross as a bear" when a photographer attempted pictures. The next year, however, on June 19, 1904, the two mated. Six months and four days later, on December 23, two cubs were born (see chap. 3).

In all, Monarch lived nearly twenty-two years in San Francisco and was a continuing attraction to thousands of visitors in Golden Gate Park. By May, 1911, he had become decrepit and was killed. His skin was mounted by Vernon Shephard, a San Francisco taxidermist, who used much of the skull in the preparation. The skeleton was buried but was later recovered, cleaned, and placed in the Museum of Vertebrate Zoölogy at Berkeley (no. 24537). Shephard, on January 11, 1912, gave from memory the following measurements of Monarch: weight, 1,127 pounds; tip of nose to end of tail, 7 ft. 4 in.; height at shoulder, 48 inches; pads of feet: front, 14 in. long by 7 in. wide; hind, 12 by 6 in.; girth of neck just behind ears, 46 in. (Grinnell *et. al.*, 1937 : 89–

90). The mounted specimen, all too adequately stuffed (fig. 33), in 1952 measured: total length, 7 ft. 8½ in.; height at shoulders, 3 ft. 5¾ in.; longest foreclaw around outer curve, nearly 5 in.; length of hair on back, about 2 in.

The stuffed "Monarch" was exhibited in the Natural History Department of the M. H. de Young Memorial Museum in San Francisco for many years, then was relegated to storage behind the scenes. In October, 1953, the specimen was transferred to the California Academy of Sciences (no. 10471). Thus, appropriately, the last captive California grizzly is in the keeping of the oldest scientific organization in the state. The specimen, as seen by one of us (T.I.S.) in 1953, was in fairly good condition although one left foreclaw was missing. The pelage was a moderately light brown, possibly having faded from long exposure while exhibited; two extant photographs—one taken about 1899 by L. S. Slevin, and another in the possession of Roy Graves—suggest that the coat was decidedly dark. The skin, poisoned with mercuric bichloride, had not been damaged by insects.

# II  *Grizzly Lore*

Legends and fables tend to accumulate about any animal with which mankind is associated over a period of time. The more conspicuous the animal and the more distinctive its habits, the greater is the accumulation of lore about it. Among primitive human races, the legends and fables become so intimately woven into the fabric of the lives of the people that fact, inference, and fanciful concepts are entangled. In consequence, students of anthropology have difficulty in separating the real and the imaginary. If the contact between the newly arrived Americans and the native Indians of California had not been so disastrously brief, many more bits of Indian lore would have been recorded; as it is, only fragments of this fascinating background have survived.

In lesser degree a similar situation obtains in respect to the Spanish Californians. They delighted in storytelling, but only a small fraction of such lore has been preserved. They talked, but they did not write. Many of the glamorous tales told on the range, about campfires, or at evening gatherings in the haciendas are gone.

The grizzly was so outstanding an element in the life of Indian, Spaniard, and early American in California that it was the motif for many stories. Some of these have been told in previous chapters, and a few more will be added here. Besides the many straightforward accounts of affairs between white men and California grizzlies there are stories that depart slightly from the factual and others that are downright "tall tales." Most of them had some substance of fact but were garnished by errors

258

in observation and interpretation or were given shifts in emphasis, and some undoubtedly were improved upon as they were told and retold.

The naval officer Joseph Warren Revere, in California before the gold rush, tells us:

An old hunter once took me ... to a retired spot on the summit of a mountain, where he assured me the bears were accustomed to resort for the purpose of *dancing*. There was an old and gigantic pine, around whose base a slight hollow was regularly excavated, and the bark of the tree was completely scratched off some distance up, and the wood itself was deeply scarred. My friend assured me, that he had seen in this place a collection of bears, and had carefully approached them, keeping himself concealed, and advancing from the leeward. He solemnly asseverated, that around this tree sat the bears, and that each one of them was approached in turn by a huge old grizzly bear and led to the tree, against which they stood up and moved up and down, as if dancing. This continued until every bear had been led out by the ancient bear, and "it looked," said the hunter, "for all the world, like a lot of gals led out by a feller to dance."

Although this story seems about as tough and indigestible as Cuffee himself, ... the narrator was an honest man, and firmly believed that he had been an eye-witness to this marvellous ursine ball. The old inhabitants of California, of all degrees believe that these animals have resorts where they indulge in this fashionable amusement, and it is said that several of these ursine Tivolis exist in every district where the grizzly bear is found. I would respectfully suggest to the dancing masters to introduce, without delay, the "pas d' ours." (Revere, 1849 : 259–260.)

Available accounts of grizzly behavior provide no clue to the basis of the preceding story. Bears were wont to gather in numbers at good feeding places; the grounds under oaks shedding an abundance of acorns were well-patronized bear restaurants; but presumably the animals were more or less independent when they went off to rest or sleep.

Among other items in the "unnatural" history is a remark "that the grizzly bear is a surgeon, and, when wounded, gathers leaves of the bush called 'greasewood' [*Adenostoma*], and forces them tightly into the wound . . ." (Anon., 1857 : 816). Benjamin D. Wilson (1877, MS) in pursuing a wounded bear found the animal immersed in the mire of a marsh where only its nose showed above the surface. He said, "I have heard told by others that bears have the sagacity to seek the healing of their wounds with application of mud."

Again there is the statement that "when the bear goes into the winter quarters he contrives to stop his fundament with clay, which remains there during the whole winter, nothing passing him while asleep. In the spring the clay comes out, being first softened by a black liquid which oozes from the animal!" (Suckley and Gibbs, 1860 : 120.) This was evidently a common opinion of hunters, since it was told by men who "never had communication with each other." An explanation is possible for this belief: It is known, from recent observations on a black bear (Matson, 1954), that once a bear settles into winter dormancy there is no evacuation from the bowel. After the long period of inactivity a large dark accumulation from the lower colon is passed. This is the mass of "clay" of hunter lore.

The early Spaniards combined a bit of bona fide grizzly natural history into one of the choicest of the tales about California's big bear. Frank Latta of Bakersfield has passed it on to us.

Gervasio Ruiz carried the mail from Los Angeles to Monterey in mission times. After three trips he was worn out, although the horse was not. He tied his steed to an oak on the beach near the Cañada de los Osos and went to sleep. When he awoke he had difficulty in saddling and mounting again. On the beach there was a whale carcass with a large hole in its side where a grizzly was feeding. He rode by the whale, whereupon the grizzly took off and Ruiz' mount insisted on following it up the beach and to the chaparral. Then he returned to the beach. Another bear emerged from the whale, and in turn it was followed. In the end no less than 184 bears came out of the whale. As a consequence of following each in turn, his mount wore a trail nine feet wide and four feet deep (which was not washed out until the exceptionally high tides of 1862). Then Ruiz rode on over the mountains to visit his sister in Santa Barbara. When he dismounted he discovered he had been riding a grizzly bear. The beast had swallowed his horse, even to the hoofs and tail; only the rope tethering the horse had remained to hold the bear. Gervasio was reputedly the champion liar of San Luis Obispo County.

Man's thoughts on the methods of salvation are colored by his experiences on earth, and even the grizzly had an influence in some cases. In the 1840's an American and his Californian companion of the central coast region, both experienced hunters, had made a "drag" of deer or calf entrails and were perched with their guns at night in an oak tree over the bait. The Spaniard dozed off and fell out of the tree, but he hastily climbed back as a bear approached.

"Antonio," said I, in a whisper, "if the bear eats you, look below, as you ascend toward the gates of Paradise."
"Why must I do that, Señor?" whispered the shuddering Antonio.
"Because, like children when they go to bed in the dark, you will see the ghosts of some hundreds of grizzly bears whom you have sent to the lower world ready to lay hold upon your feet; and, thereupon, you will so move St. Peter with admiration of your quickness in climbing the gate, he will let you pass without scruple, for the sake of laughter, among the saints." (Anon., 1857 : 820.)

A few grizzlies of California became known by name because of distinctive features or because of their exploits. In 1858 there was "the big bear of Marin" (N 58); one in the Santiago Mountains of Orange County in 1875 was known as "White Face" (N 86); another southern California bear was called "Old Pinto"; and this or another was "the Ventura sheep killer" (Seton, 1929 : 73).

The most noteworthy of these bears was "Old Clubfoot," sometimes called "Old Reelfoot." Any bear that left part of a foot in a trap would thereafter have a distinctive track; "clubfoot" implies the loss of some or all claws and toes on one foot. The bear that injured J. W. Searles (see chap. 8) in the mountains of southern California, and presumably killed fourteen people, was called "Clubfoot" (Chickering, 1938 : 113–114, 117). Another of the same name killed a man at Independence Lake, Nevada County, in 1874 (N 84); and a third is said to have lived in the mountains of Monterey County. But these were not all.

"Clubfoot," half flesh-and-blood bear and half legend, has one foot in fact and the other in fiction or fantasy. "His history has already become nebulous, and . . . the glamour which is fatal to moderation of statement" has settled about his name (Chase, 1911 : 278). He grew, indeed, to be the protagonist of a tale of Sequoia-like proportions. But first let us summarize the facts and semifacts.

The locale for the most famous "Clubfoot" is toward the Oregon-California boundary. Wright *et al.* (1933 : 121–122) state that the "last" grizzly killed near Crater Lake, Oregon, in the "early 'seventies" was called "Clubfoot" according to Judge Colvig, a local informant. We mention it here merely to clear the record, for we surmise that it is not our bear; Oregon evidently had its own clubfooted grizzly.

The earliest account of a "Clubfoot" in northern California is that from the Red Bluff *News*, reprinted in the Sacramento *Daily Union* for June 21, 1889. In brief, this grizzly had been shot a few days previously near the source of Battle Creek, Tehama County, by a trapper named Hendrix.

This ferocious beast has wandered as a dreaded monarch in that section for the past twenty years and seemed to bear [!] a charmed life. Hundreds of cattle, sheep, hogs and human beings have fallen victims to his appetite during that period and many parties organized for his destruction have returned thinned in ranks and "with hair turned white in a single night, by a passing sight of the dreadful fright," which they vainly sought to destroy. The beast weighed, when dressed, 2300 pounds, which we believe is the largest animal of its species ever seen on the American continent. . . . The bear was in rather poor conditon . . . as old age had clogged his blood somewhat . . .

The principal "Clubfoot" story, however, centers in Siskiyou County, where a large grizzly lacking claws and part of a forefoot was shot in 1890. The skin was stuffed and exhibited; then the specimen disappeared. Real bear, a claw, some bullets, men, and lore are strangely intertwined. No doubt exists that there was a bear—a photograph of the stuffed specimen is extant —but the exploits of local grizzlies for a half century may have been ascribed to the one bear. George R. Schrader, of the U. S. Forest Service at Mount Shasta, California, spent much time gathering bits of the story. His letters to us and to others and an article he wrote in 1946 (Schrader, 1946), together with other materials, are the basis of the following account.

"Clubfoot" first emerged on the page of history when he was alleged to have stampeded horses of Frémont's expedition in southern Oregon during 1846; ten years later he was caught in a steel trap on the Klamath River by the Grieve brothers. When they approached with their dogs, the bear wrenched loose and fled, leaving three claws and part of the right forefoot; this loss caused his foot to turn and leave a distinctive track. Each spring he killed livestock on ranches around Humboldt Bay, and in the autumn he plundered similarly near Pilot Rock, 150 miles and three watersheds distant on the eastern tip of the Siskiyou Mountains. Stockmen posted large bounties, and many traps were set; the bear robbed them of bait but was never recaught. In 1882, a sheepherder, J. D. Williams, had his flock on a hillside; below, David Horn's cattle, led by a big bull, were grazing in a glade. A big bear, presumed to be "Clubfoot," went for a calf and downed its defending mother. The bull charged. The bear was knocked into the brush, but sprang out, battled, and finally seized the bull by the nose and broke its neck.

In April, 1890, a seventeen-year-old boy, Pearl Bean, and his older companion, William A. Wright, were hunting on Camp Creek, a tributary of the upper Klamath River. Tracks of "Clubfoot" were seen in soft snow. An hour of following the trail brought the two hunters in view of the big bear, plowing through snow on the far slope of the creek. At their first shot the grizzly charged toward them; the men fired again and again, using the old 50–70 Sharps rifle, and finally killed the bear.

The skin was removed and stuffed by amateur taxidermists in Ashland, Oregon. The mounted specimen, seen at close range by A. E. Doney, was judged to measure as follows: tip of nose to tail, 8 ft.; height at shoulder, 4 ft. 8 in.; width between ears, 12 in.; between eyes, 8 in. (Only severe overstuffing of a skin without supporting framework could achieve such dimensions.) The records of weight are unverified; Schrader had statements of 1,850 to 2,250 pounds.

Mr. Schrader saw and handled the trap reported to have caught "Clubfoot" in 1856; he was shown one of the three claws; and he saw the gun and ammunition used. He was told that nearly a quart of bullets had been taken from the carcass.

The stuffed skin was placed on exhibition, and here the "trail" of "Clubfoot" becomes both varied and diffuse: (1) He was exhibited up and down the Pacific Coast—by Bean and Wright—then sold in Portland for $500. (2) He was sold to the Native Sons Lodge in San Francisco and was destroyed in the fire of 1906—but recent lodge officers have no record. (3) He was exhibited in the Jordan Museum in San Francisco and destroyed in the fire. (4) He was purchased by a man named Jordan and exhibited in a side-show tent at the World's Columbian Exposition in Chicago in 1893, at 10¢ per admission. (5) After the Chicago episode he was exhibited by "Dr." Jordan in the principal cities of Europe. This was the end of the "Clubfoot" story until quite recently. Then, in 1949, Thomas McHenry, a veteran of World War II, stated he had seen the stuffed bear in a London museum with a plaque reading "Grizzly bear, killed near Hornbrook, California, in 1890 by Wm. A. Wright" (Wright was a great-uncle of McHenry). The young man had been in "a lot of museums," but could not remember where he had seen the bear. Inquiry by the Siskiyou County Board of Supervisors to the British Museum in London brought a reply that no such bear was in that Museum or in the London Museum. Then another overseas veteran reported seeing "Clubfoot" in a Paris museum; but when the United States Embassy inquired of the Museum National d'Histoire Naturelle (which *has* a California grizzly skeleton—see chap. 2) and the Louvre, the reply was "we have no bear." (Sacramento *Bee*, July 2, Nov. 3, Dec. 28, 1949; Feb. 4, Mar. 8, 1950.) The resting place of "Clubfoot" is still unknown.

Half a century ago, while stories about bears in California were still on the tongues of outdoorsmen and in the minds of editors, Allen Kelley (1903) wrote entertainingly, if not entirely factually, a small book on *Bears I Have Met—and Others*. The author states (p. 9) that his stories "were accumulated and written during a quarter century of intermittent wanderings and hunting on the Pacific Slope. . . . these tales illustrate many traits of the bear and at least one trait of the men who hunt him." The "Chronicles of Clubfoot" (*ibid.*, pp. 48–74) is one of the most outstanding, but is more legend or fiction than chronicle.

The most famous bear in the world was, is and will continue to be the gigantic Grizzly known variously on the Pacific Slope as "Old Brin," "Clubfoot," and "Reelfoot." He was first introduced to the public by a mining-camp editor named Townsend, who was nicknamed "Truthful James" in a spirit of playful irony. That was in the seventies. Old Brin was described as a bear of monstrous size, brindled coat, ferocious disposition and evil fame among the hunters of the Sierra. He had been caught in a steel trap and partly crippled by the loss of a toe and other mutilation of a front paw, and his clubfooted track was readily recognizable and served to identify him. Old Brin stood at least five feet high at the shoulder, weighed a ton or more and found no difficulty in carrying away a cow. He seemed to be impervious to bullets, and many hunters who took his trail never returned. A few who had met him and had the luck to escape furnished the formidable details of his description and spread his fame, with the able assistance of Truthful James and other veracious historians of the California and Nevada press.

For several years the clubfooted Grizzly ranged the Sierra Nevada from Lassen County to Mono, invulnerable, invincible and mysterious, and every old hunter in the mountains had an awesome story to tell of the ferocity and uncanny craft of the beast and of his own miraculous escape from the jaws of the bear after shooting enough lead at him to start a smelter. Old Brin was a never-failing recourse of the country editor when the foreman was insistent for copy, and those who undertook to preserve the fame of his exploits in their files scrupulously respected the rights of his discoverer and never permitted any vain-glorious bear hunter to kill him. As one of the early guardians of this incomparable monster, I can bear witness that it was the unwritten law of the journalistic profession that no serious harm should come to the clubfooted bear and he should invariably triumph over his enemies. It was also understood that a specially interesting episode in the career of Old Brin constituted a pre-emption claim to guardianship, and, if acknowledged by the preceding guardian, the claim could not be jumped as long as it was worked with reasonable diligence.

While Old Brin infested Sierra Valley [Plumas County] and vicinity he was my ward, and I regret to say that his conduct was tumultuous and sanguinary in the extreme. . .

Two Italian woodchoppers were slain by the bear. A party of hunters—"about a score"—from the Comstock Lode went out to slay the beast, but they merely brought back tales adding to the "Clubfoot" legend. One hunter of this group took refuge in a tree, and the bear shook the tree several times, finally so hard that the man was catapulted across a ravine—whereupon the bear departed! Again, the party was playing cards in a tent, which the bear invaded. As the hunters dashed out, the tent fell on the bear, and a kerosene stove set fire to the tent and its contents; the poker stakes, of gold and silver coin, were melted by the heat.

Kelley then picks up the core of the legend again:

So long as Old Brin was under the guardianship of his early friends, it was certain that no serious harm would come to him and that no hunter would be permitted to boast of having conquered him. But a later breed of journalistic historians, having no reverence for the traditions of the craft—and no regard for the truth, sprang up, and the slaughter of the clubfooted grizzly began. His range was extended "from Siskiyou to San Diego, from the Sierra to the sea," and he was encountered by mighty hunters in every county in California and killed in most of them.

Old Clubfoot's first fatal misadventure was in Sikiyou, where he was caught in a trap and shot by two intrepid men, who stuffed his skin and sent it to San Francisco for exhibition at a fair. He had degenerated to a mangy, yellow beast of about 500 pounds in weight, with a coat like a wornout doormat, and but for a card labeling him as "Old Reelfoot," and exploiting the prowess of his slayers, his old friends never would have known him.

Clubfoot's first reincarnation took place in Ventura, about 600 miles from the scene of his death. He appeared in a sheep camp at night, sending the herders up the tallest trees in terror, and scattering the flock all over a wide-spreading mountain.

After much search, all but some fifty-odd sheep were recovered. The superintendent of the ranch found that the trails of the missing sheep led to a hidden *cienega* of a few acres bounded on three sides by steep walls.

When the superintendent reached the entrance to this sunken meadow, an opening perhaps thirty yards wide, he noticed a well worn path across it from wall to wall, and a glance told him that the path had been beaten by a bear pacing to and fro. . . . the footprints were large and . . . one paw of the bear was malformed. Old Clubfoot without doubt. . . .

The frightened band of sheep, fleeing blindly before the bear, had been driven by chance or by design into this natural trap, and the wily old bear had mounted guard at the entrance and paced his beat. . . . When he wanted mutton he caught a fat sheep, carried it to his sentry beat and killed and ate it there . . . The grass in the cienega was thick and green, and there was enough seepage of water to furnish drink for the flock. So the provident bear had several months' [!] supply of mutton on the hoof, penned up and growing fat in his private storehouse, and his trail across the entrance was as good as a five-barred gate. . . ."

In the Ventura episode, "Clubfoot" was done in by the superintendent, who positioned an old muzzle-loading musket as a set-gun, filled it with a heavy overload of powder and a huge slug of lead, and placed a bait of fresh pork to tempt "a mutton-sated bear." "At midnight there was a muffled roar." Next day the bear was found a half mile away. "The big slug from the musket had entered his throat and traversed him from stem to stern . . . It is the tradition of the mountain that the ursine shepherd was none other than Old Clubfoot . . ."

His next recrudescence was in Old Tuolumne, where he forgot former experiences with steel traps and set his foot into the jaws of one placed in his way by vindictive cattlemen. Attached to the chain of the trap was a heavy pine chunk, and Old Clubfoot dragged the clog for many miles, leaving through the brush a trail easily followed, and lay down to rest in a thicket growing among a huddle of rocks.

Two hunters took up the trail, and the more inexperienced of the pair came close to the bear. The animal reared, and swinging trap, chain, and clog, sent the hunter over the cliffside into Hetch-Hetchy Valley. His companion pumped rifle bullets into the bear until it also fell down the rough sides of the valley.

In his old age, the big brindled bear grew weary of being killed and resurrected . . . Little, ordinary no-account bears had personated him and got themselves killed under false pretenses from one end of the Sierra to the other . . .

Kelley finally disposed of the bear by having him break into a country doctor's office, steal a bottle of chloroform, find a bucket and sponge, and commit suicide by clapping the bucket with the chloroform-soaked sponge over his head.

# 12 The California Grizzly as an Emblem

Naturalists and the many other persons who seek to preserve the native fauna regret that our most spectacular and interesting wild animal has been exterminated. The passing of the California grizzly—a unique race—is an outstanding example of the ruthless destruction of our wilderness heritage. If the hunters of seventy-five years ago had exercised restraint, there might be places today where persons who are sufficiently intrepid and curious could see and study the behavior of the greatest of bears in its native habitat. But such a situation obviously would be impractical. Because of its ferocity when aroused, the grizzly would be incompatible in the second most populous commonwealth of the United States of America.

Sorry as we are at the extermination of the California grizzly, its passing was an inevitable and necessary accompaniment of human occupancy of the land. Although none of us now can ever see the live animal, we take pride that the bear which was so much a part of our early history, carries on as the state's emblematic representative. As Chester Stock (1936:30) appropriately said:

Probably no other animal is so intimately connected with the early occupation and history of California as the bear. Extolled both in fiction and in fact, the part it has played in the lives of the pioneers and early emigrants is commemorated . . . by its emblematic representation on the Bear Flag and on the Great Seal of California.

## THE GREAT SEAL

On June 14, 1846, a party of American settlers, described by one of the band as being "very greasy" and "about as rough a looking set of men as one could well imagine" (Caughey, 1940 : 276), broke into the Sonoma estate of General Mariano Guadalupe Vallejo, who had long advocated annexation of California by the United States, took him prisoner, and prematurely proclaimed independence from Mexico. Because they had chosen the grizzly as an emblem, these rebellious and overly enthusiastic settlers were known as "Bear Men."

Three years later, after California had been acquired by the United States, "Bear Men" and General Vallejo were among the delegates to the Constitutional Convention assembled at Monterey to form a state government. One of the tasks that occasioned disagreement and oratory was the selection of a state seal. Most of the designs submitted by the delegates, none of whom were artists, were too "ludicrous" to be considered. Fortunately, Major Robert Shelden Garnett, U. S. Army, " 'a gentleman of modest demeanor, [who] excelled in the use of his pencil,' heard of the impasse and executed a seal embodying the emblems discussed by the convention." (Kilian, 1949 : 7.) In his design—under an arc of thirty-one stars (the number of states when California was added to the Union)—there were prominent representations of a miner (who has been unkindly called a gravedigger), of the Goddess Minerva, and of a grizzly eating grapes. The bear took the place of the lion, unicorn, or eagle of the usual seals and coats of arms. (Bowman, 1950 : 167.)

With typical frontier gallantry, none of the delegates objected to Minerva; but the grizzly, enthusiastically extolled and defended by the "Bear Men," became the center of a storm of controversy. Jacob R. Snyder, one of the participants in the insurrection at Sonoma, was especially vociferous in his praises of the bear. (Soulé *et al.*, 1855 : 805.) O. M. Wozencraft, however, who had been the first to suggest adoption of a seal, was more interested in advertising the riches of the land than its fauna, and he proposed to strike out the feasting grizzly in favor of bags of gold and bales of merchandise (Bowman, 1950 : 157).

This suggestion was not popular. Then, General Vallejo, still smarting from his incarceration during the Sonoma revolt, rose to let it be known that he would not accept an unrestrained bear. He argued that if the beast were to remain on the seal it must be firmly secured by a *reata* in the hands of a Spanish Californian vaquero. (Hittell, 1898, 2 : 773.)

When the matter came to a vote, the grizzly triumphed over the General by a vote of 21 to 16, and to this day he stands unrestrained by the side of Minerva. In fact, he has moved into a more prominent position than that allowed him by the Convention of 1849.

The description for design of the seal, as entered in the Journal of the Convention (see Bowman, 1950 : 157–158) reads:

Around the bevel of the ring are represented thirty-one stars, being the number of states of which the union will consist upon the admission of California.

The foreground figure represents the Goddess Minerva having sprung full grown from the brain of Jupiter. She is introduced as a type of the political birth of the State of California without having gone through the probation of a Territory. At her feet crouches a grisley bear feeding upon clusters from a grapevine emblematic of the peculiar characteristics of the country. A miner is engaged with a rocker and bowl at his side, illustrating the golden wealth of the Sacramento upon whose waters are seen shipping typical of commercial greatness and the snow-clad peaks of the Sierra Nevada make up the background while above is the Greek motto "Eureka" (I have found it) applying either to the principle involved in the admission of the State or the success of the miner at work.

This phrasing, written by Caleb Lyon three days before the design was presented to the Convention, is still in use except for some changes in punctuation and spelling, and in the replacement of "crouches" by "stands." Lyon evidently used the former term to indicate position rather than posture, for, as Bowman (*ibid.*, p. 158) pointed out, "naturalists seem unable to determine the posture of a crouching bear especially when eating."

The Great Seal was used first by the governor on December 5, 1849—and shortly afterward by the secretary of state. It is the mark of official authority and its pattern might be assumed to be permanent. Yet no less than four major designs were employed from the time of its original adoption in 1849 until the seal was standardized in 1937.

The original design shows the bear with lowered head turned slightly to the left and munching a cluster of grapes. The margin of the seal cuts off the lower half of his front legs and most of his hind ones. In the second version, which appeared in 1883, the bear has turned his head to the right, giving a better

The Great Seal of the State of California, 1937 version, as corrected.

view of his profile. In the 1891, or third version, having forgotten the instructions entered in the Journal of the Constitutional Convention that he should be "feeding upon clusters from a grapevine," the bear has stopped eating the fruit, to raise his head. Apparently, also, something in the distance has attracted his attention, because he has moved forward so that most of his feet are in view; it has been suggested that he is trying to sniff at the miner's rocker. At any rate, this version of the seal was the basis for the fourth design, adopted in 1937 (*Calif. Stats.*, 1937 : 1197, chap. 380, sec. 2; *Government Code*, sec. 400). On all seals the grizzly is a dumpy figure, poorly representative of his race.

Besides the four patterns used by the governor and the secretary of state down through the years, many purported replicas but actual variants have been employed by the state printer. This diversity came to attention in 1936 when the State Employees' Association, wishing to have blotters printed with a replica of the seal, found that practically every department had a version differing from that in the office of the secretary of state. Indeed, as early as 1855, the commission of a captain in the National Guard bore two seals, one with the bear sleeping soundly and the other with the bear standing!

These discrepancies prompted the San Francisco *Recorder* of February 4, 1937, to remark:

> What is the bear doing? Well that depends on the version of the Seal that you happen to study. In some versions the rascal is lying down, apparently sound asleep—hibernating al fresco, so to speak—in others he is standing up. In some he appears to be smiling, in others growling. And in all versions he looks something like a cross between a wolf and a boar, though he is probably a grizzly.
>
> We have consulted the California Blue Books of various years, and find a delightful impartiality in the matter of this bear on the Seal. The preference seems to incline toward drawing the dear old beast "couchant," as the pundits of heraldry would put it, but in a great many examples we have found him "statant," that is to say, standing at gaze, though what he is gazing at remains a mystery . . .

Finally, in 1937 the state legislature decided to standardize the seal and commissioned an artist to draw it "for the last time" (see p. 271). After its adoption, this "definitive" version was stamped on official documents and thousands of school textbooks before a ten-year-old boy discovered that it had only thirty instead of thirty-one stars. (Kilian, 1949 : 7).

## THE BEAR FLAG

The American settlers who revolted against Mexican rule at Sonoma in 1846 had to fashion a banner to emblazon their cause. They could not use "Old Glory," for the United States was not yet at war with Mexico. But the insurgents were determined to show their ties to it by including a star and a red stripe on their flag. Then a proposal that a representation of the California grizzly be added was accepted with enthusiasm. The rebels felt that the strongest animal in the country was a fitting symbol with which to identify themselves. As one of the participants, Benjamin Dewell, many years later told the historian Hubert Howe Bancroft, "A bear always stands its ground, and as long as the stars shine, we stand for the cause." Accordingly, grizzly, star, red stripe, and the words "California Republic" were incorporated into the crude, home-made flag that was run up the pole at Sonoma on June 14, 1846, replacing the eagle of Mexico. John B. Weller, later governor of the state, referred to it as the "grizzly bear flag" in a letter dated September 8, 1851. (Hussey, 1952 : 215.)

Apparently the original flag was made of white cotton with a strip of red flannel at the lower border. The figures and lettering were painted with blackberry juice, brick dust, and oil (*ibid.*, p. 209); the work was supposedly that of William Todd, a brother of Mrs. Abraham Lincoln (Los Angeles *Express*, Jan. 11, 1878). His rendition of the bear was so crude that nearly everyone maintained that it resembled a pig. Todd, a Midwesterner, reputedly had artistic abilities and could have done a better job of portraying a grizzly. Ed Mannion (Sonoma *Index Tribune*, April 27, 1953) has recently suggested that perhaps, out of loyalty to his prairie roots, the painter cleverly sabotaged the bear in favor of the hog! Regardless of the porcine aspect of his sketch, the insurrectionists definitely considered it to be a representation of a grizzly and proudly called themselves "Bear Men."

Many persons were involved in the incident at Sonoma. Their recollections and writings might be expected to yield accounts of the making of the flag that would be in substantial agreement. Unfortunately, the opposite is true. Starting soon after the revolt and continuing down through the years, there has been a barrage of charges and countercharges concerned more with petty jealousies than with revealing or getting at the truth. Part of the controversy has raged over whether the original bear had been drawn *rampant*—which in the language of heraldry means upright on its hind legs with arms extended— or *passant*, on all fours.

Two flags, each reputed to be the original, were in the possession of the Society of California Pioneers until burned in the San Francisco fire of April 18, 1906. One flag depicted the rampant bear; the other the passant (see p. 275). Finally, the society, on the basis of studies made by a special committee in 1887, issued a statement that the one showing the bear on all fours was truly the "original" (Hussey, 1952 : 208). Even this action did not stop the controversy. As late as 1953, a proponent of the rampant bear delivered a speech in its favor (Santa Barbara *News-Press*, February, 1953).

A recent investigation of the matter by John Hussey has revealed heretofore unpublished documents and letters which permit "the movements of the original Bear Flag to be traced, almost from the day it was hauled down from the staff in Sonoma" by United States troops "until it was placed in the halls of the Society of California Pioneers in 1855" (Hussey, 1952 : 207). The commander of the troops, Lieutenant Joseph W. Revere, is said to have given the Bear Flag to John Elliott Montgomery. Montgomery then wrote to his mother shortly before he met his untimely death:

> U.S. Ship Portsmouth   Ancherage Yerba buena Cove
> Bay of San Francisco   Coast of California Oct. 20t. '46

My Dear Mother,
    It is with great pleasure that I seize upon an opportunity to write to you . . . I shall proceed to give you an account of the grand drama that has been acted in California since our anchoring in this Bay on the 1st of June last, to wit: On the 14th of June a revolution broke out on the part of the Americans & other Foreign residents against the Californian government & a party of 34 men surprised & took the interior town & fortress of Sonoma making Gen. Don Mariano Guadalupe Vallejo & his brother Don Salvadore Vallejo & Col. Don Victor Prudon all distinguished Mexican army officers prisoners & hoisted a flag of their own manufacture of this fashion. A white field with a red border on the lower edge a Grizzly bear in the center with a star in the upper corner the whole composed of [a] piece of white cotton with a Stripe of red Flannel the white

colloured with *black berry juice Brick Dust & oil* such was the first standard of liberty ever raised in California. (*Ibid.,* p. 209.)

With the letter, Montgomery included a drawing of the flag showing the bear on all fours.

From this and other evidence Hussey concluded that "the flag [with passant grizzly] which was displayed by the Pioneers as the 'original' Bear Flag until it was destroyed in the conflagration of 1906, and which served as the general model for the

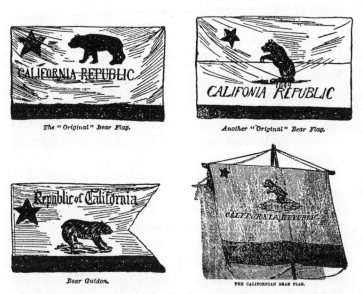

The " Original" Bear Flag.

Another "Original" Bear Flag.

Bear Guidon.

THE CALIFORNIAN BEAR FLAG.

Some early manifestations of the California Bear Flag. Two "originals" and a guidon, all lost in the San Francisco fire of 1906 (San Francisco *Call*, Sept. 7, 1890). *Lower right.*—The earliest discovered portrayal of the flag in print (*Hutching's California Magazine*, November, 1860).

present State flag, was indeed the standard which flew over the Sonoma plaza during the Bear Flag Revolt" (*ibid.,* p. 207).

Before that flag was burned, an exact replica was made for the purpose of raising it at Sonoma on the fiftieth anniversary of the Revolt. This duplicate banner is now on public display at the Sonoma Mission Historical Monument.

In 1911, the California State Legislature formally adopted

the Bear Flag as the state's emblem, using the following words (now in the *Government Code*):

The Bear Flag is the State Flag of California. Its length is one and one-half times its width; the upper five-sixths of the width thereof is a white field and the lower sixth a red stripe; there appears in the white field in the upper left corner a single red star, and at the bottom of the white field the words "California Republic," and in the center of the white field a California Grizzly Bear upon a grass plat, in the position of walking towards the left of the white field. The bear is dark brown in color and its length one-third of the length of the flag. (*Stats.*, 1911, chap. 9.)

In 1943, display of the Bear Flag beside or below the flag of the United States was prescribed by law for state buildings and institutions.

Because the act of 1911 did not specify an actual design for the grizzly but merely said that that the animal was to be in the position of walking, different flag manufacturers produced differing versions; the bear on many of these again reminded the public of a pig. The most common rendering, however, portrayed a hungry, wolflike animal. This was based on a photostat of a bear illustration that the Emerson Manufacturing Company of San Francisco had obtained from the California State Library in 1928. It is one of several forms of an early drawing or painting of the California grizzly by Charles Nahl.

The variable appearance of the bear on flags from different sources was a continual source of disturbance to Robert Stewart, Senior Buyer, Purchasing Division, Sacramento, and largely through his efforts, state-wide attention was directed to the fact that there was no officially recognized state flag. In an effort to have a single design adopted as the official flag bear, Fred W. Links, Assistant Director of Finance, in 1952 requested our aid in obtaining a portrayal of the California grizzly that would be ursine rather than porcine or lupine, and Don Greame Kelley of the California Academy of Sciences was commissioned to draw such a "portrait" and also a colored reference standard of the entire flag. It was felt that the general pose of the bear should follow the Nahl representation, since that has been on the flag for many years (see back cover; also Storer and Tevis, 1953).

When the new design was submitted to the legislature, Senator Earl Desmond of Sacramento said in a jocular mood that the drawing showed the bear with four claws on the left paw and three on the right and asked, "How come?" Senator Jesse Mayo of Angels Camp replied, "Must have stuck the right one in a trap."

This light touch aside, the revised version was approved. Senate Bill 1014 of the 1953 session incorporated the new illustration together with specifications for dimensions and colors to be used in the production of flags. The bill was passed by the legislature and signed by Governor Earl Warren on June 14, 1953 (*Stats.*, 1953, chap. 1140; *Government Code*, sec. 420).

## STATE ANIMAL

The act prescribing the details for the flag also made the California grizzly bear the "state animal" (sec. 425). This action rectified a curious legal oversight—previous laws had designated the valley quail as the state bird, the California golden trout as the state fish, and the California poppy (*Eschscholtzia*) as the state flower; but the bear had been overlooked.

## BEAR REPLICAS

The oldest stone replicas of grizzlies still extant in California are two that were carved under the supervision of the padres at Mission Santa Barbara in 1808 (fig. 34). One spouted water through its mouth from a fountain of Moorish design into a huge laundry basin; the other carried off the overflow. When Duhaut-Cilly visited Santa Barbara in 1827 he was surprised to find so sophisticated a structure as the fountain and laundry trough:

In front of the building [Mission], in the middle of a large plaza, running water gushes from a full fountain. In art it is imperfect, but still it caused so much the more surprise, the less we expected to find in this land, so distant from the comforts of Europe, an ornament or species of luxury which among ourselves is reserved only for the palaces of the more wealthy. The clear, sparkling water of this fountain, first rises more than eight feet above the ground and then falls in wide sprays over a series of stone basins that form an octangular pyramid. It then falls into the basin and when this is full, the water spouts forth through the mouth of a bear carved from stone, and finally it flows into a neat lavatory where the Indian girls and the young Californians were standing and washing. (Engelhardt, 1923 : 150–151.)

Today the water no longer flows from the mouths of the bears. One statue is so worn by the ravages of time as to be almost unrecognizable; the other, although headless, has front paws with distinctly grizzly-like claws.

Among decorative insignia using the California grizzly are several metal plaques. One, of bronze, about ten and a half inches long and delicately sculptured (fig. 35), was purchased at a secondhand store on McAllister Street in San Francisco some years ago by Mr. Francis P. Farquhar. Another, of lead alloy and bronzed, once in the collection of Miss B. A. Bowman of San Francisco, was purchased by one of us (T.I.S.) in 1954 from an antique dealer. The third, of iron, of the same size as the other two but cruder in detail, is in the museum of the Shasta State Historical Monument in the town of Shasta. The fourth plaque was cast at the Union Iron Works in San Francisco during 1882 and is in the office of that company. The plaques are of identical design and the posture of the bear somewhat resembles that in the Nahl illustrations. Other such plaques are reported to be in existence, but no information has come to us about the origin, manufacture, or date of any of them.

Other uses of the bear in decorative insignia are seen in the badge and membership certificate of the Society of California Pioneers and in the California souvenir used at the Columbian World's Fair at Chicago in 1893. "Golden" grizzlies grace the entrance to the County Building of the State Fair Grounds in Sacramento.

## NATIVE SONS OF THE GOLDEN WEST

A patriotic impulse among persons born in California led in 1875 to the formation of a society known as the Native Sons of the Golden West. After the society's first parade, held in San Francisco that same year, the grizzly was adopted as the insignia of the organization. The parade was witnessed by Major William Downie, an early pioneer of the state:

The Native Sons made their debut in a new role; they carried a handsome silk American flag, lent them by a patriotic citizen, and as an emblem they exhibited a stuffed bear, which had been found in a deserted room in Anthony's Hall. The bear was rather the worse for moths; it was a cub about three feet long and had been used as one of the insignia of a disbanded club, but it answered the purpose, and few, who today see the bear emblem upon the breast of a Native Son, would think that it originated with a musty, old straw-stuffed cub that had been discarded by its rightful owners. (Downie, 1893 : 359–360.)

In 1907, the *Grizzly Bear*, the official publication of the Native Sons, was first issued. The title on the cover was decorated with the head of a grizzly, and for a number of issues the cover showed a grizzly in various poses—fishing, sailing, and even bearing a platter with a Thanksgiving turkey. The pin of the order also contains the bear motif.

The N.S.G.W. soon began to hold their chief celebrations on Admission Day (September 9). A decorative arch erected in San Francisco at Market and Stockton streets for celebration of "the glorious 9th" in 1890 had, above portraits of Sutter and Frémont and a scene of overland travel, a huge replica of the grizzly, flanked by an Indian and a miner (fig. 36).

## UNIVERSITY OF CALIFORNIA

The bear that serves as the emblem of the University of California is the grizzly. This is highly appropriate because the principal opponent of the University's athletic teams is Stanford, whose symbol is the red Indian. When the University expanded to include the Los Angeles Campus, the southern student representatives became the "Bruins."

## PLACE NAMES

The grizzly's name was used often as California was explored in detail and settled. It marks nearly two hundred place names designating topographic features, waters, and settlements (Gudde, 1949 : 136). There is a Grizzly Springs in Lake County, a settlement named Grizzly in Plumas County near Beckwith, a Grizzly Bluff in Humboldt County, and a Grizzly Flats in El Dorado County. Trinity County has its Grizzly Mountain and Alameda County its Grizzly Peak back of Berkeley. Bear Valley in the San Bernardino Mountains and Cañada de los Osos in San Luis Obispo County both attest the former abundance of the big bears in those places. There are at least twenty-two Grizzly creeks in the state.

In 1852, there was a Grizzly-Bear House on the road to Nevada City. When Borthwick (1857 : 175) made inquiries about his route, he

found that the first habitation I should reach was a ranch called the Grizzly-Bear House. . . . There could be no mistake about it, for a strip of canvass, on which "The Grizzly-Bear house" was painted in letters a foot and a half high, was stretched along the front of the cabin over the door; and that there might be no doubt as to the meaning of this announcement, the idea was further impressed upon one by the skin of an enormous grizzly bear, which, spread out upon the wall, seemed to be taking the whole house into its embrace.

## THE OVERLAND MONTHLY

Ralph Waldo Emerson (1904 : 277–280), at the end of his essay on "Courage," wrote a mediocre poem about an encounter between the dauntless early-day bear hunter George Nidever and a California grizzly. But the grizzly's most convincing claim for immortality in the annals of literature is as the symbol for the magazine *Overland Monthly*.

This journal was started in 1868 at San Francisco for "the development of the country"—particularly in literary matters. The editor was Bret Harte. The emblem of the journal, on cover and title page, throughout most of its life, was a grizzly crossing a railroad track. In 1871 the magazine published a whimsical summation of the animal's character by a writer signed "URSUS," under the heading "Grizzly Papers":

Despite his faults—which heaven forbid I should excuse—the Grizzly Bear is as gentlemanly a brute as you shall find in a morning's ramble. He has that loose-jointed largeness of bulk, that shambling carelessness of stride, that comic honesty of expression, which are so intimately associated with our recollection of the late Mr. Lincoln, and to which that gentleman was probably as much indebted for his *sobriquet* of Honest Abe, as to any merely moral qualities he may have possessed. To be seen at his best, the Grizzly must be seen at home ... In his native wild, the grizzly is much given to the arts of peace, and seldom takes the war-path against his fellow-man, unless rudely disturbed in some philosophical meditation; in which case, it must be confessed, he usually devours the intruder while getting his scattered faculties well in hand. But the major part of his time is passed in peaceful pursuits, among which the pursuit of the California Indian holds an honorable place. To nose about for the edible nut and drag it from beneath the dead leaf, to spoil the acquisitive bee, to nip the nimble Digger as he flies—these are the humble triumphs of his unambitious life. (Anon, 1871 : 92.)

As the *Overland Monthly* achieved national success and prestige, its emblem became famous, and the editors were able to boast (1894, July, advertisements, p. 6):

WHEN YOU SEE A GRIZZLY WHAT DO YOU THINK OF? WHY, OF THE OVERLAND MONTHLY. The ONLY Literary Magazine on the Pacific Coast. Do you know the History of the OVERLAND'S GRIZZLY? No! Here it is.

Our world-renowned Bear Trade Mark had its origin in a sketch by the pioneer artist, Charles Nahl, of the famous old bear "Samson," owned by "Grizzly Adams." A cut of this sketch happened to be on the first check book used by the first publisher for OVERLAND business. He suggested to Bret Harte that the cut would make a good vignette for the cover of the new magazine, and after thinking it over, Bret Harte, in honor of the nearly finished transcontinental railroad, with his pencil traced the few lines that make the railroad track, and the "Overland Bear" was complete. (See tailpiece.)

In an editorial, Harte (1868 : 99) explained why he had selected the grizzly:

The bear who adorns the cover may be "an ill-favored" beast whom "women cannot abide," but he is honest withal. Take him if you please as a symbol of local primitive barbarism. He is crossing the track of the Pacific Railroad, and has paused a moment to look at the coming engine of civilization and progress—which moves like a good many other engines of civilization and progress with a prodigious shrieking and puffing—and apparently recognizes his rival and his doom ... As a cub he is playful and boisterous, and I have often thought was not a bad symbol of our San Francisco climate. Look at him well, for he is passing away. Fifty years and he will be as extinct as the dodo or dinornis.

Harte was indeed a prophet!

## BEGINNINGS OF A MYTH

Before the grizzly passed entirely from the California scene, Charles Howard Shinn (1890 : 130–131) wrote an article suggesting that, although the great beast was doomed, it might perhaps live on in the minds of men like the Lion of England and the Winged Bull of Assyria:

> A great many persons have told stories about grizzlies and about pioneers. But there is an aspect in which the grizzly and the pioneer may be said to represent the beginnings of a chapter of national folklore, or the first halting steps towards the development of a noble myth. . . .
>
> In the course of time . . . it may be that the two giant shadows of the past, the Argonaut and his grizzly, will loom up over the Sierras, as Hercules and his Nemean lion in the legends of the Greeks. . . .
>
> When the last grizzly has perished, when the old race of miners is . . . far lost in tradition . . . when the great Californian valleys and all the shining slopes of the long, parallel mountain ranges beside the Pacific are clothed with continuous gardens and orchards, and mighty and populous cities grow from the villages of to-day, there ought to be a background of sublime fable to inspire poet, artist, and sculptor.
>
> It is the first step towards a myth that always proves the most difficult. Already, the world over, men have come to know the old cañon-keeper and forest dweller as "the grizzly," not the grizzly bear. He . . . is on the way to still further separation from other bears, and other creatures of the high order that furnish noble subjects for art. Sometimes, I am sure, an American Thorwaldsen will know how to hew a Sierra grizzly out of some gray cliff of Rocklin granite, and there it will remain while the world endures, supreme as the Lion of Lucerne. . . . Perhaps in the day of battle, a thousand years hence, in some wild Sierra pass, the free men of the mountains, changing the course of history, and broadening the California myth to a world myth, will make the American Grizzly for all time such a name as the Lion of England, or the ancient Winged Bull of Assyria.
>
> The Pacific Coast . . . has already adopted the grizzly in its common speech. . . . A man is said to be "as strong as a grizzly," or as dreadful when aroused, or as much of a boss, or "a regular grizzly of a fellow." It is not a light phrase; it goes deep down to the roots of the matter . . .
>
> Again, the grizzly stories that frontiersmen tell have all the unconscious dignity of their subject; they rise at times to the height

of an epic of the Sierras, and they possess a singular vitality. . . . The grizzly has somehow impressed himself irrevocably upon the imagination of the man of the Pacific Coast, and this in a way that the black and brown bears have never yet done to any people. In the delightful German tales Bruin is a good-natured, stupid fellow, whom one cannot but like even while smiling over his adventures. . . . But the grizzly stands apart, so different in his very nature, and so impressive in every aspect that another long step towards the creation of a noble and satisfactory myth appears to have been taken by the pioneers, the true myth-builders and makers of literature in their log cabins; by their winter fires. How long a step has thus been gained we shall know better when the grizzly is gone from the Sierras.

The California grizzly
of the *Overland Monthly*.

# Appendixes

## APPENDIX A

### KNOWN SPECIMENS OF CALIFORNIA GRIZZLIES

In a zoölogical museum having materials for research and reference use, a scientific specimen of a mammal typically comprises the skin and the skull (cranium plus jaws, with all teeth), and sometimes the body skeleton. Each specimen is tagged with precise locality, date of capture, sex (as determined from the carcass), and the collector's name. Certain standard dimensions are recorded before the animal is skinned or the skull removed; these are: total length from tip of nose to end of last tail vertebra; tail length (vertebrae, excluding hair at end); length of hind foot from heel to tip of longest claw; and length of ear from crown of head. The weight is sometimes recorded. For large mammals, other features may be recorded, such as chest girth, height at shoulder, and weight (total or after evisceration). All these details are essential or at least helpful in subsequent study for identifying the specimen, making comparisons with other specimens, learning the variation within a population and between different populations, describing species new to science, and determining the geographic range of the species, its season of activity, and other matters. Large mammals are often photographed after death, and notes may be added on the color of the eyes and claws.

The relics of California grizzlies are disappointingly few and imperfect. So far as we know, there is not a complete "scientific specimen" with full data and measurements in any museum. There is one skin and skull (USNM 156594) and one skin as a rug with skull (MVZ 46918). Of the last captive, "Monarch" (see chap. 10), the California Academy of Sciences has the mounted skin (reportedly with skull inside) and the Museum of Vertebrate Zoölogy at the University of California has the body skeleton. The accompanying table lists all "specimens," of whatever sort, catalogued in museums or otherwise. In summary, there are parts of about 66 animals, including 1 skull and 1 cranium known to us in private ownership, as follows: skin and skull, 2; skin only, 9; skull only, 34; cranium only, 11; jaws (one or both), 5; complete skeleton, 1; body skeleton, 2; individual tooth, 2; claws (one or two), 2. Few of these catalogued items are satisfactory as scientific specimens. A number of the skulls, crania, and jaws are weathered, and some lack all or part of the teeth.

The reasons for this lack of good museum material are simple. During the years following the gold rush, when these bears were being killed for their flesh or because they were hazards to man or livestock, there were no scientific museums in the West. A few specimens were shipped to Washington and Philadelphia. Pelts and skulls were occasionally kept by private individuals, but through the years the skins probably became worn or moth-eaten and the owners gradually lost interest in the skulls. Finally, a half century later, zoölogists became interested and a number of such relics reached museums. The San Luis Obispo *Tribune* of March 13, 1953, in its column "Fifty Years Ago" mentions an advertisement placed by Miss Annie M. Alexander of Oakland (who later endowed the Museum of Vertebrate Zoölogy at Berkeley) offering five dollars each for grizzly skulls in good condition. Many years ago, C. H. Merriam began actively to search for and purchase grizzly relics. The quest still continues, but the lapse of time and the effects of weather make it unlikely that any substantial addition will be made to the scant series of known specimens.

Of the National Museum collection in Washington, Dr. David H. Johnson, in the Division of Mammals, wrote (letter, Aug. 14, 1952): "The early records were not well kept, and the specimens have never been adequately labeled. . . . The early catalog entries are sometimes contradictory, and so many different people have made annotations on the labels or in the catalogs through the years that it is now impossible to pick out the original information. . . . some specimens have been given away or exchanged, or have simply disappeared."

Of nine California specimens (3 skins with skull, 2 separate skins, and 4 skulls) mentioned by Baird (1857), only one skin and three skulls were on hand in 1952.

### Known Museum Specimens of California Grizzly

Museum of Vertebrate Zoölogy, University of California, Berkeley

    16379 Indian Shell Mound near Mayfield, Santa Clara Co. N. C. Nelson. Picked up in December, 1911. *Jaw only.*

    16615 Tejon [or San Emigdio] Mountains, Kern Co. Received from W. S. Tevis. Collected between 1893 and 1896. *Skin only*, as floor rug with turned bone teeth in head.

16616 Same data and type of specimen as no. 16615.

24537 Mount Gleason, Los Angeles Co. Collected about October, 1889, by Allen Kelley. This is "Monarch" (see chap. 10) *Body skeleton only*. (Mounted skin in California Academy of Sciences, no. 10471; fig. 33.)

27928 Buck Camp, 16 mi. NE of Wawona, Madera Co. R. S. Wellman and Jim Duncan, Oct. 21, 1887. *Skin only*, obtained April or May, 1918.

28007 Mountains 5 mi. E of Point Gorda, Monterey Co. L. S. Neville and Mr. Plaskett. Picked up in 1918. *Skull only*.

31826 E of Crescent Lake, elevation about 8,500 ft., Madera Co. (about 12 mi. S of Yosemite Valley and 10 mi. E of Wawona). Received from Mrs. J. S. Washburn. Killed about 1895. *Skin only*, without claws.

35388 East end of Frazier Mountain, elevation 7,000 ft., Ventura Co. V. C. De Lapp. Picked up in 1924. *Skull only*, broken.

46910 Head of San Gregorio Creek, near Kings Mountain, San Mateo Co. S. Hardy and Chase Littlejohn. Picked up about 1922. *Skull only*.

46918 Near head of Tujunga Canyon, San Gabriel Mountains, Los Angeles Co. W. L. Richardson. Killed May 16, 1894. *Skull and skin*, as a rug.

84215 San Benito River, 10 mi. SE of San Benito, San Benito Co. Carl J. Bleifus. Picked up Aug. 13, 1933. *Cranium only*.

84216 Locality unknown, probably San Benito Co. Carl J. Bleifus and Lew Smith. *One tooth*.

91034 Probably Monterey Co. Obtained by E. R. Hall. No date. *Skull only*.

91035 Upper end of Oso Canyon, Santa Barbara Co. Alonzo G. Yates. Collected before 1896. *Cranium only*.

91036 Probably Santa Barbara Co. Obtained by E. R. Hall. No date. *Lower jaw only*.

97901 Moffitt Field, Sunnyvale, Santa Clara Co. W. C. Russell. Picked up in February, 1942. *Part of lower jaw*.

101053 Redwood City Wharf, San Mateo Co. Chase Littlejohn, 1858. *One claw*.

101161 Corral Hollow, 1½ mi. E of Tesla, Contra Costa Co. S. B. Benson. Picked up July 10, 1943. *One lower molar*.

[24408 2 mi. NE of Sunland, Los Angeles Co. Shot by Cornelius B. Johnson, Oct. 28, 1916. *Skull and part skeleton*, ♀ adult. Long considered one of the last native grizzlies killed in

California, this bear actually was one that had escaped from the zoo in Griffith Park, Los Angeles. It was not known to be a California grizzly, and the skull is not like others taken in the state.]

Museum of Paleontology, University of California, Berkeley

Acc. 1298 Seventeen-Mile Drive, Carmel, Monterey Co. (Locality V-5228, late Pleistocene to Recent). July 24, 1952. Ray C. Roberts. *Mandible and two cervical vertebrae.* Beneath beach sand.

California Academy of Sciences, San Francisco

5567 Marina Postoffice, Monterey Co. William McGowan. Received Nov. 15, 1946. *Skull and part skeleton.*

9377 2½ mi. SE of Mayfield, Santa Clara Co. (in Indian mound). Received Jan. 25, 1948. *Part of skull.*

10471 Mount Gleason, Los Angeles Co. Collected by Allen Kelley about Oct., 1889. Male. *Mounted skin* (skull inside?) of "Monarch" (see chap. 10 and fig. 33). Body skeleton is MVZ 24537.

Los Angeles County Museum

M1000 Near Mount Hamilton, Santa Clara Co. G. G. Johnson. Received in 1916. *Two claws.*

Academy of Natural Sciences, Philadelphia, Pa.

961 California. Received March, 1861. *Skin only,* juvenile female.

986 California. Received April, 1861. *Skin only.*

2524 California. No other data. *Skull only.*

2525 California. No other data. *Skull only.*

2792 California. No other data. *Skull only.*

Museum of Comparative Zoölogy, Harvard College, Cambridge, Mass.

41998 Probably near San Francisco. W. O. Potter. Received May 15, 1862. *Skull only.*

Carnegie Museum, Pittsburgh, Pa.

20989 Sierra Nevada. Dr. Hiram De Puy. *Skull only.*

Peabody Museum of Natural History, Yale University, New Haven, Conn.

01143 Mission San Jose, Alameda Co. M. Overacker. Received by Yale Expedition of 1871. *Skull only.*

United States National Museum, Washington, D.C. (Including Biological Surveys Collection)

1219 Probably Monterey. Lt. W. P. Trowbridge. No date; catalogued Nov. 9, 1854. *Skull*, middle-aged adult, sex unknown; length,[1] 367 mm. (Baird [1857 : 224] measurement 14.10 in.=358 mm. Specimen originally numbered 1218 and so reported by Baird. Note on label: "Certainly from W. Coast U. S. possibly San Francisco or San Diego.")

1220 Probably Monterey. Lt. W. P. Trowbridge. No date; catalogued Nov. 9, 1854. *Skull*, juvenile; length 225± mm. Note in catalogue: "Certainly from W. Coast U. S. possibly San Francisco or San Diego." (Baird's no. 1220 was designated skull and skin, and stated to be largest skin from California, 6 ft. long [Baird, 1857, 218, 220]; his number and that of the catalogue seemingly pertain to different individuals.)

1444 Calaveras Co. Dr. [J. S.] Newberry. November, 1855. *Stuffed skin of six-month-old cub*, 3½ ft. long. Data from Baird, 1857 : 225; no locality or date on label or in catalogue.

2037 Perhaps San Francisco. Dr. Newberry. No date. *Skull*, subadult or young adult; length 347 mm. (Note on label queries accuracy of locality; catalogue entry 2037 is for a raccoon from Wisconsin. Baird, 1857 : 224, lists this number as from San Francisco.)

3536 Fort Tejon. J. Xantus (no. 1172.) Catalogued March 25, 1859. *Skull*, moderately old adult; length 346 mm.

3537 Fort Tejon. J. Xantus (no. 1171). Catalogued March 25, 1859. *Skull*, old adult male; length 349 mm.

3538 Fort Tejon. J. Xantus (no. 1319). Catalogued March 25, 1859. *Cranium only*, young adult male; length 312 mm.

3630 Monterey. A. S. Taylor. No date; catalogued Oct. 7, 1859. *Skull*, moderately old adult; length 378 mm.

3837 Sacramento River, probably between Colusa and Sacramento. Collected in autumn of 1841. Catalogued June 8, 1860. *Skull*, old adult, weathered; length 370 mm.

4161 California. E. Samuels. No date; catalogued July 6, 1860. *Cranium only*, young adult; length 347 mm. (Baird, 1857 : 225, listed as a skull, no. 3100, taken at Petaluma by Samuels.)

---

[1] Condylobasilar length in this and succeeding USNM specimens.

6905 Monterey. C. A. Canfield. No date; catalogued in April, 1866. *Skull*, young adult; length 380 mm.

7401 Monterey. C. A. Canfield. No date; catalogued Nov. 19, 1867. *Skull*, young adult; length 338 mm.

11809 California. No other data. *Stuffed skin of cub.*

15421 California. F. V. Hayden. No date; catalogued March 29, 1876. *Cranium only*, adult, weathered and broken; length 315± mm.

15671 Near Havilah, Kern Co. (southern Sierra Nevada). J. T. Rothrock and H. W. Henshaw (no. 953). October, 1875. *Skull*, middle-aged adult; length 338 mm.

15682 San Fernando Mission. G. M. Wheeler. September, 1875. *Skull*, subadult; length 303 mm.

16624 Baird, Shasta Co. Livingston Stone. No date; catalogued Feb. 10, 1882. *Skull*, old adult, weathered; length 352 mm.

156594 Trabuco Canyon, Santa Ana Mountains, San Diego Co. Andrew Joplin. January, 1908. *Skin and skull*, old female; length 312 mm. Reputedly mate of no. 160155.

160155 Los Biacitos, head of San Onofre Canyon, Santa Ana Mountains, San Diego Co. Henry A. Stewart. August, 1900, or 1901. *Skull*, old adult; length 375 mm.

178735 Beswick, near mouth of Shovel Creek, Siskiyou Co. Charles F. Edson. No date; catalogued May 13, 1912. *Skull*, young adult male; length 353 mm.

180880 2 mi. S of Mayfield, Santa Clara Co. (in Ponce Mound, 6 ft. 9 in. below surface). Harold Heath. 1912. *Right mandible only*, young adult.

206624 Dobbins Creek Canyon, Trinity and Humboldt counties. Thomas Murphy. No date; received 1914. *Cranium only*, old, weathered; length 370 mm.

206625 Near Long Valley, Mendocino Co. Jim Farley. No date. *Cranium only*, old adult, weathered and broken, no teeth; length 350± mm.

223401 Lassen Peak Canyon, 10 mi. from Blocksburg, Humboldt Co. Received from Ephraim Hoover. Reportedly killed in 1880. *Cranium only*, adult, weathered, no teeth; length 357 mm.

228225 Isabel Valley, Mount Hamilton, Santa Clara Co. Received from Frank Hubbard. Reportedly killed about 1882. *Cranium only*, adult, weathered, no teeth; length 321 mm.

228226 Isabel Valley, Mount Hamilton, Santa Clara Co. Received from Frank Hubbard. Reportedly killed about 1882. *Cranium only*, adult, weathered; length 365 mm.

Leningrad [St. Petersburg] Academy of Science, Leningrad, U.S.S.R.
  1514–1519  California. Wasnesensky. *Skulls only* (?): one 1841; one 1843; four juveniles, August, 1848 (letter, E. R. Hall, Feb. 2, 1953; notes from longhand catalogue seen by him at the Museum Aug. 10, 1937).

Museum National d'Histoire Naturelle, Paris, France
  A2754 Monterey. Surgeon Neboux. Collected between Oct. 18, and Nov. 14, 1837. *Complete skeleton*, mounted, old male; in 1954 lacks only sternum, several teeth, tail vertebrae, and claws. Total length, 1766 mm.; skull, 401 mm. (fig. 10).

## Uncatalogued Specimens

Pacific Grove Museum, Pacific Grove, California
  —— Carmel Valley, Monterey Co. Mrs. Herbert Hoover. Obtained probably between 1894 and 1898. *Skull*.
  —— No data. *Skull*.

Albert Kennedy, Clear Lake Oaks, Lake Co., California
  —— Long Valley, Lake Co. Hiram Kennedy. Probably in 1860's. *Cranium*.

Lloyd P. Tevis, Jr., Carmel, California
  —— 9 mi. from Carmel, George G. Moore ranch. W. S. Tevis. 1885. *Skull*.

# APPENDIX B

## LAST RECORDS OF GRIZZLIES IN CALIFORNIA (BY COUNTIES)

*Northwest Coast*

1879  Trinity (N 90)
1868  Humboldt (Grinnell, 1938)
1875  Mendocino (Grinnell, 1938)
1868  Sonoma (N 71)
1888  Marin (Van Atta, 1946)

*Central Coast*

1879   San Mateo (Grinnell,
           1938)
1886   Santa Cruz (Welch, 1931)
1866   Alameda (N 67)
1879   Santa Clara (N 91)
1886   Monterey (Grinnell,
           1938)
1878   San Benito (N 89)
1912(?) Santa Barbara (Grinnell,
           1938)

*Sierra Nevada*

1902   Siskiyou (Schrader,
           1946)
1868   Shasta (N 73)
1884   Lassen (Townsend, 1887)
1874   Sierra (N 85)
1874   Nevada (N 84)
1902–1903  Amador (Storer,
           field notes)
1875   Calaveras (A. K. Fisher,
           1920)
1911(?) Mariposa (Grinnell and
           Storer, 1924)
1922   Fresno (Merriam, 1925)
1924(?) Tulare (Fry, 1924)

*Central Valley*

1889   Tehama (N 99)
1870   Butte (N 74)
1862   Colusa (Grinnell,
           1938)
1864   Yolo (N 63)
1865   San Joaquin (N 64)
1861   Merced (N 60)
1898   Kern (Grinnell,
           1938)

*Southern California*

1882[+]   Ventura
1897   Los Angeles (Grinnell,
           1938)
1868   San Bernardino (N 69)
1895   Riverside (Grinnell,
           1938)
1903   Orange (N 101)
1908   San Diego (USNM
           156594)

# APPENDIX C

## INDIAN NAMES FOR BEARS; TRIBES AND TRIBAL TERRITORIES

Dr. C. Hart Merriam (see Introduction) spent much time visiting old Indians of various tribes in California to learn of their *rancheria* sites, tribal boundaries, modes of life, and speech. Equipped with illustrations and samples of local animals and plants, he recorded many of their words for biological elements of their environment. From his manuscript "vocabularies" now in the Department of Anthropology of the University of California at Berkeley we have copied some names used for bears. His system for transliteration of Indian words differed somewhat from those used by other anthropologists, but the examples tabulated suffice to show that the tribes and their subdivisions had their own distinctive names for the bears and that these differed from place to place. The areas for which Dr. Merriam recorded no specific words for one or another kind of bear we infer lacked the animals. The Pass Cahuilla had a name for the black bear, although the species was not native closer than the Tehachapi region; this may indicate that they, like other tribes in country that contained only grizzlies, sometimes had captive black bears which they probably obtained as cubs.

## TRIBAL NAMES FOR BEARS

| Tribe | Locality | A Bear | Grizzly Bear | Black Bear |
|-------|----------|--------|--------------|------------|
| Chem-e-we'-ve | Colorado River | No word | No word | No word |
| | Twenty-nine Palms | No word | No word | No word |
| Cahuilla | near Banning | Hoo'-naht | Hoo-naht | Ho-nut; To-ran-ko |
| | | Hoon'-wit | Hoon'-wit | No word |
| Eastern Mono | Long Pine | | No word | Wăh-zut-nah (small one) |
| | Owens Valley | Pah-hah'-bitch | Pah-hah'-bitch | Tŭ-trah'-bit |
| | Mono Lake | Pah-roo-ah' | Too-hwid (white) | Wăh-zut-nah; also Too-hwid |
| | | | Too-hah'-give-dă (white and bad) | |
| | Bridgeport | Pah-roo-ah | Pe-jas'-soo-e (red paint) | Pah-roo'-ah |
| Yokuts | Tejon Ranch | Mut-teep' | Pah-roo-it | Pa-e-taht'-se |
| | | | Mut-teep' | |
| | Bakersfield | No'-ho'; | No'-ho'; | Nŏ-ŏh; Mŏl-loi |
| | | Naw'-ah-ah' | Naw'-ă-ah' | |
| | Tulare Lake | No'-ho | Oo-lŏw'e; Mo'-lo-e | No word |
| | Kaweah River foothills | No-hah | Nă-hah'-ah; Doo-hoon' | Duo'-kahn |
| Maidu | Cosumnes River | Kah'-pah' | Tah'-kah-pah' | Em-mŭl' |
| | Deer Creek | Kap'pă; Yu'-we | Kah'-pah | Em'-mool |
| | Big Meadows | Mŭ-deh | Pah'-no | Bul-lah'-li |
| Sinkyone | Eel River | No'-nā | Sahs'; Shash | No-nā'; Tah'-nā; Tah-hă-nā |
| Mattole | Mattole River | No'-ne | Ĭ ͨʰ'-le-ge | No'-ne |
| Hupa | Trinity River | Sah'ts; Sah'ahts | Mĕ-chā-e-s | Sah'-ahts |

NAMES OF CALIFORNIA INDIAN TRIBES MENTIONED IN
CHAPTER 4 AND THEIR TERRITORIES

*Atsugewi.*—Part of the drainage of Hat Creek in the northeast

*Cahuilla.*—Coachella Valley and vicinity in the Colorado Desert region

*Chumash.*—Coastal southern California near Santa Barbara

*Costanoan.*—Southern shores of San Francisco Bay south to Point Sur and the Salinas River

*Cupeño.*—Headwaters of the San Luis Rey River in the south

*Diegueño.*—Coastal southern California near San Diego

*Gabrielino.*—Coastal southern California near Los Angeles

*Hupa.*—Lower part of Trinity River in the Klamath Mountains

*Juaneño.*—Coastal southern California near San Juan Capistrano

*Karok.*—Middle part of Klamath River in the extreme northwest

*Kato.*—Uppermost drainage of South Fork of Eel River in the northern Coast Ranges

*Kawaiisu.*—Tehachapi Mountains at southern end of Sierra Nevada

*Klamath.*—Subdivision of the Lutuami of southeastern Oregon

*Luiseño.*—Western foothills of the Santa Rosa and San Jacinto Mountains in southern California

*Lutuami.*—Southeastern Oregon and northern Modoc region

*Maidu.*—Northern Sierra Nevada and eastern half of Sacramento Valley

*Mattole.*—Drainages of the Bear and Mattole rivers in the northern Coast Ranges

*Miwok.*—Central coast from the Golden Gate north to Bodega Bay; part of the basin of Clear Lake; part of the deltas of the San Joaquin and Sacramento rivers; and the western slope of the Sierra Nevada from Cosumnes River to Fresno River

*Mono.*—Drainages of the upper San Joaquin, Kings, and Kaweah rivers of the southern Sierra Nevada and arid eastern base of that range

*Nisenan.*—Subdivision of the Maidu

*Nomkensus.*—Drainage of Klamath River

*Nomlaki.*—Subdivision of the Wintun

*Paiute (Owens Valley).*—Along eastern base of the southern Sierra Nevada

*Patwin.*—Subdivision of the Wintun

*Pomo.*—Drainage of Russian River, adjacent coastal lands to the west, and part of the basin of Clear Lake, all in the northern Coast Ranges

*Salinan.*—Headwaters of Salinas River and adjacent Santa Lucia Mountains in the southern Coast Ranges

*Shasta.*—Drainage of Shasta and Scott rivers and adjoining part of Klamath drainage in the extreme north

*Sinkyone.*—South Fork of Eel River (except upper drainage) and part of adjacent coast in the northern Coast Ranges

*Tachi.*—Subdivision of the Yokut

*Wailaki.*—Part of middle drainage of Eel River in the northern Coast Ranges

*Wappo.*—Part of Russian and Napa rivers and adjacent uplands in the northern Coast Ranges

*Wintun.*—Western half of Sacramento Valley to crest of the northern Coast Ranges

*Wiyot.*—Humboldt Bay and lower drainage of Mad and Eel rivers in the northwest

*Yokuts.*—Valley of the San Joaquin River and much of the adjacent foothill territory from Fresno River south to Bakersfield

*Yuki.*—Upper drainage of Eel River in the northern Coast Ranges

*Yurok.*—Lower part of Klamath River and adjoining coastal lands in the extreme northwest

# APPENDIX D

## SOURCES OF INFORMATION

Search of the literature has shown no substantial early account of the natural history of the grizzly, in California or elsewhere. The dearth of precise information is due to the rushing course of events and the status of biological study in the years when these bears were numerous. No early writer had much time or facility for field investigation, and the relatively slow means of travel and communication precluded gathering detailed factual information. Since there is no adequate firsthand account, we have constructed our story of the California grizzly from the following sources.

### EARLY ACCOUNTS

The first records, merely of "bears" but actually grizzlies, are in the diaries and reports of the early Spaniards: Father Ascensión of the Vizcaíno expedition in 1602–1603; Costanso, Fages, and Crespi with Portolá in 1769; and Font of the Anza party in 1776 (see our chap. 5).

The several exploring parties of other nationalities that touched California, from Kotzebue in 1816 to the *Venus* expedition in 1837, each reported a bit about bears. The first overland trapper-explorers, such as Jedediah Smith and John Work, who traversed inland areas unknown to the Spanish Californians, met grizzlies and made notes about them. The earliest Americans to settle in California—William Heath Davis, Benjamin Wilson, George Nidever, and their like—left reminiscences which recorded encounters with bears. It was the Americans of the gold rush, however, who left the largest record of the big bear of California.

## NEWSPAPERS

A significant part of the early chronicles on grizzlies and Americans is in the newspapers of the 1850's and later, where some of the first citizens reported their actual experiences while the circumstances were vividly in mind. Like the occasional cache of gold dust hidden by a departing miner, this accumulation has been dormant in library files for a century. Discovered by the senior author some years ago, these contemporary stories were the stimulus toward preparing the present volume. The Sacramento *Daily Union*, the San Francisco *Daily Alta California* and *Daily Evening Bulletin*, and other newspapers contain many articles detailing experiences that were variously benign, injurious, or fatal to pioneer Californians (see list following Bibliography). The subject catalogue in the California Department of the State Library made it easy to find these newspaper articles. Other, shorter references are now appearing in historical columns of contemporary rural papers, under their headings of "Seventy-five Years Ago" or "One Hundred Years Ago." The number of articles actually found in each decade serves as a crude index to the relative numbers of grizzlies: there were between forty and fifty articles in the years 1851–1860, fifteen in 1861–1870, and sixteen in 1871–1879.

A fortunate circumstance in evaluating early literature was the original distribution of bears. The black bear was not present coastwise south of Sonoma County or south of the Tehachapi region, and did not inhabit the Great Central Valley. Early American activities—except in the Mother Lode and some northern mountainous counties—were in territory that contained only grizzlies; any reference to bears in those areas meant the larger species.

## SCIENTIFIC LITERATURE

The relations between Indians and grizzlies (our chap. 4) are set forth mainly in the writings of anthropologists, usually as incidental items amid much other subject matter on the first human inhabitants.

Scientific literature on the grizzly of California is meager. The most notable early item is the illustration of a mounted skeleton (fig. 10), from Monterey, obtained by the French *Venus* expedition of 1837 (see our chap. 2 and App. A). The reports of the Wilkes Expedition of 1841 (Cassin, 1858) contain little material about the bear. There is more in the Pacific Railroad Reports, particularly in the account by Spencer F. Baird (1857), which is the first straightforward scientific article on this animal. Thereafter short notes were the rule, and not until 1937 was there any extensive study of the grizzly—that by Grinnell, Dixon, and Linsdale in *The Fur-bearing Mammals of California* (pp. 65–94). These authors digested the accounts of Hittell (1860), W. H. Brewer (Farquhar, 1930), and Jedediah Smith (Sullivan, 1934), and added miscellaneous material, oral and written, from various sources.

## PRINCIPAL AUTHORS

Joseph Grinnell (1877–1939) was a naturalist and zoölogist in California from the 1890's onward. He was Director of the Museum of Vertebrate Zoölogy in the University of California at Berkeley from 1908 until his death. He was a keen, persevering student of the California fauna, on which he published extensively. As a young man he had near encounters with the grizzly (see chap. 1), probably the last California biologist to do so. He used every opportunity to obtain specimens and scientific information, particularly about the big bear and other scarce or vanishing species. Besides the coöperative volume just mentioned, he published, in 1938, a modest essay setting forth his findings and conclusions about the grizzly in California.

Ernest Thompson Seton (1860–1946) was an artist and naturalist, the author and illustrator of many books and popular magazine articles for young people about animals. One of his small volumes is *Monarch the Big Bear of Tallac* (1904). This is a children's story, in which a few facts and names of California grizzlies are mingled with much fiction and some fancy. Seton wrote more technical articles on birds and mammals for scientific journals and was the author of two large technical works on mammals: *Life Histories of Northern Animals* (2 vols., 1909) and *Lives of Game Animals* (4 vols., 1925–1928). In some respects these stand as the original edition and a greatly expanded revision of the same work. The rather extensive chapter on the grizzly (*Lives*, 2 : 5–91, 1926, reissued in 1929) pools data from the entire North American range. It is a mixture of information, not all equally reliable. For our account we have culled various details that seemed trustworthy.

The best-known and oldest account of the grizzly in California and the West is that by Theodore H. Hittell (1830–1917), *The Adventures of James Capen Adams, Mountaineer and Grizzly Bear*

*Hunter, of California* (see our chap. 9). The author graduated from Yale University in 1849, came to California in 1855, and was in newspaper work until 1860 on the San Francisco *Daily Evening Bulletin*. He practiced law from 1862 until 1906, became an attorney of distinction, twice summarized the general laws of the state, and wrote a four-volume *History of California* (1885–1897).

A scarcer volume—in lighter vein—is that by Allen Kelley (1855–1916), *Bears I Have Met—and Others* (1903). The author was variously a newspaper reporter and a hunter. Certain of his accounts have considerable factual basis, but others are dubious. He negotiated the transport of "Monarch" to San Francisco in 1889, wrote a newspaper report of that experience, and later brought together material for his book.

Joaquin Miller (1841?–1913), termed "poet of the Sierras" by his generation of Californians, wrote a volume of *True Bear Stories* (1900), which was erroneous in both title and contents. The account of "Monarch" was uncritical and contained many errors, according to Kelley.

Outside California, several persons have written books on the grizzly—some about the animal elsewhere in the United States and others of it in Alaska. One by William H. Wright (1856–1934), *The Grizzly Bear*, carries as subtitle "The narrative of a hunter-naturalist, historical, scientific and adventurous" (1909). Wright lived afield for months in Montana and British Columbia. For about twelve years he was an active hunter, then for an equal period became an observer and photographer. The book has much on grizzly habits. Wright and a companion, J. B. Kerfoot, pioneered with flash photographs of grizzlies at dusk when lighting for this purpose was the flare of burning magnesium powder. A reminiscent volume on grizzly hunting in New Mexico during the 1880's and early 1900's is that by Montague Stevens, *Meet Mr. Grizzly* (1943).

The volume by Theodore Roosevelt (1858–1919), *Hunting the Grisly* (1893), contains two chapters on the animals: "Old Ephraim, the Grisly Bear," dealing with habits, and "Hunting the Grisly," telling mainly of Roosevelt's own experiences in the Dakota-Wyoming region during the 1880's. Besides being a vigorous and successful hunter, Roosevelt was accurate and judicious as a naturalist-observer.

Finally, a naturalist of the Rockies, Enos A. Mills (1870–1922), wrote in 1919 *The Grizzly, Our Greatest Wild Animal*. For thirty years he lived in the wilds and studied grizzlies at close range but never carried or used firearms. His is a simple, straightforward account of the animal's habits. We have used occasional quotations from the Mills book by way of comparison with the more fragmentary record of grizzly traits in California.

## ILLUSTRATIONS AND ILLUSTRATORS

Many persons sought to portray the California grizzly, and they used various media—oil paints, water colors, lithography, crayon, pen and ink, and even photography. The majority of these renderings are no more than mediocre, but a few have special merit. Some of both types are reproduced in this volume.

Of all contemporary artists who tried to picture the living bear only one succeeded. He was Charles Christian Nahl (1818–1878), who had come to California in 1850 with his mother and his half brother, Hugo Wilhelm Arthur Nahl. Through four centuries the ancestral Nahls had been artists in Germany—painters, sculptors, or engravers. Charles and Hugo were of the sixth generation. Charles Nahl had a brief trial at mining and a year as an artist in Sacramento. In 1852 he moved to San Francisco and lived there for the remainder of his life. His skill was varied: drawings for wood engravings, etchings, lithographs, daguerrotypes, photographs, and oil paintings are known from his hand. Several of his paintings survive, along with a miscellany of his other work, but many originals probably were lost in the San Francisco fire of 1906. The ancestry and the work of the brothers Nahl in California are summarized in an essay by Eugen Neuhaus (1936 : 295–305).

Two original Nahl illustrations of a live California grizzly are in existence. One, owned by Mrs. Virgil (Edna C.) Nahl of Corte Madera, Marin County, is painted on glass. The bear faces to the right, the claws on all four feet are pale, and a strip of grass lies before the right hind foot. The side of the bear is predominantly light. The legend at the top, in script, reads "California Animals"; centered below the figure, in capitals, is the name "GRIZZLY BEAR." Lettering at the bottom reads, at left: "Drawn from Nature in Adams Menagerie, S.F."; at right: "Published by Nahl Brothers, 79 Broadway, San Francisco"; at center: "Entered according to Act of Congress in the year 1855 by Nahl Bros. in the Clerk's Office of the United States District Court."

The second painting, about 7 by 12 inches, now in Colton Hall, the city hall of Monterey, California, has the bear headed to the left; only the claws of the left front foot are pale; no grass is shown; and the body is dark (fig. 2). Appended is the following manuscript statement:

This drawing is by *Charles C. Nahl* of San Francisco, in the period between 1850–1856. The original from which the grizzly bear of the California State Seal and nearly every representation of the grizzly were copied.

From my own personal knowledge this drawing has remained in the Nahl family ever since to the presentation date June 21st, 1931.

[Signed] PERHAM W. NAHL
Prof. of Art University of California

Title to authorship of this painting is slightly clouded by a remark in the Neuhaus article (1936 : 302) which states: "His daughter, Augusta Nahl Allen, of Portland, Oregon, is the owner of a drawing of a Grizzly Bear by Hugo Wilhelm Arthur Nahl from which the great seal of the State of California was made." Professor Neuhaus, in answer to an inquiry, replied (letter, Feb. 28, 1953) that "the notation on the drawing in Monterey apparently stems from a divergency of claims within the Nahl family. . . . either claim would be difficult to authenticate at this late date, but at least it seems safe to say that the original . . . was by one of the two Nahls in question."

The claim regarding the seal, however, cannot be correct, because the seal was designed and used (chap. 12) before the Nahl family arrived in California. Also the copyright notation on the first figure is confusing because that item is not included in a listing of copyrights issued at San Francisco from 1851 to 1856 (Coulter, 1943).

The earliest published illustration of a grizzly by Charles Nahl is in *Hutching's Illustrated California Magazine* of September, 1856 (Vol. 1, No. 3, p. 106) as a woodcut inscribed "C. Nahl–D. Van Vleck Sc." The bear faces to the right, has pale claws, and light areas on the sides; it closely resembles the first painting, of 1855. The cut accompanies a three-page anonymous article on "The Grizzly Bear of California." Two years later the cut reappeared (Vol. 3, No. 1, p. 5) with figures of other animals similarly signed.

Horace H. Allen of Los Angeles, a grandson of Arthur Nahl, has several original unsigned pictures of mammals—wolf, fox, mink, weasel, seal, and so forth—"probably all made at about the same time." Mr. Allen states that the Colton Hall picture, together with a number of paintings and sketches by Charles and Arthur Nahl, were shipped in the late 1920's from San Francisco to his mother's home in Portland (letter, Nov. 9, 1953). The authorship of the Colton Hall figure remains in doubt.

The design for the grizzly used on the State Flag from about 1928 until 1953 by manufacturers in producing flags ordered on state requisitions is based on a photostat of an illustration in the possession of the California Department of the State Library. That photograph is of the 1855 Charles Nahl drawing (see chap. 12).

The Nahl brothers were associated as engraver-publishers in San Francisco. Several of their products, signed by Charles Nahl as artist, include illustrations of grizzlies. The membership certificate of the Society of California Pioneers has a fine recumbent grizzly at the upper left; that of the San Francisco Volunteer Fire Department has "the arms of the City, a phoenix and a shield, supported by two grizzly bears." Charles Nahl also drew the originals of the twelve illustrations in Hittell's 1860 volume on Grizzly Adams (see our chap. 9); those of Adams and "Ben Franklin" and of "Samson" are particularly noteworthy and were doubtless made from life in Adams' museum. The picture of Adams' menagerie (fig. 26) is ascribed to Nahl along with others. The emblem of the *Overland Monthly* is another example of his work. It is possible that he made still other grizzly illustrations that are hidden in private possession or were lost in the 1906 San Francisco fire.

The naturalist-artist Ernest Thompson Seton made many sketches and paintings of animals, both wild and captive, among which there is one of 'Monarch" made in San Francisco in 1901, and several of his cubs drawn on March 18, 1905 (p. 67).

Two oil paintings of California grizzly subjects are worthy of mention. The first, by James Walker (1818–1889), titled "Roping the Bear at Santa Margarita Rancho" (fig. 15), reportedly was painted about 1876 from notes made by the artist in the 1840's. It shows six gentlemen in decorative Spanish costume on their excited horses, with several *reatas* around the bear's neck. In this respect the portrayal differs from descriptions of bear roping (chap. 5) which emphasize the importance of lassoing the legs of the bear to hold it at bay. The painting, about 24 by 42 inches, is in the possession of the California Historical Society, in San Francisco.

The second is a great painting, 54¾ by 88 inches, "The Return from the Bear Hunt" by Karl Wilhelm Hahn (1829–1887), now hanging in the M. H. de Young Memorial Museum, San Francisco (frontispiece). The artist was born in Ebersbach and died in Dresden; after training at art academics in Germany, he spent the years 1867–1882 in the United States, principally in California. Paintings by Hahn were exhibited in San Francisco in 1874 and 1876 and, posthumously, at the expositions there in 1893 and 1940. The M. H. de Young Museum has seven paintings by this artist, including the one mentioned and a smaller and very poor one (21 by 30 inches), "Bear Hunting," made in 1875 or 1878.

The superbly detailed "Return" was painted in 1882 presumably near The Geysers, Sonoma County (statement of Henry Trost, Aug. 1, 1952). The plant cover shown is representative of that region, where grizzlies were once abundant.

The earliest known portrayal of a California grizzly is that by Ludovik Choris, who accompanied Otto von Kotzebue on the *Rurik*. While the vessel was in San Francisco Bay in 1816, on October 22 hunters brought in a small bear, which served the artist as a model. The rather mediocre drawing (our fig. 12) appears in Choris' *Voyage pittoresque* (1882, Pt. 3, pl. V), a volume issued at Paris in 1822. Baron Georges Cuvier, the distinguished French comparative anatomist, prepared a brief text account to accompany the Choris illustration (Choris, 1822; Mahr, 1932 : 41, 125, 447).

Nearly three decades elapsed before the next picture of grizzlies from California. In the summer of 1850 three cubs were shipped to the Zoological Gardens in London. An account of their travels and a woodcut of the trio appeared in the *Illustrated London News* of October 12, 1850. The first pictures of a bear-and-bull fight are four crayon sketches (fig. 18) by J. D. Borthwick of an affair at Moquelumne Hill in the early 1850's (Borthwick, 1857 : 296). In succeeding years other illustrations of the big bear appeared, but few were of any real merit. Some were in *Vischer's Pictorial* (1870), including the one of Adams' menagerie. A steel engraving of three mounted men with guns and *reatas* shows a bear at bay against a rock. Copyrighted in 1873, it is marked "F.O.C. Darley, Fecit . . . Francis Holl, Sculp." Joseph Warren Revere, the naval officer, in 1849 produced a "Design for the Arms of California" showing a grizzly holding a draped American flag above a horse and bull.

The pioneer photographs of the California grizzly are two of "Bob," a yearling in the Kent family of Chicago, made in 1869 (fig. 30). The slow wet-plate emulsions of those years were scarcely suited to recording the characteristics of a live grizzly, yet the prints give some idea of posture in a young one. Many persons must have taken pictures of "Monarch" (chap. 10), the long-time captive in San Francisco, but only three have come to our attention. One reproduced by Kelley (1903 : 46), which gives evidence of having been retouched, is the only photograph we have seen of the animal on all fours. Roy Graves has two, showing the bear standing and prone. None of the photographs is dated.

Finally, there is one illustration of a mounted grizzly skeleton (our fig. 10), that of a bear captured in 1837 at Monterey. The illustration was published in 1846 in the report of the *Venus* expedition.

## ACKNOWLEDGMENTS

In their search for subject matter the authors of any historical study must depend heavily on many persons and institutions. We are especially grateful for aid from the following: California Academy of Sciences: Don G. Kelley and Robert T. Orr; California Historical Society: J. de T. Abajian and Edna M. Parratt; California State Library: Eudora Garoutte, Mabel R. Gillis, Allan R. Ottley, and Caroline Wenzel; Chas. Scribner's Sons, for use of material in *Adventures of James C. Adams* by Hittell; Henry E. Huntington Library: Leslie R. Bliss; Museum National d'Histoire Naturelle, Paris: J. Millot, for information on and measurements of the Monterey skeleton; Santa Barbara Museum of Natural History: A. S. Cogshill,

Margaret C. Irwin, and Phil C. Orr; United States Fish and Wildlife Service: Viola Schantz, for map of grizzly range; United States National Museum, Division of Mammals: David H. Johnson, for list of specimens and use of collection; United States National Park Service: Lowell Sumner, for use of Dixon photographs; University of California (Berkeley), Department of Anthropology, for access to the Merriam vocabularies; University of California, Bancroft Library: John Barr Tompkins; University of California, Museum of Vertebrate Zoölogy: Seth B. Benson and Alden H. Miller, for use of specimens and for access to correspondence files; Horace H. Allen, for information on Nahl illustrations; Charles L. Camp, for references; Henry B. Collins, for photograph of a trap; Eileen L. Donohue, for transcription of many newspaper articles; Roy D. Graves, for photographs; Alex and Albert Kennedy, for views of trap and skull and for data; William Kent, Jr., for story and photograph of "Bob"; Fred W. Links, for data on the Bear Flag; Mayo O'Donnell, for stories from Monterey; Francis P. Farquhar, for references, especially to the *New York Weekly;* Anne B. Fisher, for notes on Monterey County reminiscences; Joseph Schoeninger, Jr., for stories from Monterey; Arthur Woodward, for miscellaneous notes and references; and Josephine Zane, for miscellaneous notes.

# Bibliography

ABBOTT, C. G.
  1935. Bears in San Diego County, California. Journal of Mammalogy, 16 : 149–151.

ALLEN, G. M.
  1938a. The mammals of China and Mongolia, *in* Natural History of Central Asia. New York, American Museum of Natural History, Vol. 11, pt. 1. xxv+620 pp., 9 pls., 23 figs.
  1938b. Zoological results of the second Dolan expedition to western China . . . Mammals. Academy of Natural Sciences, Philadelphia, Proceedings, 90 : 261–294.
  1942. Extinct and vanishing mammals of the Western Hemisphere . . . Lancaster, Pa., American Committee for International Wildlife Protection, Special Publication 11. xv+620 pp., 24 illus.

ALLEN, J. A.
  1876. The American bisons, living and extinct. Geological Survey of Kentucky, Memoirs, Vol. 1, pt. 2. ix+246 pp., map, 12 pls.

ANDERSON, R. M.
  1946. Catalogue of Canadian Recent mammals. Canada, National Museum, Bulletin 102. v+238 pp., 1 map.

[ANGEL, MYRON]
  1883. History of San Luis Obispo County, California . . . Oakland, Thompson & West. viii+9–385 pp., illus.

Anon. [L. MASCALL]
  1590. A booke of engines and traps . . . London, John Wolfe, *Appended to* A booke of fishing with hooke and line, 1599, pp. 51–93(?).

Anon.
  1850a. Bears from California. Illustrated London News, 17 (Sept. 7): 210.
  1850b. Grisly bears, in the garden of the Zoological Society, Regent's Park. Illustrated London News, 17 (Oct. 12): 298, 300, 1 fig.
  1856. The grizzly bear. Hutching's Illustrated California Magazine, 1 : 106–108; illus.
  1857. The grizzly bear of California. Harper's New Monthly Magazine, 15 : 816–826.
  1858. California animals. Hutching's Illustrated California Magazine, 3 : 1–6; 9 figs.
  1859. The life of Joaquin Murieta, the brigand chief of California. San Francisco, California Police Gazette. 71 pp., illus.

Anon. [TAYLOR, A. S., trans.]
  1860*a*. Journal of a mission-founding expedition north of San Francisco, in 1823 [by José Altimira, June 25–July 6]. Hutching's California Magazine, 5 : 58–62; 115–118.
Anon.
  1860*b*. *Review of* T. H. Hittell, Adventures of James Capen Adams, mountaineer and grizzly bear hunter, of California. Hutching's California Magazine, 5 : 103–114, 4 figs.
  1861. The coast rangers: A chronicle of events in California, III— Hunting adventures. Harper's New Monthly Magazine, 23 : 593–606.
Anon. [signed URSUS]
  1871. Grizzly papers: No. 1. Overland Monthly, 6 : 92–96.
Anon.
  1877. The grizzly bear. Pacific Rural Press, 14 (July 7): 1, 1 fig.
  1880. History of Sonoma County . . . San Francisco, Alley, Bowen. xv+16–717 pp., illus.
  1881. History of Napa and Lake counties, California . . . San Francisco, Slocum, Bowen. xvi+600+291 pp., illus.
  1894. When you see a grizzly what do you think of? Why, of the Overland Monthly. Overland Monthly, 2d series, 24 : 5–8 of adv.
  1932. Joaquin Murieta, the brigand chief of California. Notes by Francis P. Farquhar. San Francisco, Grabhorn Press. x+116+4 pp.
  1936. Continuation of the annals of San Francisco [extracted from the daily papers of the time, June, 1854]. California Historical Society Quarterly, 15 : 163–186.
  1949. The Oregon and California letters of Bradford Ripley Allan. California Historical Society Quarterly, 28 : 199–232.
ANTHONY, H. E.
  1928. Field book of North American mammals. New York, G. P. Putnam. xxv+625 pp., 48 pls., 150 figs., 1 map.
ASCENSIÓN, ANTONIO DE LA
  *ca.* 1611. Relacion de la jornada que hizo el General Sevastian Vizcayno al descubrimiento de las Californias . . . MS. 248 pp. (This is translated in Wagner, 1929.)
AUDUBON, J. J., and JOHN BACHMAN
  1856. The quadrupeds of North America. New York. 3 vols. Vol. 3. v+348 pp., pls. 101–155.
AYERS, J. J.
  1922. Gold and sunshine; reminiscences of early California. Boston, R. G. Badger. xiv+11–359 pp.
BACHMAN, J. H.
  1943. The diary of a "used-up" miner. Edited by J. S. Van Nostrand. California Historical Society Quarterly, 22 : 67–83.
BAILEY, VERNON
  1905. Biological survey of Texas. U. S. Department of Agriculture, Biological Survey, North American Fauna, no. 25. 222 pp., 16 pls., 24 figs.

1936. The mammals and life zones of Oregon. U. S. Department of Agriculture, Biological Survey, North American Fauna, no. 55. 416 pp., 51 pls., 1 map, 102 figs.

BAIRD, S. F.
1857. Mammals [of North America], *in* Reports of explorations and surveys ... for a railroad from the Mississippi River to the Pacific Ocean. . . . Washington. Vol. 8, pt. 1. xix–xlviii+757 pp., 60 pls.

BAKER, A. B.
1912. Notes on animals now, or recently living in the National Zoölogical Park. Smithsonian Miscellaneous Collections, 59(9). 3 pp., 1 pl.

BANCROFT, H. H.
1884–1890. History of California. San Francisco. 7 vols. Vol. 1, 1884 : lxxxviii+744 pp.; vol. 2, 1886 : xvi+795 pp.; vol. 6, 1888 : xi+787 pp.
1888. California pastoral. San Francisco. vi+808 pp.

BANDINI, DON ARTURO
1893. Big game in the West. Californian Illustrated Magazine, 4(2) : 314–320.

BARNES, C. T.
1922. Mammals of Utah. University of Utah Bulletin, 12(15). 166+10 pp., 32 figs. (maps).

BARNUM, P. T.
1875. Struggles and triumphs; or, Forty years' recollections of P. T. Barnum. Buffalo, N.Y. vii+2+[13]–874 pp., 33 figs.

BARRETT, A. S.
1917. Pomo bear doctors. University of California Publications in American Archaeology and Ethnology, 12 : 443–465, pl. 7.

BARTLETT, J. R.
1854. Personal narrative of exploration and incidents in Texas . . . California . . . connected with the United States and Mexican Boundary Commission during ... 1850 ... '53. New York. 2 vols. Vol. 2. xvii+624 pp., illus.

BEALS, R. L.
1933. Ethnology of the Nisenan. University of California Publications in American Archaeology and Ethnology, 31 : 333–413.

BEATTIE, G. W., and H. P. BEATTIE
1939. Heritage of the Valley: San Bernardino's first century. San Pasqual, Calif., San Pasqual Press. xxv+459 pp., illus.

BEECHEY, F. W.
1831. Narrative of a voyage to the Pacific and Beering's Strait . . . in His Majesty's ship Blossom . . . in the years 1825, 26, 27, 28 . . . London. 2 vols. Vol. 2. iv+452 pp.

BELL, HORACE
1927. Reminiscences of a ranger; or, Early times in Southern California. Santa Barbara, Wallace Hebbard. 16+499 pp., illus.
1930. On the Old West Coast ... New York, William Morrow. xiv+336 pp., illus.

BENSON, S. B.
1944. A "specimen" of grizzly bear from Alameda County, California. California Fish and Game, 30 : 98–100, fig. 43.

BIDWELL, JOHN
1897. Echoes of the past. Chico, Calif. 91 pp.

BLAKE, W. P.
1857. Geological Report, *in* Reports of explorations and surveys . . . for a railroad from the Mississippi River to the Pacific Ocean . . . Washington. Vol. 5, pt. 2. xviii+328+xi pp., 22 pls.

BODDAM-WHETHAM, J. W.
1874. Western wanderings: A record of travel in the evening land. London. xii+364 pp.

BOLTON, H. E.
1926. Historical memoirs of New California by Fray Francisco Palóu, O.F.M. Translated into English from the manuscript in the archives of Mexico. Berkeley, University of California Press. 4 vols. Vol. 2. xii+390 pp.
1927. Fray Juan Crespi, missionary explorer on the Pacific Coast, 1769–1774. Berkeley, University of California Press. lxiv+402 pp.
1930. Anza's California expeditions. Berkeley, University of California Press. 5 vols.
1931. In the south San Joaquin ahead of Garces. California Historical Society Quarterly, 10 : 210–219.

BONESTELL, C. L.
1920. A Woodside reminiscence as told by Grizzly Ryder. San Francisco, privately printed. 16 pp., 2 pl.

BORTHWICK, J. D.
1857. Three years in California. Edinburgh. vi+384 pp., illus.
1948. [Reprint.] 3 years in California. Oakland, Calif., Biobooks. 10+318 pp., illus., map.

BOUCARD, A[DOLPHE]
1894. Travels of a naturalist. London. viii+ii+204 pp., front.

BOWMAN, J. N.
1950. The Great Seal of California, *in* California Blue Book [Sacramento], 1950 : 157–171, 7 figs.

BRACE, C. L.
1869. The new West; or, California in 1867–1868. New York. xii+13–373 pp.

BRACKENRIDGE, H. M.
1814. Views of Louisiana; together with a journal of a voyage up the Missouri River, in 1811. Pittsburgh. 304 pp.

BROWNE, J. R.
1862. A dangerous journey. Harper's New Monthly Magazine, 24 : 741–756; 25 : 6–19, illus.
1864. Crusoe's island: A ramble in the footsteps of Alexander Selkirk, with sketches of adventure in California and Washoe. New York. vii+9–436 pp.

BRUFF, J. G.
1949. Gold rush. [The journals, drawings, and other papers of J. Goldsborough Bruff.] Edited by G. W. Read and Ruth Gaines. New York, Columbia University Press. lxxii+794 pp.

BRYANT, EDWIN
1858. What I saw in California; being the journal of a tour . . . in the years 1846, 1847. 2d ed. New York. 455 pp.

BUEL, J. W.
1882. Metropolitan life unveiled; or, The mysteries and miseries of America's great cities . . . St. Louis. 606 pp., illus.
1887. Sea and land. Philadelphia. 800 pp., illus.

BUFFUM, E. G.
1850. Six months in the gold mines; from a journal of three years' residence in Upper and Lower California, 1847–8–9. Philadelphia. xxiv+26–172 pp.

BUNNELL, L. H.
1892. Discovery of the Yosemite, and the Indian war of 1851, which led to that event. 3d ed. New York. 338 pp.

BURGHDUFF, A. E.
1935. Black bears released in southern California. California Fish and Game, 21 : 83–84.

CABRERA, ANGEL, AND JOSÉ YEPES
1940. Mamíferos sud-americanos (vida, costumbres, y description). Buenos Aires, Compania Argentina de Editores. 370 pp., 78 col. pls., map.

CAHALANE, V. H.
1952. Wildlife resources of the National Park system . . . 1951. U. S. Department of the Interior, National Park Service. 135 pp. (mimeographed).

CALIFORNIA WORLD'S FAIR COMMISSION
1894. Final report of the California World's Fair Commisssion [in Chicago] . . . Sacramento. vii+240 pp.

CAMP, C. L.
1937. William Alexander Trubody and the overland pioneers of 1847. California Historical Society Quarterly, 16 : 122–143.
1952. Earth song, a prologue to history. Berkeley and Los Angeles, University of California Press. 8+127 pp., illus.

CANFIELD, C. L. (ed.)
1906. Diary of a forty-niner [Alfred T. Jackson]. New York, Morgan Shepard. xiii+5+231 pp.

CARLETON, M. T.
1938. The Byrnes of Berkeley. California Historical Society Quarterly 17 : 41–49.

CARTER, C. F.
1929. Duhaut-Cilly's account of California in the years 1827–28. California Historical Society Quarterly, 8 : 131–166, 214–250, 306–336.

CASSIN, JOHN
  1858. Mammalogy and Ornithology, *in* Charles Wilkes, United States Exploring Expedition during the years 1838 . . . 1842. Philadelphia. Vol. 8. viii+466 pp.

CAUGHEY, J. W.
  1940. California. New York, Prentice-Hall. xiv+680 pp.

CHAMBERLAIN, N. D.
  1936. The call of gold: True tales on the gold road to Yosemite. Mariposa, Calif., Gazette Press. xii+183 pp., illus.

CHASE, J. S.
  1911. Yosemite trails. Boston, Houghton Mifflin. x+354 pp., illus.

CHEVER, E. E.
  1870. The Indians of California. American Naturalist, 4 : 17–148.

CHICKERING, A. L.
  1938. Bandits, borax, and bears, a trip to Searles Lake in 1874. (Translated from the French of Edmond Leuba.) California Historical Society Quarterly, 17 : 99–117.

CHORIS, LOUIS [LUDOVIK]
  1822. Voyage pittoresque autour du monde . . . Paris. vi+149 pp., 103 pls., maps.

CHRISTMAN, ENOS
  1930. One man's gold; the letters & journal of a forty-niner. Edited by Florence Morrow Christman. New York, McGraw-Hill, Whittlesey House. xiii+278 pp.

CLELAND, R. G.
  1929. Pathfinders. Los Angeles, Powell. 8+452 pp., maps.
  1940. The place called Sespe: The history of a California ranch. Chicago, Lakeside Press. vi+120 pp.
  1941. The cattle on a thousand hills: Southern California, 1850–1870. San Marino, Huntington Library. xiv+327 pp.
  1944. From wilderness to empire: A history of California, 1542–1900. New York, Knopf. xii+388 pp.

CLYMAN, JAMES
  1926. James Clyman, his diaries and reminiscences. Edited by Charles L. Camp. California Historical Society Quarterly, 5 : 109–138; 255–282.

COCKRUM, E. L.
  1952. Mammals of Kansas. University of Kansas Museum of Natural History, Publications, Vol. 7, no. 1. 303 pp., 73 figs.

COLLINS, CARVEL (ed.)
  1949. Sam Ward in the gold rush. Stanford University Press. x+189 pp.

COLTON, WALTER
  1850. Three years in California [1846–1848]. New York. 456 pp., illus.
  1948. [Reprint] Three years in California. Oakland, Calif., Biobooks. xv+261 pp., illus.

COOLIDGE, DANE
  1939. Old California cowboys. New York, Dutton. 158 pp.

[COOPER, J. G.]
  1868. Zoology, *in* T. F. Cronise, The natural wealth of California, pp. 434–501.

CORDUA, THEODOR
  1933. The memoirs of Theodor Cordua, the pioneer of New Mecklenburg in the Sacramento Valley. Translated by Erwin G. Gudde. California Historical Society Quarterly, 12 : 279–311.

CORNELIUS, BROTHER
  1942. Keith, old master of California. New York, Putnam. xix+631 pp., 200 illus.

COUES, ELLIOTT
  1893. History of the expedition under the command of Lewis and Clark . . . New York, 4 vols.

COULTER, E. M.
  1943. California copyrights, 1851–1856. California Historical Society Quarterly, 22 : 27–40.

COUTURIER, M. A. J.
  1954. L'ours brun, Ursus arctos L. Grenoble, Isére, France, Marcel Couturier. xi+905 pp., 82 pls., 49 figs.

COWAN, R. E., and R. G. COWAN
  1933. A bibliography of the history of California. San Francisco, J. H. Nash. v+704 pp.

CRITES, A. S.
  1952. A hunter's tale of the great outdoors. Los Angeles, Ward Ritchie. 229 pp., illus.

CRONISE, T. F.
  1868. The natural wealth of California . . . San Francisco, xvi+696 pp.

CULLIMORE, CLARENCE
  1941. Old adobes of forgotten Fort Tejon. Bakersfield, Kern County Historical Society. 88 pp., 19 figs.

CURTIS, E. S.
  1907–1930. The North American Indian; being a series of volumes picturing and describing the Indians of the United States, and Alaska. Seattle, Curtis. 20 vol. Vol. 14. xiv+284 pp.

DALQUEST, W. W.
  1948. Mammals of Washington. University of Kansas Museum of Natural History, Publications, 2 : 1–144, 140 figs.

DAVIS, W. H.
  1889. Sixty years in California. San Francisco. xxii+639 pp.
  1929. Seventy-five years in California. San Francisco, John Howell. xxxii+422 pp., illus.

DAY, MRS. F. H.
  1859. Sketches of the early settlers of California: George C. Yount. The Hesperian, 2(1) : 1–6.

[DELANO, ALONZO]
  1853. Pen knife sketches; or, Chips of the Old Block. Sacramento. 112 pp., illus.

DIXON, JOSEPH
    1916. Does the grizzly bear still exist in California? California Fish and
        Game, 2 : 65–69.

DIXON, R. B.
    1905. The northern Maidu. American Museum of Natural History,
        Bulletin 17 : 119–346.
    1907. The Shasta. American Museum of Natural History, Bulletin
        17 : 381–498, pls. 59–72, figs. 68–118.

DOLLMAN, J. G., and J. B. BURLACE
    1928. Rowland Ward's records of big game . . . 9th ed. London, Row-
        land Ward. xiii+523 pp., illus.

DOWNIE, WILLIAM
    1893. Hunting for gold: Reminiscences of personal experience and
        research in the early days of the Pacific Coast from Alaska to
        Panama. San Francisco. 407 pp.

DRIVER, H. E.
    1936. Wappo ethnography. University of California Publications in
        American Archaeology and Ethnology, 36 : 179–220.

DU BOIS, C. G.
    1908. The religion of the Luiseño Indians of southern California. Uni-
        versity of California Publications in American Archaeology and
        Ethnology, 8 : 69–186.

DU BOIS, CORA
    1935. Wintu ethnography. University of California Publications in
        American Archaeology and Ethnology, 36 : 1–148.

DU BOIS, CORA, and DOROTHY DEMETRACOPOULOU
    1931. Wintu myths. University of California Publications in American
        Archaeology and Ethnology, 28 : 279–403.

DUHAUT-CILLY, A. [AUGUSTE BERNARD]
    1834. Voyage autour du monde principalement a la Californie et aux
        Iles Sandwich . . . 1826 . . . 1829. Paris. (See C. F. Carter, 1929.)

EDGAR, W. F.
    1893. Historical notes of old land marks in California. . . . Historical
        Society of Southern California Publications, 3 : 22–30.

ELLERMAN, J. R., and T. C. S. MORRISON-SCOTT
    1951. Checklist of Palaearctic and Indian mammals, 1758 to 1946. Lon-
        don, British Museum. 4+810 pp., 1 map.

ELLISON, W. H., and FRANCIS PRICE (eds.)
    1953. The life and adventures in California of Don Agustín Janssens
        1834–1856. San Marino, Huntington Library. xi+165 pp., illus.

ELLSWORTH, R. S.
    1931. Reminiscences of a hunter and collector. California Fish and
        Game, 17 : 87–88.

EMERSON, R. W.
    1904. Society and solitude. Boston, Houghton Mifflin. 451 pp.

ENGELHARDT, ZEPHYRIN
    1913. The missions and missionaries of California. San Francisco, James H. Barry Co. 4 vols. Vol. 3, pt. 2. xviii+663 pp.
    1923. Santa Barbara Mission. San Francisco, James H. Barry. xviii+470 pp.
    1924. San Francisco or Mission Dolores. Chicago, Franciscan Herald Press. xv+432 pp.
    1930. San Buenaventura, the mission by the sea. Santa Barbara, Calif., Mission Santa Barbara. ix+166 pp.
    1932. Mission Santa Ines, Virgen y Martir, and its ecclesiastical seminary. Santa Barbara, Santa Barbara Mission. ix+194 pp.
    1933. Mission San Luis Obispo in the Valley of the Bears. Santa Barbara, Mission Santa Barbara. x+213 pp.
    1934. Mission San Carlos Borromeo (Carmelo): The father of the missions. Santa Barbara, Mission Santa Barbara. xii+264 pp.

ERDBRINK, D. P.
    1953. A review of fossil and recent bears of the Old World. Deventer, Holland, Jan de Lange. 2 vols. xii+597 pp., 22 pls., 61 figs., 1 map.

EVANS, A. S.
    1873. Á la California. Sketches of life in the Golden State. San Francisco. 379 pp., illus. Reissued 1889.

FAGES, PEDRO
    1937. A historical, political, and natural description of California by Pedro Fages, soldier of Spain. Translated by H. I. Priestley. Berkeley, University of California Press. xi+83 pp.

FAIRFIELD, A. M.
    1916. Fairfield's pioneer history of Lassen County California . . . San Francisco, Crocker. xxii+506 pp.

FARNHAM, T. J.
    1849. Life, adventures and travels in California. New York. 468 pp., illus.
    1947. [Reprint:] Travels in California. Oakland, Calif., Biobooks. xv+166 pp.

FARQUHAR, F. P.
    1930. Up and down California in 1860–1864. [Letters of W. H. Brewer, edited by Francis P. Farquhar.] New Haven, Yale University Press. xxx+601 pp., 33 pls.
    1947. The grizzly bear hunter [Adams] of California, a bibliographic essay, *in* Essays for Henry R. Wagner. San Francisco, Grabhorn, pp. 27–42.

FERRY, PHILIP
    1945. The passing of the California grizzly. Westways, 37(10, pt. 1) : 14.

FISHER, A. K.
    1920. In Memoriam: Lyman Belding. Auk, 37 : 33–45, 1 fig.

FISHER, ANNE B.
    1945. The Salinas, upside-down river. New York, Farrar & Rinehart. xiii+316 pp., illus.

FLINT, TIMOTHY (ed.)
1930. The personal narrative of James O. Pattie of Kentucky. Chicago, Lakeside Press. xliii+428 pp. (A reprint; 1st ed., Cincinnati. 1831.)

FLOWER, S. S.
1931. Contributions to our knowledge of the duration of life in vertebrate animals. V, Mammals. Zoological Society of London, Proceedings, 1931 : 145–234.

FRÉMONT, J. C.
1887. Memoirs of my life . . . Chicago. xix+655 pp., illus.

FREMONT, COL. [J. C.], and MAJ. [W. H.] EMORY
1849. Notes of travel in California . . . From the official reports of Col. Fremont and Maj. Emory. New York. 29+83 pp.

FROWD, J. G. PLAYER
1872. Six months in California. London. 3+164 pp.

FRY, WALTER
1924. The California grizzly. Sequoia National Park, Historic Series, Nature Guide Service, Bulletin 2, Dec. 4, 1924. 4 pp. (mimeographed).

G., W. [W. R. GARNER]
1847a. Bear-hunting in California, *in* Littell's Living Age, 14 (171) : 370–371.

1847b. Letters from California (1846). The North American, Philadelphia. 1 folio sheet. [Reprint, 1924:] Letters from California, 1846. The Magazine of History (Tarrytown, N.Y.), 26(3) : 153–205.

GARNETT, PORTER
1913. San Francisco one hundred years ago. Translated from the French of Louis Choris. San Francisco, Robertson. v+20 pp., illus.

GARTH, T. R.
1953. Atsugewi ethnography. University of California Publications: Anthropological Records, 14(2) : 128–212.

GAYTON, A. H.
1948. Yokuts and Western Mono ethnography. II: Northern foothill Yokuts and Western Mono. University of California Publications: Anthropological Records, 10(2) : 142–301.

GEOFFROY SAINT-HILAIRE, I. [ISADORE?]
1846. Atlas de zoologie, *accompanying* A. A. Du Petit-Thouars, Voyage autour du monde sur la frégate La Vénus, pendant les années 1836–1839. Paris, Gide, 1840–1864. 70 pl. Pl. V.

1852. Mammifères, *in* A. A. Du Petit-Thouars, Voyage autour du monde sur la frégate La Vénus . . . Vol. 5(?)

GIBBS, GEORGE
1853. Journal of the expedition of Colonel Redick M'Kee . . . through northwestern California . . . in 1851, *in* H. R. Schoolcraft, Archives of aboriginal knowledge . . . history . . . of the Indian Tribes of the United States (Philadelphia), 3 : 99–177.

GIDNEY, C. M., BENJAMIN BROOKS, and E. M. SHERIDAN
1917. History of Santa Barbara, San Luis Obispo and Ventura counties, California. Chicago, Lewis Publishing Co. 2 vols. Vol. 1 : xvii+ 486 pp.; vol. 2 : 487–873 pp.

GIFFEN, H. S., and ARTHUR WOODWARD
1942. The story of El Tejon. Los Angeles, Dawson. 146 pp.

GIFFORD, E. W.
1932. The Northfork Mono. University of California Publications in American Archaeology and Ethnology, 31 : 15–65.
1947. California shell artifacts. University of California Publications: Anthropological Records, 9(1): iii+132 pp.

GIFFORD, E. W., and G. H. BLOCK
1930. Californian Indian nights entertainments . . . Glendale, Calif., Clark. 323 pp.

GILBERT, F. T.
1879. The illustrated atlas and history of Yolo County, Calif. San Francisco. iv+105 pp., illus.

GODDARD, P. E.
1903. Life and culture of the Hupa. University of California Publications in American Archaeology and Ethnology, 1 : 1–88.

GOLDSCHMIDT, WALTER
1951. Nomlaki ethnography. University of California Publications in American Archaeology and Ethnology, 42 : viii+303–443.

GRINNELL, JOSEPH
1933. Review of the recent mammal fauna of California. University of California Publications in Zoölogy, 40 : 71–234.
1938. California's grizzly bears. Sierra Club Bulletin, 23(2) : 70–81.

GRINNELL, JOSEPH, J. S. DIXON, and J. M. LINSDALE
1937. Fur-bearing mammals of California, their natural history, systematic status, and relations to man. Berkeley, University of California Press. 2 vols. Vol. 1 : xii+375 pp., 7 pls., 138 figs; vol. 2 : xiv+377 pp., pls. 8–13, figs. 139–345.

GRINNELL, JOSEPH, and T. I. STORER
1924. Animal life in the Yosemite. Berkeley, University of California Press. xviii+752 pp., 60 pls., 2 maps, 65 figs.

GRINNELL, JOSEPH, and H. S. SWARTH
1913. An account of the birds and mammals of the San Jacinto area of southern California . . . University of California Publications in Zoölogy, 10 : 197–406, 5 pls., 3 figs.

GUDDE, E. G.
1949. California place names. Berkeley and Los Angeles, University of California Press. xxviii+431 pp.

GUTHRIE, WILLIAM
1815. A new geographical, historical, and commercial grammar; and present state of the several kingdoms of the world. 2d American ed. improved. Philadelphia, Johnson & Warner. 2 vols. Vol. 2. Zoology, by George Ord.
1894. [Reprint; see Rhoads, 1894.]

HAFEN, L. R.
  1953. Colorado mountain men. Colorado Magazine, 30 : 14–28.
HALL, E. R.
  1928. A new race of black bear from Vancouver Island, British Colum-
    bia, with remarks on other northwest coast forms of Euarctos.
    University of California Publications in Zoölogy, 30 : 231–242,
    pls. 12, 13.
  1939. The grizzly bear of California. California Fish and Game, 25 :
    237–244, figs. 90–92.
  1946. Mammals of Nevada. Berkeley and Los Angeles, University of
    California Press. xi+710 pp., 11 pls., 485 figs.
[HARTE, BRET]
  1868. Editorial comment. Overland Monthly, 1(1) : 99–100.
HARTE, BRET
  1875. Baby Sylvester, *in* Tales of the Argonauts and other sketches.
    Boston. pp. 173–198.
HASTINGS, L. W.
  1845. The emigrant's guide to Oregon and California containing . . .
    a description of California . . . Cincinnati. 152 pp.
HEARNE, SAMUEL
  1795. A journey from Prince of Wales's Fort, in Hudson's Bay to the
    Northern Ocean . . . London. xliv+458 pp., illus.
HEERMANN, A. L.
  1859. Report upon the birds collected on the survey [near the 35th
    and 32d parallels], *in* Reports of explorations . . . for a railroad
    from the Mississippi River to the Pacific Ocean . . . Washington.
    Vol. 10, pt. 4, Zoological Report [no. 2], pp. 29–80, 7 pls.
HELPER, H. R.
  1855. The land of gold. Reality versus fiction. Baltimore. xii+300 pp.
HENSHAW, H. W.
  1876. Notes on the mammals taken and observed in California in 1875
    by H. W. Henshaw, *in* G. M. Wheeler, Annual report on the
    geographical surveys west of the one hundredth meridian; annual
    report of the Chief of Engineers for 1876, App. JJ : 305–312.
HERRICK, B. F.
  1946. Grade-school grizzly. California Historical Society Quarterly,
    25 : 178–181.
HESS, C. N.
  1930. Moccasin Jim. Touring Topics, 22(5) : 28–31, 52, 4 illus.
HICKIE, PAUL
  1952. Inventory of big-game animals in the United States. U. S. Fish
    and Wildlife Service, Wildlife Leaflet 348. 2 pp.
HITTEL[L], J. S.
  1863. The resources of California . . . San Francisco. xvi+464 pp.
HITTELL, T. H.
  1860. The adventures of James Capen Adams, mountaineer and grizzly
    bear hunter, of California. San Francisco, Towne and Bacon.
    vi+378 pp., 12 figs. Also 1860, Boston, Crosby, Nichols, Lee and
    Co. Reprinted 1861, 1867.

1885; 1898. History of California. San Francisco. 4 vols.

1911. The adventures of James Capen Adams, mountaineer and grizzly bear hunter of California. New York, Scribner. xiii+373 pp., 12 illus.

HOBBS, JAMES

1875. Wild life in the Far West; personal adventures of a border mountain man. Hartford, Conn. 488 pp., 33 figs.

HOLT, CATHARINE

1946. Shasta ethnography. University of California Publications: Anthropological records, 3(4) : 298–349.

HOOVER, M. B., H. E. RENSCH, and E. G. RENSCH

1948. Historic spots in California. [2d ed.] Stanford University Press. xiv+411 pp.

HUSSEY, J. A.

1952. New light on the original bear flag. California Historical Society Quarterly, 31 : 205–217.

HUTCHINGS, J. M.

1870. Scenes of wonder and curiosity in California. New York. 292 pp., more than 100 figs.

INGERSOLL, L. A.

1904. Ingersoll's century annals of San Bernardino County 1769 to 1904 . . . Los Angeles, Ingersoll. xxii+887 pp., illus.

ISBELL, F. A.

1948. Mining & hunting in the Far West, 1852–1870. Introduction by Nathan Van Patten. Burlingame, Calif., W. P. Wreden. 13+36 pp.

JACKSON, H. H. T.

1944a. Big-game resources of the United States. U. S. Department of the Interior, Fish and Wildlife Service, Research Report 8 : 1–56, 31 figs.

1944b. Conserving endangered wildlife species. Wisconsin Academy of Sciences, Arts and Letters, 35 : 61–89, 2 figs., 4 maps.

JACKSON, J. J.

1945. Early history of Trinity County. The Mountaineer (Rotary Club, Weaverville, Calif.), 7, nos. 16–19.

JOHNSON, T. T.

1849. Sights in the gold region, and scenes by the way. New York. xii+278 pp.

KELLEY, ALLEN

1903. Bears I have met—and others. Philadelphia, Drexel Biddle. 209 pp., 16 illus.

KELLY, WILLIAM

1851. An excursion to California over the . . . Sierra Nevada. London. 2 vols. Vol. 2. viii+334 pp.

KENT, G. F.

1941. Life in California in 1849, as described in the "journal" of George F. Kent [edited by J. W. Caughey]. California Historical Society Quarterly, 20 : 26–46.

KILIAN, C. H.

1949. The story of the Great Seal. What's Doing, August, 1949 : 7.

KINGSLEY, HOMER
   1920. Roping grizzlies. Overland Monthly, n.s., 97(6) : 22–24. 1 fig.
      [of black bear!].

KIP, W. I.
   1921. A California pilgrimage. Fresno, Calif., Louis C. Sanford, pri-
      vately printed. 64 pp., illus.

KOTZEBUE, OTTO VON
   1821a. A voyage of discovery, into the South Sea and Beering's Straits
      . . . in the years 1815–1818 . . . in the ship Rurick . . . Translated
      by H. E. Lloyd. London. 3 vols. Vol. 1 : xvi+358 pp., 6 illus.;
      vol. 3 : 1+442 pp., 3 illus.
   1821b. Voyage of discovery in the South Sea, and to Behring's Straits,
      . . . in the years 1815, 16, 17, and 18, in the ship Rurick. London,
      Sir Richard Phillips and Co. *In* Richard Phillips, New voyages
      and travels, London, 1819–1823, vol. 6, nos. 1 and 2.

KROEBER, A. L.
   1907. The religion of the Indians of California. University of Cali-
      fornia Publications in American Archaeology and Ethnology,
      4 : 319–356.
   1916. California place names of Indian origin. University of California
      Publications in American Archaeology and Ethnology, 12 : 31–
      69.
   1922. Elements of culture in native California. University of California
      Publications in American Archaeology and Ethnology, 13 : 259–
      328.
   1925. Handbook of the Indians of California. U. S. Bureau of Ameri-
      can Ethnology, Bulletin 78. xviii+995 pp., 83 pls., 78 figs.
   1929. The valley Nisenan. University of California Publications in
      American Archaeology and Ethnology, 24 : 253–290.
   1932. The Patwin and their neighbors. University of California Publi-
      cations in American Archaeology and Ethnology, 29 : 253–423.

LAHONTAN, L. A., BARON DE
   1703. New voyages to North America. . . . London. 2 vols. in 1.

LAWRENCE, BARBARA
   1944. Skull of a California grizzly. California Fish and Game, 30 : 98.

LEONARD, ZENAS
   1904. Leonard's narrative: Adventures of Zenas Leonard, fur trader
      and trapper, 1831–1836. Edited by W. F. Wagner. Cleveland,
      Burrows. 317 pp., 6 figs. Reprinted from ed. of 1839.

LETTS, J. M.
   1852. California illustrated: including a description of the Panama and
      Nicaragua routes by a returned Californian. New York. vii+224
      pp.

LILIENCRANTZ, H. T.
   1948. Grizzly bears and reatas in early California. Western Horseman,
      13(5) : 12–13, 49–52.

LOEB, E. M.
  1926. Pomo folkways. University of California Publications in American Archaeology and Ethnology, 19 : 149–405.
  1932. The western Kuksu cult. University of California Publications in American Archaeology and Ethnology, 33 : 1–137.
  1933. The eastern Kuksu cult. University of California Publications in American Archaeology and Ethnology, 33 : 139–232.

LYMAN, C. S.
  1924. Around the Horn to the Sandwich Islands and California, 1845–1850. Edited by F. J. Teggart. New Haven, Yale University Press. xviii+328 pp., illus.

MCALLISTER, M. H.
  1919. Game conditions in southern California thirty-five years ago. California Fish and Game, 5 : 172–173.

MCCAULEY, JAMES
  1910. How a grizzly stopped berrying. Grizzly Bear, 6(3) : 5.

MCCLELLAN, R. G.
  1872. The golden state: A history of the region west of the Rocky Mountains; embracing California . . . Philadelphia. [ii]+15–685 pp., illus.

MCGROARTY, J. S.
  1911. California its history and romance. Los Angeles, Grafton Publishing Co. 393 pp.

MACKIE, W. W.
  1915. Antelope in Kern County. California Fish and Game, 15 : 285.

MCKITTRICK, M. M.
  1944. Vallejo, son of California. Portland, Binfords & Mort. 377 pp.

MADDEN, H. M.
  1949. Xántus, Hungarian naturalist in the pioneer west. Palo Alto, Books of the West. 312 pp., 5 figs.

MAHR, A. C.
  1932. The visit of the "Rurik" to San Francisco in 1816. Stanford University Press, 194 pp., illus.

MALONEY, A. B. (ed.)
  1943, 1944. Fur brigade to the Bonaventura: John Work's California expedition of 1823–33 for the Hudson's Bay Company. California Historical Society Quarterly, 22 : 193–222, 323–348; 23 : 19–40, 123–146.

MARRYAT, F. S.
  1855. Mountains and molehills; or, Recollections of a burnt journal. London. x+443 pp.
  1952. [Reprint:] Mountains and Molehills. Stanford University Press. xiv+393 pp.

MARTIN, O. F.
  1887. Bears. Overland Monthly, ser. 2, 10 : 33–50.

MAZET, H. S.
  1946. Mightiest hunter of them all. Saturday Evening Post, 219 (Oct 19) : 38, 105–108, 110, 1 fig.

MERRIAM, C. H.
1896. Preliminary synopsis of the American bears. Biological Societv
of Washington, Proceedings, 10 : 65–83, pls. 4–6, 17 figs.
1899. Results of a biological survey of Mount Shasta California. U. S.
Department of Agriculture, Biological Survey, North American
Fauna, 16 : 179 pp., 5 pls., 46 figs.
1914. Descriptions of thirty apparently new grizzly and brown bears
from North America. Biological Society of Washington, Proceedings, 27 : 173–196.
1918. Review of the grizzly and big brown bears of North America
(Genus Ursus) . . . U. S. Department of Agriculture, Biological
Survey, North American Fauna, 41 : 1–136, 16 pls.
1922. Distribution of grizzly bears in [the] U. S. Outdoor Life, 50
(6) : 405–406, 1 map.
1925. Recent evidence of grizzly bear in the Sierras. Sierra Club, Cir-
cular 12 : 1–4.

MEYER, CARL
1938. Bound for Sacramento. Travel-pictures of a returned wanderer.
Translated by Ruth Frey Axe. Claremont, Calif., Saunders Stu-
dio Press. xii+2+282 pp.

MILLER, G. S.
1912. Catalogue of the mammals of western Europe. London, British
Museum. xv+1019 pp., 213 figs.
1924. List of North American Recent mammals, 1923. U. S. National
Museum, Bulletin 128. xvi+673 pp.

MILLER, JOAQUIN
1900. True bear stories. Chicago, Rand McNally. 6+259 pp., 6 pls.

MASON, J. A.
1912. The ethnology of the Salinan Indians. University of California
Publications in American Archaeology and Ethnology, 10 : 97–
240, pls. 21–37.

MATSON, J. R.
1954. Observations on the dormant phase of a female black bear. Journal of Mammalogy, 35 : 28–35, 1 pl.

MILLS, E. A.
1919. The grizzly our greatest wild animal. Boston, Houghton Mifflin.
xiii+289 pp., illus.

MORA, JO
1949. Californios, the saga of the hard-riding vaqueros, America's first
cowboys. Garden City, N.Y., Doubleday. 175 pp., illus.

MUIR, JOHN
1894. Picturesque California: the Rocky Mountains and the Pacific
Slope. Edited by John Muir. New York. 508 pp., illus.
1898. Among the animals of the Yosemite. Atlantic Monthly, 82 : 617–
622.
1917. Our national parks. Boston, Houghton Mifflin. xii+391 pp., illus.
1938. John of the mountains; the unpublished journals of John Muir.
Edited by L. M. Wolfe. Boston, Houghton Mifflin. xii+459 pp.,
illus.

NAHL, P. W.
  1894. Souvenir, early days in California. San Francisco. 32 pp., illus.
NELSON, E. W.
  1918. Wild animals [mammals] of North America. Washington, National Geographic Society. 228 pp., illus.
NEUHAUS, EUGEN
  1936. Charles Christian Nahl, the painter of California pioneer life. California Historical Society Quarterly, 15 : 295–305, 1 fig.
NEWBERRY, J. S.
  1857. Report upon the zoölogy of the route [Sacramento Valley to Columbia River], *in* Reports of explorations and surveys . . . for a railroad from the Mississippi River to the Pacific Ocean. Vol. 6, pt. iv, no. 2, pp. 35–110, pls. 26–29.
NEWMARK, HARRIS
  1916. Sixty years in southern California, 1853–1913 . . . Edited by M. H. and M. R. Newmark. New York, Knickerbocker. xxviii+732 pp., 150 figs.
  1926. Sixty years in southern California, 1853–1913 . . . Edited by M. H. and M. R. Newmark. New York, Knickerbocker. xxxiii+ 732 pp., 172 figs.
  1930. Sixty years in southern California, 1853–1913 . . . Edited by M. H. and M. R. Newmark. Boston, Houghton Mifflin. xxxv+744 pp., 182 illus.
NIDEVER, GEORGE
  1937. The life and adventures of George Nidever. Edited by W. H. Ellison. Berkeley, University of California Press. xi+128 pp.
NORDHOFF, CHARLES
  1875. California for health, pleasure, and residence . . . New York. 255 pp., illus.
O'BRIEN, ROBERT
  1951. California called them. New York, McGraw-Hill. xv+251 pp., illus.
ORD, GEORGE
  1815. [See Guthrie, 1815.]
O[RR], P. C.
  1942. The Ojai expedition. Santa Barbara Museum of Natural History, Museum Leaflet, 17(7) : 79–82.
ORR, R. T.
  1950. Additional records of *Ursus californicus*. Journal of Mammalogy, 31 : 362–363.
PALLISER, JOHN
  1859. The solitary hunter: Sporting adventures in the prairies. [2d ed.] London. xvi+234 pp., illus.
PALOU, FRANCISCO
  1913. Francisco Palou's Life and apostolic labors of the venerable Father Junípero Serra, founder of the Franciscan missions of California. Introduction and notes by George Wharton James. Translated by C. Scott Williams. Pasadena, G. W. James. xxxiv+338 pp.

PATTIE, J. O.
   1831. [See Flint, 1930, and Thwaites, 1905.]
PAULSON, L. L.
   1874. Handbook and directory of Napa, Lake, Sonoma and Mendo-
      cino counties ... San Francisco. xvi+296 pp., illus.
PENNOYER, A. S. (comp. and ed.)
   1938. This was California; a collection of woodcuts and engravings
      of historical events, human achievements and trivialities from
      pioneer days to the gay nineties. New York, Putnam. 224 pp.,
      illus.
PERLOT, J.-N.
   1897. Vie et aventures d'un enfant de l'Ardenne, autobiographie par
      J.-N. Perlot. Arlon. 545+x pp.
PINKERTON, JOHN
   1812. A general collection of the best and most interesting voyages
      and travels ... London. Vol. 13, North America. 876 pp.
POTTER, E. G.
   1945. Columbia—"Gem of the southern mines." California Historical
      Society Quarterly, 24 : 267–270.
POWERS, STEPHEN
   1872. Afoot and alone ... adventures ... in southern California ...
      Hartford, Conn. xvi+17–327 pp.
   1877. Tribes of California. Contributions to North American Ethnol-
      ogy, Vol. 3. 635 pp.
PRICE, W. W.
   1894. Notes on a collection of mammals from the Sierra Nevada Moun-
      tains. Zoe, 4 : 315–332.
PRIESTLEY, H. I.
   1937. A historical, political, and natural description of California by
      Pedro Fages, soldier of Spain. Berkeley, University of California
      Press. xi+83 pp.
RAUSCH, ROBERT
   1953. On the status of some Arctic mammals. Arctic, 6 : 91–148, 17
      figs.
READING, P. B.
   1930a. [Letter of Feb. 7, 1844, from Monterey to his brother.] Society
      of California Pioneers, Quarterly, 7(3) : 143–147.
   1930b. Journal of Pierson Barton Reading ... in 1843. Society of Cali-
      fornia Pioneers, Quarterly, 7(3) : 148–198.
REID, H. A.
   1895. History of Pasadena. Pasadena, Calif. 675 pp., illus.
REID, HUGO
   1926. The Indians of Los Angeles County. Los Angeles, privately
      printed. 70 pp.
REVERE, J. W.
   1849. A tour of duty in California. New York. vi+305 pp.
   1947. [Reprint:] Naval duty in California. Oakland, Calif., Biobooks.
      15+245 pp., illus.

RHOADS, S. N.
  1894. A reprint of the North American zoology, by George Ord. . . . the part originally compiled by Mr. Ord for Johnson & Warner and first published in their second American edition of Guthrie's geography, in 1815. . . . To which is added an appendix on the more important scientific and historic questions involved. By Samuel N. Rhoads. Haddonsfield, N.J., published by the editor. x+1 l., 290–362+52 pp. of appendix+53–90 pp. of index.

ROBINSON, ALFRED
  1846. Life in California. New York. xii+341 pp.
  1947. [Reprint:] Life in California. Oakland, Biobooks. xviii+147 pp., illus.

ROJAS, A. R.
  1953. California vaquero. Fresno, Calif., Academy Library Guild. 125 pp., illus.

ROOSEVELT, THEODORE
  1893. The wilderness hunter. Part II. Hunting the grisly. New York. 279 pp.

SANCHEZ, NELLIE VAN DE GRIFT
  1928. Memoirs of a merchant, being the recollections of life and customs in pastoral California by José Arnaz, trader and ranchero. Translated and edited by Nellie van de Grift Sanchez. Touring Topics, 20(9) : 14–19, 47–48.
  1929. Spanish Arcadia. San Francisco, Powell. 399 pp.

SCHRADER, G. R.
  1946. The grizzly bear of California. Siskiyou County Historical Society, 1(1) : 15–18.

SETON, E. T.
  1904. Monarch, the big bear of Tallac. New York, Scribner. 214 pp., illus.
  1909. Life histories of northern animals: An account of the mammals of Manitoba. New York, Scribner. 2 vols. Vol. 2. xii+647–1267 pp., pls. 47–100, figs. 183–267, maps 39–68.
  1925–1928. Lives of game animals [mammals of North America]. 4 vols. New York, Doubleday, Page.
  1929. [Reissue:] Lives of game animals. New York, Doubleday, Doran. 4 vols. in 8. Vol. 2. xvii+746 pp., 101 pls., 27 figs., 13 maps.

SHINN, C. H.
  1890. Californiana. Grizzly and pioneer. Century Magazine, 41(1) : 130–131.

SIMPSON, G. G.
  1945. The principles of classification and a classification of mammals. American Museum of Natural History Bulletin 85. 350 pp.

SIMPSON, L. B. (trans.)
  1938. California in 1792: The expedition of José Longinos Martínez. San Marino, Huntington Library. xiii+111 pp.

SKINNER, M. P.
  1925. Bears in the Yellowstone. Chicago, McClurg. 10+158 pp., illus.
  1936. Birth and early life of grizzly bears. Outdoor America, n.s. 1(5) :
       4–5, 9; (6) 4–5, 14, 4 figs.

SMITH, D. E., and F. J. TEGGART (eds.)
  1909. Diary of Gaspar de Portola during the California expedition of
       1769–1770. Academy of Pacific Coast History, Publications, 1(3) :
       30–59.

SOULÉ, FRANK, J. H. GIHON, and JAMES NISBET (comps.)
  1855. The annals of San Francisco . . . New York. 822 pp., illus.

SPARKMAN, P. S.
  1908. The culture of the Luiseño Indians. University of California
       Publications in American Archaeology and Ethnology, 8 : 187–
       234.

SPIER, LESLIE
  1930. Klamath ethnography. University of California Publications in
       American Archaeology and Ethnology, 30 : 1–338.

SPURR, G. G.
  1881. The land of gold: A tale of '49, illustrative of early pioneer life
       in California, and founded upon fact. Boston. ix+271 pp.

STEPHENS, FRANK
  1906. California mammals. San Diego, West Coast Publishing Co. 351
       pp., illus.

STEVENS, MONTAGUE
  1943. Meet Mr. Grizzly: A saga on the passing of the grizzly. Al-
       buquerque, University of New Mexico Press. x+281 pp., illus.

STEWARD, J. H.
  1933. Ethnography of the Owens Valley Paiute. University of Cali-
       fornia Publications in American Archaeology and Ethnology,
       33 : 233–350.

STILLMAN, J. D. B.
  1877. Seeking the golden fleece . . . pioneer life in California . . . San
       Francisco. 352 pp., illus.

STOCK, CHESTER
  1936. Ursus; or, The past of the California bears. Westways, 28(11) :
       30.

STONE, LIVINGSTON
  1883. Habits of the panther in California. American Naturalist, 17 :
       1188–1190.

STORER, T. I., and L. P. TEVIS, JR.
  1953. The California grizzly—emblem of the golden state. Pacific Dis-
       covery, 6(4) : 22–27, 9 figs.

STRONG, W. D.
  1926. Indian records of California carnivores. Journal of Mammalogy,
       7 : 59–60.
  1929. Aboriginal society in southern California. University of Cali-
       fornia Publications in American Archaeology and Ethnology,
       Vol. 26. x+358 pp., 7 maps.

SUCKLEY, GEORGE, and GEORGE GIBBS
1860. Report upon the mammals collected on the survey [near the 47th and 49th parallels], *in* Reports of explorations . . . for a railroad from the Mississippi River to the Pacific Ocean . . . Washington. Vol. 12, Book II, Zoological Report [no. 2, chap. iv], pp. 107–139, 5 pls.

SULLIVAN, M. S.
1934. The travels of Jedediah Smith: A documentary outline including the journal of the great American pathfinder. Santa Ana, Calif., Fine Arts Press. 16+195 pp., map, 12 illus.

SUMNER, LOWELL, and J. S. DIXON
1953. Birds and mammals of the Sierra Nevada, with records from Sequoia and Kings Canyon National Parks. Berkeley and Los Angeles, University of California Press. xvii+484 pp., illus.

SYMONS, T. W.
1878. . . . Report of . . . [Corps of Engineers] party no. 1, California Section, Field Season of 1877. Secretary of War, Report, 1877. Vol. 2, pt. 3, pp. 1535–1542.

T. B. M.
1907. A bear steak. The Grizzly Bear, 1(5) : 30–31.

TAYLOR, BAYARD
1850. El Dorado; or, Adventures in the path of empire . . . New York. 2 vols. Vol. 2. 247 pp., illus.
1951. [Reprint:] El Dorado . . . Oakland, Calif., Biobooks. xviii+135 pp., illus.

TAYLOR, WILLIAM
1858. California life illustrated. New York. 348 pp., illus.

TEGGART, F. J.
1911. The Portola expedition of 1769–1770: Diary of Miguel Costanso. Academy of Pacific Coast History, Publications, 2(4) : 1–167.

THOMAS, D. J.
1856. The charter and ordinances of the City of Sacramento. Sacramento.

THORNTON, J. Q.
1855. Oregon and California in 1848. New York. 2 vols. Vol. 2. ix+2+ 13–379 pp., 6 figs.

THWAITES, R. G. (ed.)
1905. Early western travels, 1748–1846. Cleveland, Clark, 1904–1907. 32 vols. Vol. 18. Pattie's Personal Narrative, 1824–1830. [Reprint of original edition, Cincinnati, 1831.] 379 pp., 6 illus.

TINSLEY, H. G.
1902. Grizzly bear lore. Outing, 41 : 154–162, illus.

TORCHIANA, H. A. VAN COENEN
1933. Story of the Mission Santa Cruz. San Francisco, Paul Elder. xix+460 pp.

TOWNSEND, C. H.
1887. Field-notes on the mammals, birds and reptiles of northern California. U. S. National Museum, Proceedings, 10 : 159–241.

TYSON, J. L.
  1850.  Diary of a physician in California . . . New York. 92 pp.
UMFREVILLE, EDWARD
  1790.  The present state of Hudson's Bay . . . London. vii+230 pp.
U. S. DECENNIAL CENSUS
  1931.  Fifteenth census of the United States: 1930. Washington, Gov-
          ernment Printing Office. Vol. I. Population . . . iv+1268 pp.
VACHELL, H. A.
  1901.  Life and sport on the Pacific Coast. New York, Dodd Mead.
          x+393 pp.
VALLEJO, GUADALUPE
  1890.  Ranch and mission days in Alta California. Century Magazine,
          41(2) : 183–192.
VAN ATTA, C. E.
  1946.  Notes on the former presence of grizzly and black bears in
          Marin County, California. California Fish and Game, 32 : 27–29.
VICTOR, F. F.
  1872.  All over Oregon and Washington. San Francisco. vi+7–368 pp.
VISCHER, EDWARD
  1870.  Vischer's Pictorial of California . . . scenery, life, traffic and
          customs. San Francisco. 16 pp., 199 photos.
WAGNER, H. R.
  1929.  Spanish voyages to the northwest coast of America in the six-
          teenth century. San Francisco, California Historical Society.
          viii+571 pp., 20 pls.
WELCH, W. R.
  1931.  Game reminiscences of yesteryears. California Fish and Game,
          17 : 255–263.
WHEAT, C. I.
  1942.  The maps of the California gold region, 1848–1857. San Fran-
          cisco, Grabhorn. xlii+153 pp., 26 maps.
WHITE, K. A.
  1930.  A Yankee trader in the gold rush. The letters of Franklin A.
          Buck. Boston, Houghton Mifflin. viii+294 pp., 8 illus.
WILKES, CHARLES
  1844.  Narrative of the United States Exploring Expedition during the
          years 1838 . . . 1842. Philadelphia. 5 vols. and atlas. Vol. 5. xv+
          591 pp., illus.
WILSON, B. D.
  1877.  Observations on early days in California and New Mexico. MS
          in Bancroft Library.
WINSHIP, G. P.
  1896.  The Coronado Expedition, 1540–1542. U. S. Bureau of [Ameri-
          can] Ethnology, Annual Report, 14(2) : 329–613, pls. 38–80.
WISTAR, I. J.
  1937.  Autobiography of Isaac Jones Wistar 1827–1905. Philadelphia,
          Wistar Institute of Anatomy and Biology. vii+528 pp., 8 illus.,
          1 map.

WOLLE, M. S.
  1953. The bonanza trail ghost towns and mining camps of the West. Bloomington, Indiana University Press. xvi+510 pp., illus.
WOOD, L. K.
  1932. Discovery of Humboldt Bay, a narrative. Society of California Pioneers, Quarterly, 9(1) : 40–64.
WOOD, R. F.
  1954. The life and death of Peter Lebec. Fresno, Calif., Academy Library Guild. 78 pp.
WRIGHT, G. M., J. S. DIXON, and B. H. THOMPSON
  1933. A preliminary survey of faunal relations in national parks. U. S. Department of the Interior, National Park Service, Fauna Series, Vol. 1. iv+157 pp., 56 figs.
WRIGHT, W. H.
  1909. The grizzly bear, the narrative of a hunter-naturalist, historic, scientific and adventurous. New York, Scribner. x+274 pp., illus.
XÁNTUS, JÁNOS
  1860a. Utazás Kalifornia déli részeiben. Pest. 10+191 pp., 16 figs.
  1860b. Travels in the southern part of California . . . [Translation by Edgar H. Yolland of the foregoing book.] Typescript, Bancroft Library. 9+304 pp.
YOUNT, G. C.
  1923. The chronicles of George C. Yount, California pioneer of 1826. California Historical Society Quarterly, 2 : 3–66.

# EARLY NEWSPAPER ARTICLES ON THE CALIFORNIA GRIZZLY

References in the text are indicated thus: (N 24)

1. San Francisco *Californian*, March 15, 1848.
2. *Ibid.*, April 26, 1848.
3. Sacramento, *Placer Times*, February 9, 1850.
4. San Francisco *Daily Alta California*, February 20, 1850.
5. *Ibid.*, February 7, 1850.
6. *Ibid.*, June 24, 1850.
7. *Ibid.*, September 11, 1850.
8. *Ibid.*, October 4, 1850.
9. *Ibid.*, October 16, 1850.
10. Sacramento *Transcript*, March 12, 1851.
11. San Francisco *Daily Alta California*, November 25, 1850 (from Sacramento *Times*).
12. *Ibid.*, April 12, 1851.
13. *Ibid.*, June 9, 1851 (from Sacramento *Transcript*).
14. *Ibid.*, September 27, 1851.
15. *Ibid.*, September 16, 1851.
16. *Ibid.*, October 15, 1851 (from Sacramento *Union*).

17. *Ibid.*, October 12, 1851 (from Sacramento *Union*).
18. *Ibid.*, November 14, 1851 (from Stockton *Journal*).
19. Sacramento *Daily Union*, February 26, 1858 (from Placerville *Index*).
20. San Francisco *Daily Alta California*, April 26, 1852 (from *San Joaquin News* through *California Chronicle*).
21. San Francisco *Daily Alta California*, May 19, 1852 (from El Dorado *News*).
22. *Ibid.*, August 1, 1852 (from *San Joaquin News* through Stockton *Journal*).
23. San Francisco *Daily Alta California*, April 26, 1853.
24. *Ibid.*, August 7, 1853 (from San Diego letter).
25. *Ibid.*, June 14, 1854.
26. *Ibid.*, June 18, 1854 (from Grass Valley *Telegraph*)
27. *Ibid.*, June 6, 1854 (from Alameda *Express*).
28. *Ibid.*, July 19, 1854.
29. Sacramento *Steamer Union*, October 14, 1854.
30. San Francisco *Daily Alta California*, August 6, 1855 (from *Southern Californian*).
31. *Ibid.*, September 27, 1855 (from Sacramento *Union*)
32. *Ibid.*, January 7, 1856.
33. San Francisco *California Chronicle*, January 23, 1856 (from Mariposa *Chronicle*).
34. San Francisco *Daily Alta California*, March 19, 1856.
35. San Francisco *Evening Bulletin*, April 2, 1856 (from Sacramento *Union*).
36. San Francisco *Herald*, August 10, 1856.
37. San Francisco *Daily Alta California*, August 6, 1856.
38. *Ibid.*, September 27, 1856.
39. *Ibid.*, September 10, 1856.
40. San Francisco *Daily Evening Bulletin*, October 10, 1856.
41. *Ibid.*, November 1, 1856 (from Mariposa *Democrat*).
42. *Ibid.*, November 12, 1856.
43. San Francisco *Daily Alta California*, November 28, 1856 (from *Nevada Democrat*).
44. *Ibid.*, April 12, 1857.
45. San Francisco *Daily Evening Bulletin*, April 25, 1857.
46. San Francisco *Daily Alta California*, May 19, 1857.
47. *Ibid.*, September 22, 1857.
48. San Francisco *Evening Bulletin*, September 22, 1857 (from Shasta *Courier*).
49. *Ibid.*, September 18, 1857.
50. San Francisco *Daily Alta California*, November 22, 1857
51. San Francisco *Herald*, March 28, 1858 (from Sonora *Democrat*).
52. San Francisco *Daily Alta California*, June 7, 1858.
53. *Ibid.*, July 4, 1858.
54. *Ibid.*, November 13, 1858.
55. *Ibid.*, December 15, 1858 (from Shasta *Courier*).
56. San Francisco *Daily Evening Bulletin*, September 8, 1859 (from Nevada *Democrat*).

57. Sacramento *Daily Union*, May 29, 1860 (from San Juan *Press*).
58. Sacramento *Daily Union*, June 18, 1860.
59. San Francisco *Daily Alta California*, July 25, 1860 (from Alameda *Herald*.)
60. *Ibid.*, January 18, 1861 (from Mariposa *Gazette*).
61. Sacramento *Daily Union*, May 6, 1862.
62. San Francisco *Daily Alta California*, February 22, 1863.
63. Sacramento *Daily Union*, January 6, 1864.
64. *Ibid.*, August 29, 1865.
65. *Ibid.*, July 26, 1866.
66. *Ibid.*, August 22, 1866.
67. *Ibid.*, October 4, 1866.
68. *Ibid.*, November 24, 1866.
69. San Francisco *Daily Alta California*, August 28, 1868.
70. Sacramento *Daily Union*, August 10, 1868.
71. San Francisco *Daily Alta California*, September 14, 1868.
72. *Ibid.*, September 22, 1868.
73. *Ibid.*, September 25, 1868.
74. Sacramento *Daily Union*, November 22, 1870.
75. San Diego *Union*, November 1, 1871.
76. *Ibid.*, November 2, 1871.
77. *Ibid.*, June 21, 1872.
78. *Ibid.*, July 23, 1872.
79. *Ibid.*, August 19, 1873.
80. *Ibid.*, October 31, 1873.
81. Sacramento *Daily Union*, November 11, 1875 (from Watsonville *Pajaronian*).
82. San Diego *Union*, September 20, 1874 (from San Bernardino *Guardian*).
83. *Ibid.*, November 21, 1874.
84. San Francisco *Daily Alta California*, December 10, 1874.
85. *Ibid.*, December 13, 1874 (from Downieville *Messenger*).
86. *Ibid.*, April 15, 1875 (from Anaheim *Gazette*)
87. *Ibid.*, March 25, 1877.
88. San Diego *Union*, May 15, 1877.
89. San Francisco *Daily Alta California*, June 28, 1878.
90. *Ibid.*, January 6, 1879 (from *Trinity Journal*)
91. *Ibid.*, July 31, 1879.
92. *Ibid.*, September 2, 1881.
93. San Francisco *Call Bulletin*, September 5, 1882 (from Ventura *Signal*).
94. *Ibid.*, December 22, 1882.
95. Sacramento *Daily Record-Union*, May 22, 1884.
96. *Ibid.*, May 21, 1884.
97. Santa Barbara [*Weekly*] *Herald*, November 25, 1886.
98. San Francisco *Examiner*, October 23, 1888.
99. Sacramento *Daily Union*, June 21, 1889 (from Red Bluff *News*)
100. *Ibid.*, August 11, 1889.
101. San Francisco *Chronicle*, January 14, 1903
102. *Ibid.*, September 29, 1907

# Index

Where not otherwise indicated, entries in Index pertain to California Grizzly.

331